Springer Earth System Sciences

More information about this series at http://www.springer.com/series/10178

Gabriela Catalina Cusminsky
Emiliana Bernasconi · Graciela Andrea Concheyro
Editors

Advances in South American Micropaleontology

Selected Papers of the 11th Argentine Paleontological Congress

Springer

Editors
Gabriela Catalina Cusminsky
INIBIOMA-CONICET
Universidad Nacional del Comahue
San Carlos de Bariloche, Argentina

Emiliana Bernasconi
INIBIOMA-CONICET
Universidad Nacional del Comahue
San Carlos de Bariloche, Argentina

Graciela Andrea Concheyro
UBA-CONICET Instituto de Estudios Andinos (IDEAN)
Unversidad de Buenos Aires
Ciudad Autónoma de Buenos Aires
Argentina

ISSN 2197-9596 ISSN 2197-960X (electronic)
Springer Earth System Sciences
ISBN 978-3-030-02118-4 ISBN 978-3-030-02119-1 (eBook)
https://doi.org/10.1007/978-3-030-02119-1

Library of Congress Control Number: 2018958356

This Springer imprint is published by the registered company Springer Nature Switzerland AG
The registered company address is: Gewerbestrasse 11, 6330 Cham, Switzerland

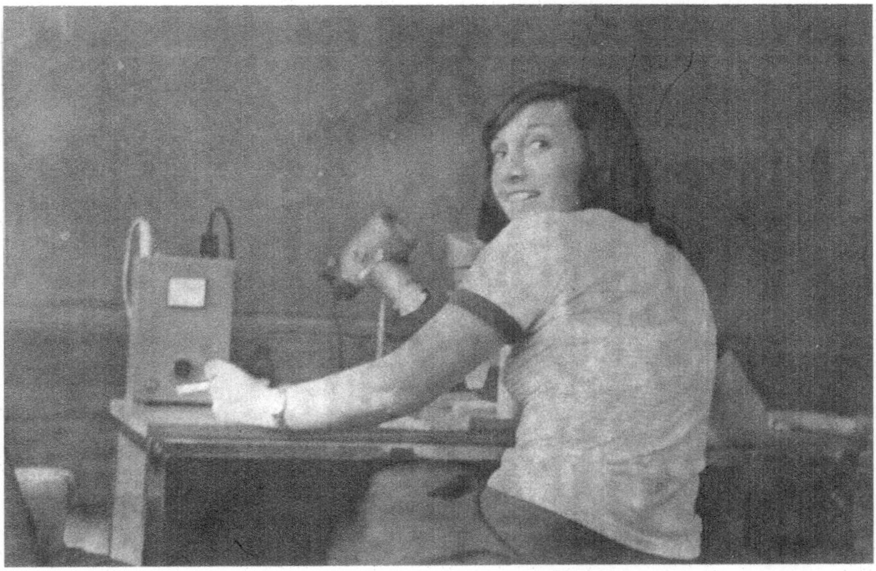

Sara Ballent In memoriam, who pioneered the Symposia of Micropaleontology of South America and Antarctica. Her enthusiasm encouraged several Argentinean micropaleontologists to follow her steps and patiently continue her work in this discipline.

Preface

Micropaleontology is a very well established discipline in Europe, America, Africa, and Asia, and it is a valuable tool in basic and applied geology and paleobiology. It also complements interdisciplinary studies on climate change and pollution and forensic studies apart from the traditional uses in biostratigraphy, paleoenvironments, and paleoecology.

The first South American and Antarctic Micropaleontology Symposium took place in Corrientes, Argentina, in 2002, during the VII Congreso Argentino de Paleontología y Bioestratigrafía. Since then, both meetings have been associated.

The present book, *Advances in South American Micropaleontology: Selected Papers of the 11th Argentine Paleontological Congress*, intends to provide an update on the research carried out by South American micropaleontologist in the fields of microfossils, such as foraminifera, ostracods, calcareous nannofossils, and diatoms, present in the geological record.

Their research in these fields was presented and discussed in the IV Simposio de Micropaleontología de America del Sur y Antártida in the forum of the XI Congreso de la Asociación Paleontológica Argentina (CAPA 2016) which took place in General Roca, Río Negro, Argentina, from October 17 to 21, 2016. The symposium was well attended by Brazilian, Colombian, Ecuadorian, and Argentinean scientists. There were 23 presentations and posters.

The editors acknowledge and thank Dr. J. Rabassa for the invitation to publish some of the research presented at the IV Simposio of Micropaleontología de America del Sur y Antártida in a Springer Special Volume in the Springer Earth System Science book series and for his support. Thanks are also expressed to those researchers from several countries that helped us as reviewers, and their opinion and contributions improved the original manuscripts.

The book consists of nine chapters. The chapters are written by single authors or interdisciplinary group of authors, highlighting the team approach to research.

The chapters follow a stratigraphic order, and we summarize them below.

The first chapter is entitled "New Finding of Regular Echinoid Elements and Microfossils from the Pilmatué Member, Agrio Formation (Early Cretaceous), Neuquén Basin, Argentina," *by A. Caramés, S. Adamonis, A. Concheyro and M. Remírez.* This chapter presents new findings of regular echinoid elements from Upper Valanginian claystones, marlstones, and mudstones recovered near the base of the Pilmatué Member, Agrio Formation, and points out the presence of non-cidaroid regular echinoids belonging to the "diademataceans." They have been found in fine sediments associated with quiet marine environments. The productive horizons also contain a rare to very abundant assemblage of benthic foraminifera, few ostracods, and resistant calcareous nannofossils.

The second chapter is named "Lower Cretaceous Benthic Foraminifers and Ostracods of the Middle Magdalena Basin, Colombia: Biostratigraphic and Paleoenvironmental Inferences," *by G. D. Patarroyo, D. Ceolin, G. Fauth, G. A. Torres, C. P. Cárdenas, and D. A. Rincón.* This chapter comprises the biostratigraphic framework and depositional environments of the Lower Cretaceous Rosablanca Formation (Middle Magdalena Basin), studying two wells drilled in this formation. This study provides the first evidence of Early Cretaceous ostracods in Colombia. Despite the moderate recovery of microfossils, the lower part of this unit contains several ostracods. In contrast, the upper part of the Rosablanca Formation contains benthic foraminifera which indicate a transitional to normal marine environment and are Valanginian to Barremian in age.

The third chapter is entitled "Turonian–Santonian Calcareous Nannofossil Biozonation for La Luna Formation, Middle Magdalena Valley Basin, Northern Colombia," *by J. P. Pérez Panera, F. J. Parra, R. E. Navarrete, C. Sánchez, D. Daza, M. Rodriguez, and M. Prince.* A well site biostratigraphic study based on calcareous nannofossils of Chuira-2 ST-1 well, through the Umir and La Luna formations in the Middle Magdalena Valley Basin (MMVB), north of Colombia, is presented herein. Biostratigraphic analyses allowed to determine six local calcareous nannofossil interval biozones spanning the Early Turonian to Late Santonian and also recognize an inverse fault traversed by this well.

The fourth chapter is related with Chapter "Turonian–Santonian Calcareous Nannofossil Biozonation for La Luna Formation, Middle Magdalena Valley Basin, Northern Colombia," entitled "Turonian-Campanian Foraminifera Zonation for the La Luna and Lower Umir Formations, Middle Magdalena Valley Basin, Northern Colombia," *by R. E. Navarrete-Parra, F. Parra, J. P. Pérez Panera, D. Daza R., C. Sánchez G., M. Prince S., and M. Rodriguez.* This chapter presents a new biozonation using planktonic foraminifera, palynomorphs, and calcareous nannofossils of the Chuira-2ST well, Middle Magdalena Valley Basin, Colombia, focusing their analysis in the latest Early Campanian marly and sandy-shale deposits of the Umir Formation and the limestone and shale deposits of the La Luna Formation. Fossiliferous levels with foraminifera are discussed and correlated with palynomorphs and calcareous nannofossils. Complementary data provided by Chuira-2 and other exploration wells and outcrop sections are also considered.

The fifth chapter corresponds to "Paleoenvironmental Interpretation of Late Miocene Deposits, from El Cajón Valley (Catamarca Province, Argentina) Based on Calcareous Microfossils," *by B. Zamudio and S. Morton*. The Neogene sedimentary sequence from El Cajón Valley, Catamarca Province, Argentina, includes the alloformations Peñas Azules, Totoral, and Playa del Zorro, being the latter unit studied herein, at the Río Totoral. Ostracods, foraminifera, gastropods, and bivalves were found in green and yellow-green fine sandstones and mudstones, interbedded with gypsum. Ostracods were dominant in all the associations, and they were represented by valves and carapaces of juveniles and adults. In many cases, they showed signs of compression and fragmentation; meanwhile, foraminifera were found in only one horizon. Microfossils from the lower section could characterize brackish waters, and the presence of *Ammonia tepida* could indicate an increase in the salinity conditions, probably resulting from dry periods. Up to section, monospecific associations of ostracods, mainly composed of *Cyprideis*, would suggest drier conditions and shallower water for these levels.

The sixth chapter is named "Late Pleistocene Diatoms of the Lower Basin from the Quequén Salado River, Argentina," *by M. A. Espinosa and E. Beilinson*. Here, diatom assemblages from *Cueva del Tigre* were studied in the lower basin of the Quequén Salado River, Buenos Aires Province, Argentina, with the aim of reconstructing the environmental history of south-eastern Buenos Aires during the Late Cenozoic. The studied sedimentary succession consisted of coarse sandstone and clayed siltstones, and diatom analyses were performed. As a result, 74 diatoms taxa were recognized and grouped according to salt tolerance and life form. Cluster analysis allowed dividing the sedimentary sequence into two diatom zones. The comparison between fossil and modern samples through NMDS analyses was also defined.

The seventh chapter is "New Micropaleontological Record (Foraminifera and Ostracoda) from the Late Glacial and Holocene in Salinas del Bebedero, San Luis, Argentina: A Paleoenvironmental Interpretation," *by L. Calvo Marcilese, D. Sabina D'Ambrosio, L. D. Rojo, and G. C. Cusminsky*. Few studies have been conducted related to foraminifera and Ostracoda in endorheic environments of Argentina. In core SB III from Salinas del Bebedero, a preliminary report based on 17 productive micropaleontological samples (ca. 12600–3600 yr BP) is analyzed. Some samples contained foraminifera which indicate typical marginal marine microfaunal assemblages, and ostracods, poorly represented, with the exception of six levels. It is noteworthy to note that ostracods and foraminifers were not recorded coexisting, with the exception of one sample. The high abundance observed in foraminiferal assemblages is outstanding and indicates shallow marginal marine environments, whereas the ostracod characterizes lacustrine environments. These new studies increase the knowledge about the typical marine microfauna living in continental environments and demonstrate their potential implications in paleoenvironmental interpretation.

The eighth chapter is named "Calcareous Microfossils (Ostracoda and Foraminifera) from the Holocene (MIS 1) at the Salado Basin (Arroyo San Miguel Site, Pila, Buenos Aires Province, Argentina)," *by N. A. Ramos,*

A. P. Carignano, G. C. Cusminsky, and E. Fucks. A sedimentary section attributable to the Marine Isotopic Stage 1 (MIS 1) (Canal de Las Escobas Formation) was identified at the study site, between the Members Gorch and Puente Las Gaviotas of the Luján Formation. The ostracods recovered include common species of estuarine environments with wide salinity tolerance. The foraminifera are also typical of oligo–mesohaline conditions, and this low diversity association is indicative of environments with marine influence. Bivalve shells were dated obtaining a radiocarbon age of 5580 ± 110 years BP (6278–6511 cal. years BP), suggesting that Arroyo San Miguel site represents the most internal locality assignable to the Middle Holocene (MIS 1) transgression.

The ninth and last chapter is "Modern Benthic Foraminifera "Phylum Foraminifera (D'Orbigny 1826)" of the Panama Bight: A Census Report Based on Thanatocoenoses from the Continental Slope," *by A. M. Ballesteros.* This research compiles the published reports of the foraminiferal assemblages in the Panama Bight (Eastern Equatorial Pacific) and adds new information from samples located in the continental margin of Panama Bight, with the purpose of illustrating and updating taxonomic list and exploring the presence and richness of the species along this region. The faunal composition corresponds zoogeographically to the Panamanian Province, and some foraminifera reported are indicators of environments with high flow of organic matter and high productivity.

Finally, the publication of this book will certainly reinforce the link between scientists working in micropaleontology in South America and it will hopefully give access to colleagues from other countries unfamiliar with some of the publications in Spanish.

In the era of globalization, this book provides the base for the reader to be aware of the active research issues in Argentina and South America and gives the opportunity for future cooperation research by members in different countries.

San Carlos de Bariloche, Argentina — Gabriela Catalina Cusminsky
San Carlos de Bariloche, Argentina — Emiliana Bernasconi
Ciudad Autónoma de Buenos Aires, Argentina — Graciela Andrea Concheyro

Participants Symposium CAPA 2016

Photograph of some participants of the IV Symposium of Micropaleontology of South America and Antarctica, October 19, 2016, General Roca Rio Negro Province, Argentina.

From left to right **1** Juan Pablo Pérez Panera, **2** Angélica Ballesteros Prada, **3** Emiliana Bernasconi, **4** Germán David Patarroyo, **5** Ana Paula Carignano, **6** Melina Márquez, **7** Lydia Calvo-Marcilese, **8** Laura Ferrero, **9** Gabriela Cusminsky, **10** Nicolás Ramos, **11** Daiane Ceolin, **12** Rosa Navarrete Parra, **13** Francisco Parra, **14** Andrea Concheyro, **15** Marcela Espinosa

Contents

New Finding of Regular Echinoid Elements and Microfossils from the Pilmatué Member, Agrio Formation (Early Cretaceous), Neuquén Basin, Argentina

Andrea Caramés[1(✉)], Susana Adamonis[1,2], Andrea Concheyro[1,2,3], and Mariano Remírez[4]

[1] IDEAN—Instituto de Estudios Geológicos Andinos "Don Pablo Groeber", CONICET, Ciudad Autónoma de Buenos Aires, Argentina
{carames, suad, andrea}@gl.fcen.uba.ar

[2] Departamento de Ciencias Geológicas, Facultad de Ciencias Exactas y Naturales, Universidad de Buenos Aires, Ciudad Autónoma de Buenos Aires, Argentina

[3] Instituto Antártico Argentino, San Martín, Buenos Aires, Argentina

[4] Facultad de Ciencias Naturales y Museo, CIG—Centro de Investigaciones Geológicas, CONICET, Universidad Nacional de La Plata, La Plata, Buenos Aires, Argentina
mremirez@cig.museo.unlp.edu.ar

Abstract. This contribution presents new findings of regular echinoid elements from Upper Valanginian claystones, marlstones, and mudstones recovered near the base of the Pilmatué Member, Agrio Formation, in the Puerta Curaco and Área 3000 sections, Province of Neuquén. The recovered material includes isolated elements of Aristotle's lantern, spines, ambulacral and interambulacral plates obtained from washed residues of micropaleontological samples. Their taxonomy covers several possible regular echinoids included within the Acroechinoidea. Considering the tubercles, pore-pairs, type of plates, and ambulacral plating style, these fragments could be assigned to the "diademataceans". The spines and the rotulae cannot be attributed to a specific taxon because they are found in several different regular echinoid clades. The morphology of the hemi-pyramids is common to different regulars; this fact coupled with their poor preservation prevents a confident assignment to a higher taxonomic level. Wedge-shaped teeth, typical of irregular echinoids, cannot be linked to any of the other recovered remains. These fossiliferous horizons also contain a rare to very abundant assemblage of benthic foraminifera dominated by *Epistomina loncochensis* Ballent, few ostracods, and resistant calcareous nannofossils. The presence of *E. loncochensis* indicates quiet marine environments and scarcity of oxygen in the water/bottom interface. Claystone composition of the level rich in echinoid remains suggests a detrital input from shallower environments by means of currents, explaining the disarticulation of the echinoids.

Keywords: Echinoids · Early Cretaceous · Late Valanginian · Agrio Formation Neuquén Basin · Argentina

G. C. Cusminsky et al. (eds.), *Advances in South American Micropaleontology*, Springer Earth System Sciences, https://doi.org/10.1007/978-3-030-02119-1_1

1 Introduction

With the aim to improve the biostratigraphic framework of the Lower Cretaceous strata of the Agrio Formation, a micropaleontological study was carried out in the lower levels of the Pilmatué Member of two sections located in the Curaco area, within the central sector of Neuquén Basin. Initially, the benthic foraminifera were studied, which consists of a very low diverse and poorly preserved assemblage. An unexpected large number of echinoderm remains belonging to the Echinoidea and Holothuroidea were also recovered. The first ones have inspired this contribution, whose main objective is to present and illustrate this new record from the Neuquén Basin.

Echinoids are important organisms of current marine benthic communities living from the intertidal zone to 5000 m depth. They are characterized by having a test of high-Mg calcite made of individual elements that usually interlock to a lesser or greater extent to form the test (Kroh and Nebelsick 2010). The irregular echinoids are more likely to be preserved than the regulars because they live buried within the sediment (Kier 1977). Test construction and post-mortem conditions also play an important role (Kidwell and Baumiller 1990) along with sedimentary environments (Nebelsick et al. 2015).

The specimens studied by Rodríguez (2007) from the Agrio Formation include complete coronas of both regular and irregular echinoids. The following species were recovered from the Pilmatué Member: the irregular echinoids *Clypeopygus robinaldinus* d'Orbigny, *Pygopyrina gerthi* (Weaver), and *Coenholectypus planatus* Roemer var. *numismalis* (Gabb) as well as the presence of unidentifiable regular echinoids and their spines from Barda del Cerro Marucho locality. Rodríguez (2007) also reported the presence of the regular echinoids *Leptosalenia prestensis* (Desor), *Codechinus* sp. A, and the irregular echinoids *Pygorhynchus obovatus* L. Agassiz and *Coenholectypus planatus* Roemer var. *numismalis* (Gabb) from the younger Agua de la Mula Member.

Other records of echinoids from the Agrio Formation are limited to brief reports. These include the presence of spines from the basal beds of the Pilmatué Member, further north in Cuesta del Chihuido section, south of Mendoza Province (Ballent et al. 2006). There, the spines were collected from the same levels as *Epistomina loncochensis*, benthic foraminifera originally described as a widespread biostratigraphic marker species in the lower beds of the Agrio Formation (Sagasti and Ballent 2002).

2 Geological Setting

The analyzed material was recovered from the Lower Cretaceous marine deposits of the Pilmatué Member which belongs to the Agrio Formation which itself is part of the Mendoza Group. The Agrio Formation consists of three members: the Pilmatué, Avilé and Agua de la Mula Members (Leanza and Hugo 2001). Both the Pilmatué and Agua de la Mula Members are of marine origin and mainly composed of shales and sandy limestones (packstones and wackestones). Their deposition has been related to the Pacific Ocean transgressions. These members have been interpreted as siliciclastic-carbonate successions ranging from the offshore to shoreface and are influenced by wave movements of both fair weather and storm conditions (Spalletti et al. 2011).

The Avilé Member is non-marine sandstone that has been interpreted as a lowstand wedge produced by a rapid relative sea-level drop during mid-Hauterivian times (Legarreta and Gulisano 1989) and provides a conspicuous marker horizon separating the two marine members.

Ammonoids and calcareous nannofossils have provided a detailed biostratigraphy (Aguirre-Urreta and Rawson 1997, 2012; Aguirre-Urreta et al. 2005, 2007, 2017; Bown and Concheyro 2004; Concheyro et al. 2009). According to both groups, the age of the Agrio Formation is Upper Lower Valanginian to Lower Barremian and, in particular, the Pilmatué Member is assigned to the Lower Valanginian to Lower Hauterivian.

There are also few absolute ages, the first one obtained from the Pilmatué Member is a sensitive high resolution ion microprobe—SHRIMP U–Pb age of 130.0 ± 0.6 or 0.8 Ma for the tuff layer that occurs at Mina San Eduardo section within beds belonging to the *Holcoptychites agrioensis* Subzone of the *Holcoptychites neuquensis* Zone (Schwarz et al. 2016). This dated tuff layer provides an estimated minimum age for the lower boundary of the Hauterivian in the Neuquén Basin. Recently, a new chemical abrasion isotope dilution thermal ionization mass spectrometry—CA-ID-TIMS U–Pb age of 130.39 ± 0.16 Ma has been defined at the El Portón section, located further north of San Eduardo (Aguirre-Urreta et al. 2017), and was obtained from the same *Holcoptychites agrioensis* Subzone of the *Holcoptychites neuquensis* Zone. In addition, calcareous nannofossils from El Portón indicate the CC4A Zone (Roth 1983).

3 Materials and Methods

The remnants of echinoids studied herein were recovered from samples of the Pilmatué Member of the Agrio Formation and were collected at Puerta Curaco (37° 22′ 47″ S, 69° 55′ 45″ W), and Área 3000 (37° 25′ 10″ S, 69° 57′ 4″ W) sections, exposed at the center of the Neuquén Basin (Fig. 1). The Puerta Curaco log (*ca* 557 m thick) predominantly consists of marlstones and mudstones in the lower part and claystones and siltstones in the upper part (Fig. 2), similar to that of the Área 3000 log (*ca* 600 m thick).

The lower part of these sections yielded dissociated echinoderm ossicles (echinoids, holothurians), foraminifera, few ostracods, and an impoverished assemblage of calcareous nannofossils. The most abundant echinoid remains of both sections come from claystones (samples PC009b and AT009 from the Puerta Curaco—PC, and Área 3000—AT sections) being Puerta Curaco more productive than Área 3000.

The studied samples were disaggregated using hydrogen peroxide solution (5%, 200 vols.), and the large residual fragments of non-disaggregated rock were treated using the repeated freezing-thawing technique (Hanna and Church 1928). Then, the disaggregated samples were washed over 74 μ and 297 μ sieves. The washed samples were dried, and the above-mentioned rests were picked under a Leica M165C stereoscopic microscope. The distribution of the different echinoid sclerites is given in Tables 1 and 2 along with the other benthic microfossils (holothurians spicules, foraminifera, and ostracoda). The abundance of all these microfossils is given in number of specimens and comes from approximately 40 g of washed sediment.

Fig. 1. Location map of the Puerta Curaco and Área 3000 sections in the north of Neuquén Province, Argentina

The distribution charts of calcareous nannofossils are given in Tables 3 and 4 and are based in a semiquantitative analysis, estimating specimens in two longitudinal transects of the slides. Selected specimens of echinoid sclerites are illustrated with photographs taken with Leica DFC295 camera. Specimens figured herein (Fig. 3) are housed in the micropaleontological collections of the Facultad de Ciencias Exactas y Naturales, Universidad de Buenos Aires (Argentina), under the catalog numbers LM-FCEN 3756-3773. Calcareous nannofossils were recovered using smear slide technique (Edwards 1963; Bown and Young 1998). Cover slides were fixed with Norland Optical

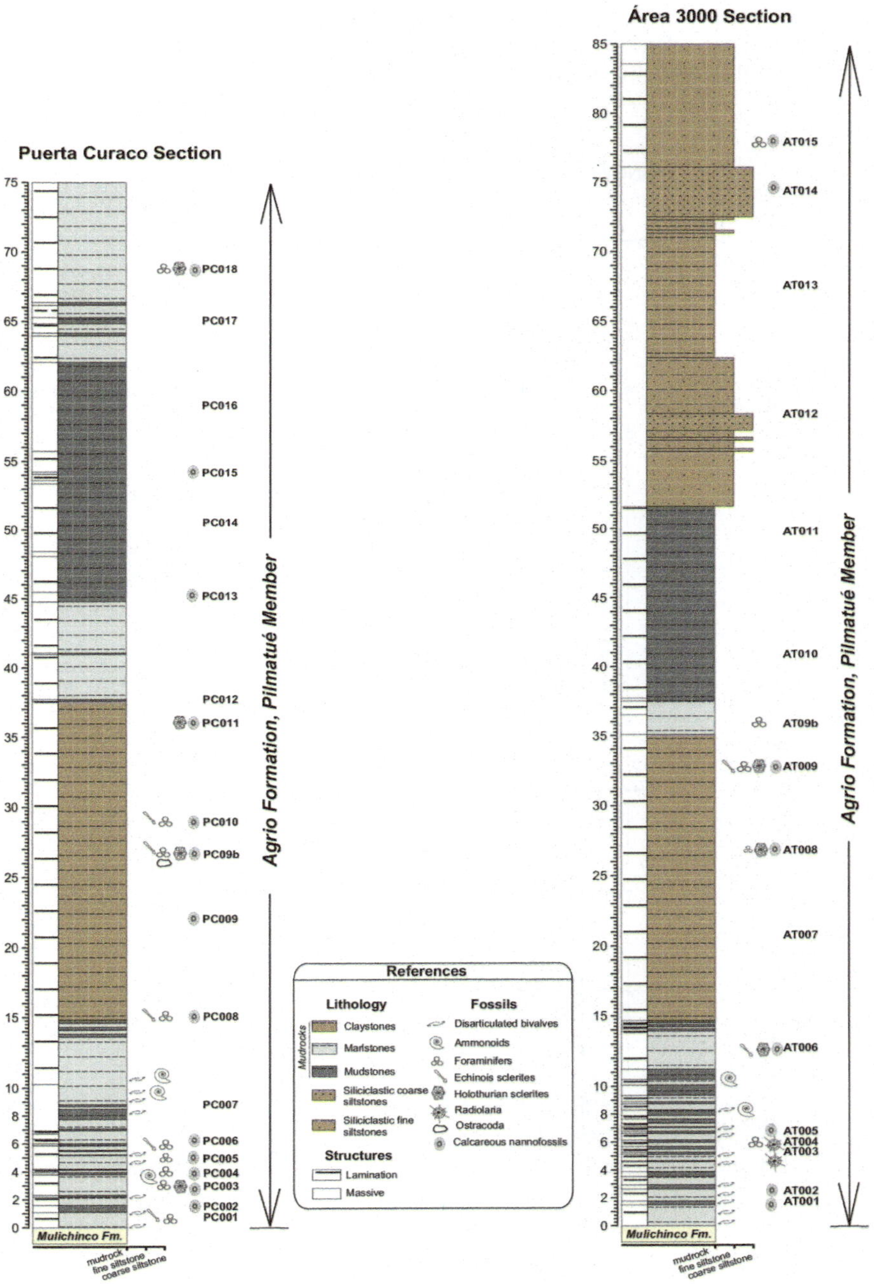

Fig. 2. Stratigraphic logs of the lower levels outcropping at the Puerta Curaco and Área 3000 sections, Neuquén Province

Table 1. Stratigraphic distribution of the echinoid ossicles together with other microfossils remains (holothurian sclerites, foraminifers, ostracods, bivalve fragments, and fish teeth) at the lower levels of the Puerta Curaco section

PUERTA CURACO SECTION	ECHINOID OSSICLES						HOLOTHURIAN REMAINS	FORAMINIFERS				OSTRACOD	BIVALVE FRAGMENTS	FISH TEETH
	Spine fragments	Fragments with tubercles	Rotulae	Hemi-pyramids	Fragments of teeth	Fragments of ambulacral plates		*Epistomina loncochensis* Ballent	*Saracenaria* sp.	*Lenticulina* sp. (juvenile)	*Lenticulina* sp.	*Sondagella colchesterensis* Valicenti and Stephens		
PC018							2	1			2			
PC015														
PC013														
PC011							1			1				
PC010	28	2	1					33						
PC09b	117	116	83	47	29	10	308	310				2		
PC009														
PC008	1							1						2
PC007														
PC006	1							12	1					
PC005								14					5	
PC004								12						
PC003							2	70						
PC002														
PC001	1							80						

Gray color indicates barren samples

Adhesive, and nannofossil identifications were carried out through a Leica DMLP petrographic microscope under 1000x magnification adding a gypsum plate. Calcareous nannofossils slides are housed in the Área Paleontología, Facultad de Ciencias Exactas y Naturales, Universidad de Buenos Aires, under the catalog numbers BAFC-NP 4103-4123.

Table 2. Stratigraphic distribution of the echinoid ossicles together with other microfossils remains (holothurian sclerites, foraminifers, Spumellaria radiolarians, and bivalve fragments) at the lower levels of the Área 3000 section

ÁREA 3000 SECTION	ECHINOID OSSICLES					HOLOTHURIAN REMAINS	FORAMINIFERS				SPUMELLARIA RADIOLARIANS	BIVALVE FRAGMENTS
	Spine fragments	Rotulae	Fragments with tubercles	Fragments of teeth	Hemi-pyramids		*Epistomina loncochensis* Ballent	*Astacolus ambanjabensis* (EspitaliÈ and Sigal)	*Lenticulina* spp. (indet.)	*Vaginulinopsis* sp.		
AT015						7						
AT014												
AT013												
AT011												
AT09b								1		1		
AT009	21			2		24	4					
AT008						2			2			
AT007												
AT006	2	2	1	1	1	8		57				7
AT005												
AT004							4				8	
AT003											3	
AT002												
AT001												

Gray color indicates barren samples

4 Results

Echinoid remains include isolated and poorly preserved ossicles belonging to Aristotle's lantern including rotulae, hemi-pyramids, teeth, as well as spines, ambulacral and interambulacral plates. The spines, teeth, and hemi-pyramids are usually fragmented. The hemi-pyramids and plates show signs of dissolution and abrasion.

Table 3. Puerta Curaco section calcareous nannofossils range chart. Preservation of calcareous nannofossils: M, moderate; P, poor. Nannofossils abundance: A, abundant; C, common; F, frequent; R, rare

PUERTA CURACO SECTION	PRESERVATION	ABUNDANCE	*Watznaueria barnesiae*	*Watznaueria fossacincta*	*Nannoconus steinmannii*	*Nannoconus* spp.	*Micrantholithus hoschulzii*	*Calculites percenis*	*Nannoconus circularis*	*Micrantholithus obtusus*	*Eiffellithus striatus*	*Eiffellithus windii*	*Rhagodiscus asper*	*Watznaueria ovata*	*Nannoconus* cf. *N. bucheri*	*Retecapsa surirella*	*Rhagodiscus* sp.	*Zeugrhabdotus embergeri*	*Watznaueria biporta*	*Cyclagelosphaera margerelii*	*Diazomatholithus lehmannii*
PC018	M	C	C	C			R				R		R					R		C	
PC015	P	C	C	C					R									R	R	C	
PC013	M	A	A	A	R		C		F	C	R		R					F		C	
PC011	P	F	F	F			F			F								F		F	F
PC010	P	R	R	R			R										R		R		
PC09b	M	C	C	C			F				R	R	R	R					R		
PC009	M	F	R	F			R		R					R				R	R	R	
PC008	M	F	F	F														F			
PC006	M	F	F	F		R	F			F	R	R	R	R	R	R					
PC005	P	R	R	R			R														
PC004	M	R	R	R			R														
PC003	P	F	F	F	R	R		R	R												
PC001																					

Gray color indicates barren samples

4.1 Echinoid Ossicles

4.1.1 Fragments of Ambulacral Plates with Pore-Pairs

Figure 3 p, q.

Material. Puerta Curaco section: PC009b, 10 fragments. Área 3000 section: AT006, 1 fragment.

Description. Ambulacra apparently straight, plating trigeminate, ambulacral plates compounded in acrosaleniid style (two elements) bearing the primary tubercle, alternating with a simple element with secondaries only. Pore-pairs uniserially arranged, approximately equal-sized (partitioned isopores) forming oblique pore-pairs, with very narrow interporal partitions. Taking into account the attachment area breath, pore size,

Table 4. Área 3000 section calcareous nannofossils range chart. Preservation of calcareous nannofossils: M, moderate; P, poor. Nannofossils abundance: A, abundant; C, common; F, frequent; R, rare; VR; very rare

ÁREA 3000 SECTION	PRESERVATION	ABUNDANCE	*Watznaueria barnesiae*	*Watznaueria fossacincta*	*Zeugrhabdotus embergeri*	*Micrantholithus hoschulzii*	*Micrantholithus obtusus*	*Retecapsa surirella*	*Cretarhabdus conicus*	*Watznaueria ovata*	*Watznaueria biporta*	*Nannoconus cross-section*	*Nannoconus steinmannii*	*Rhagodiscus asper*	*Nannoconus cf. N. bucheri*	*Eiffellithus windii*	*Cruciellipsis cuvillieri*	*Eiffellithus striatus*	*Retecapsa angustiforata*	*Diazomatholithus lehmannii*
AT015	M	A	A	A				F		R								F	R	R
AT014	M	A	A	A	F	R	R									R	R	R		
AT009	M	C	C	C	R	F	R	R	R	R	C	R	R	R	R	R				
AT008	M	R	R	R		R														
AT007																				
AT006	M	F	F	F			R													
AT005	M	R	R	R	R	R														
AT002	P	VR	VR	VR																
AT001	P	VR	VR	VR																

Gray color indicates barren samples

and interporal partition shape, the pores are classified as partitioned isopore type P2 of Smith (1978). The attachment area is asymmetric, being more developed toward the apical pole. No neural channels are observed in perradial pores. The stereom of many plate fragments show irregular struts and varying pore size and allow it to be classified as labyrinthic stereom (Smith 1980).

Remarks. Many of these features such as the acrosaleniid style, trigeminate ambulacral plating can be found in the "diademataceans", and the families Acrosaleniidae, Emiratiidae, Hemicidaridae, and Pseudodiadematidae.

The recovered plates have been compared with the descriptions and pictures of these taxa from Smith and Kroh (2011) and assigned to the "diademataceans". This group includes the orders Micropygoida, Aspidodiadematoida, and Diadematoida although the Micropygoida can be discarded because of the non-crenulate perforate tubercles.

Echinoids belonging to the Aspidodiadematoida have acrosaleniid ambulacral plating in triads at least adorally and with some elements united by tubercle overgrowth (Smith and Kroh 2011). Taking into account these features, these plates might be related to this taxon. Diadematoida includes the family Heterodiadematidae and Diadematidae. The Heterodiadematidae are discarded because although this taxon has trigeminate ambulacral plates and uniserial pore-pairs, it has diadematid-style

(a)
(b)
(c)
(d)
(e)
(f)
(g)
(h)
(i)
(j)
(k)
(l)
(m)
(n)
(o)
(p)
(q)
(r)
(s)
(t)
(u)
(v)

◀ **Fig. 3.** Echinoid ossicles of the Pilmatué Member, Agrio Formation. **a–e** Hemi-pyramids: **a** on left-hand side: interpyramidal face wing with adaxial inner border showing comb-like indentations and striated attachment area of interpyramidal muscle; on right-hand side, abaxial face (LM-FCEN 3756); **b** interpyramidal face wing (LM-FCEN 3757); **c** detail of the interpyramidal face with a flat dorsal edge that seems to have a V-shape fosseta (LM-FCEN 3758); **d** abaxial face (LM-FCEN 3756); **e** intrapyramidal face with suture face (LM-FCEN 3756). **f–g** Rotulae: **f** oral side with the abaxial condyles to the bottom and the adaxial condyles and grooves to the top, intermediate fossa in the middle (LM-FCEN 3759); **g** aboral side, large pored, abaxial head to bottom, adaxial head to top (LM-FCEN 3759). **h–i** Tooth: **h** cross section of a wedge-shaped tooth, abaxial edge to the top (LM-FCEN 3760); **i** lateral view with the oral chewing tip to the bottom and the abaxial edge to the right (LM-FCEN 3761). **j–n** Interambulacral fragments with primary tubercles perforate and crenulate (LM-FCEN 3762-3766). **m** Interambulacral plate with a primary and secondary tubercle perforate and crenulate and two small ones, being probably miliary granules (LM-FCEN 3765). **p–q** Ambulacral plates. **p** ambulacra plates compounded in acrosaleniid style and perradial suture on right (LM-FCEN 3767); **q** two plates with pore-pairs LM-FCEN 3768). **o, r, s, t, u, v** Spines (LM-FCEN 3769-3773). **o, r, s** cross section, with some wedges dichotomized (shown in **r, s**) (LM-FCEN 3769, 3770, 3771). **t–v** lateral view of the base and part of the shaft with ribs and no cortex (LM-FCEN 3772, 3773, 3771). Scale bar: 0.2 mm

compound plates. The Diadematidae have trigeminate ambulacral plates with an enlarged element and two smaller demiplates, and thus, the plates could potentially be assigned to this family. Thus, the ambulacral plates could belong either to the Aspidodiadematoida, or to the Diadematoida, particularly to the family Diadematidae.

4.1.2 Fragments with Tubercles

Figure 3 j–n.

Material. Puerta Curaco section: PC009b, 116 fragments, PC010, 2 fragments. Área 3000 section: AT006, 1 fragment.

Description. Interambulacral plates with a large surface of the plate devoid of ornamentation. Primary and secondary tubercles of various sizes all perforate and crenulate with areolae not well defined and no scrobicular tubercles observed. Mamelons of the tubercles with a well-developed neck and a central perforation up to 1/3 of the mamelon diameter, platform broad, complete, and with a radially symmetrical development of crenulation that has more than ten nodes on the parapet. Scarce miliary granules.

Remarks. The interambulacral plates have been compared with the same taxa mentioned above for the ambulacral plates, and they match with the plates observed in “diademataceans”. The Aspidodiadematoida have interambulacral plates with a single large tubercle perforate and crenulate, while the Diadematoida have perforate and crenulate primary tubercles, without surrounding scrobiculae. The Heterodiadematidae are characterized by interambulacral plates with single primary tubercle and wide adradial and interradial granular zones, and tubercles with a small mamelon in proportion to the boss. The Diadematidae present interambulacral plates with multiple primary tubercles, or if there is only a single tubercle, these plates lack a wide adradial and interradial granular zones. The observed material could perhaps belong to the Diadematidae because of the scarce and scattered granules near the tubercle.

Some features exhibited in the following families allow them to be discarded with respect to the "diademataceans" (Kroh *pers. comm.* 2017). Members of the Acrosaleniidae have the areolae sharply delimited as the Emiratiidae, one or two large primary tubercles to each plate, and a dense secondary tuberculation. The Hemicidaridae are discarded because of a dense presence of secondary tubercles, the perforation of the mamelon is vertically elongate, the areolae are sunken and confluent, and the crenulation is different with respect to those described herein. The Pseudodiadematidae are also discarded as they possess dense granulate plate surfaces (Smith and Kroh 2011).

4.1.3 Spines

Figure 3 o, r, s, t, u, v.

Material. Puerta Curaco section: PC001, 1 spine; PC006, 1 spine; PC008, 1 spine; PC009b, 117 spines; PC0010, 28 spines. Área 3000 section: AT006, 2 spines; AT009, 21 spines.

Description. Spine fragments of various sizes corresponding to primary and secondary tubercles. The bases of the spines are about as wide as the basal part of the shaft, which is straight, cylindrical, slender and with eight or more closely set, smooth longitudinal ribs, some of which dichotomize in cross section toward the base. The center of the spines seems to be hollow, but is filled with sediment. Spine articulation is by means of a restricted-pivot system.

Remarks. According to the above-mentioned characters especially the longitudinal ribs, these spines cannot be attributed to the "diademataceans" as these taxa have a verticillate surface, with the exception of the Micropygoida. Nevertheless, the spines described herein are probably hollow. The spines of the Aspidodiadematoida possess a central lumen crossed by partitions and fine needle-like pillars and those of the Diadematoida are fragile with a sparse trabecular meshwork filling the central lumen of the spines.

The spines characterizing the Acrosaleniidae are solid, with a smooth cortex. Spines of the Emiratiidae are unknown except for those of *Tetragramma* and *Tiaromma* (subfamily Diplopodiinae) and those of *Allomma* (subfamily Emiratiinae). In the Hemicidaridae, the only spines known belong to *Cidaropsis* (Pseudocinarinae) and *Pseudocidaris* (Hemicidarinae). In the Pseudodiadematidae, the only genera whose spines are known belong to *Pseudodiadema* and *Acrocidaris* (Smith and Kroh 2011). The spines described herein are, however, closely similar in their general morphology to the spines of *Schizechinus* (Toxopneustidae, Echinacea) (Kroh 2003).

The comparison of spines described herein with those of other families shows that the morphology of the former cannot be related to one specific taxon, because they possess characters that are present in several different regular echinoids.

4.1.4 Elements of the Aristotle's Lantern

The non-cidaroid regular echinoids are characterized by gripping lanterns with strong hinge-joints, whereas the cidaroids posses socket joints (Märkel 1979). The jaw elements studied herein are compatible with those of non-cidaroid lanterns. Of the taxa considered herein, the "diademataceans" posses aulodont lanterns, while the other families (Acrosaleniidae, Emiratiidae, Hemicidaridae, and Pseudodiadematidae) have stirodont lanterns. Aulodont and stirodont lanterns have a short pyramidal suture, a more or less deep foramen magnum, and narrow epiphyses that do not meet. Aulodont lanterns posses grooved teeth, while stirodont lanterns posses keeled teeth.

Hemi-pyramids

Figure 3 a–e.

Material. Puerta Curaco section: PC009b, 49 hemi-pyramids. Área 3000 section: AT006, 1 hemi-pyramid.

Dimensions. Interpyramidal length 0.074–0.156 cm, width 0.022–0.093 cm, abaxial face 0.021–0.073 cm.

Description. Hemi-pyramids isolated, not bounded together, rectangular on abaxial view, lateral wing (interpyramidal face) with a V-shaped apical edge and horizontal ridges forming a short comb-like fringe in the adaxial border. The oral tips of the hemi-pyramids are strongly curved axially. Intrapyramidal face with a short and narrow pyramidal suture. Processus superioris, supra alveolaris, foramen magnum, and epiphyses not observed.

Remarks. The presence of isolated hemi-pyramids, not bounded together, does not allow the observation of the foramen magnum. Nevertheless, it can be inferred by the short pyramidal suture, that a deep V-shaped foramen is present which is deeper than the lateral notches for epiphyses and seems to broaden upwards. The lack of epiphyses hinders the determination of either an aulodont, stirodont, or camarodont construction. The general morphology of the described hemi-pyramids is shared by several regular taxa. This, coupled with the lack of diagnostic characters does not allow a confident assignation to lower level taxa.

Rotulae

Figure 3 f, g.

Material. Puerta Curaco section: PC009b, 85 rotulae; PC0010, 1 rotula. Área 3000 section: AT006, 2 rotulae.

Dimensions. Length 0.029–0.09 cm, width 0.016–0.038 cm, thickness 0.007–0.028 cm.

Description. Rectangular in outline, with parallel sides, dorso–ventrally flattened. Apical surface of the rotula with large pores, oral surface with a symmetrical arrangement of the articular areas that interact with the epiphyses. The abaxial condyle, the adaxial condyle and groove, and the intermediate fossa can be distinguished.

Remarks. Regarding the arrangement of the joints in non-cidaroids, the middle and inner joints are separated from one another by an area where the interior rotular muscle is attached, the middle tubercle of the epiphyses with the middle fossa of the rotula construct a sort of buffer, with the inner tubercle including the inner fossa representing gliding areas (Märkel 1979). The rotulae are highly similar to those of *Genocidaris catenata* (Temnopleuridae, Echinacea) (Kroh 2003). Moreover, the rotulae described herein are also similar to those of *Paracentrotus lividus* (Parechinidae, Echinacea) (Candia Carnevali et al. 1991).

Teeth

Figure 3 h, i.

Material. Puerta Curaco section: PC009b, 32 teeth. Área 3000 section: AT006, 1 tooth; AT009, 2 teeth.

Dimensions. In cross section, major axis is 0.038–0.046 cm, and minor axis is 0.018–0.025 cm.

Description. Isolated fragments of teeth, slightly abaxial curved and wedge-shape in cross section. Aboral end showing secondary plates.

Remarks. The cross section of the teeth somewhat resembles those of *Eodiadema*, an Early Jurassic paraphyletic or polyphyletic taxon, classified as a sister group to the Irregularia or between the Diadematidae and Pedinidae taking into account the different types of phylogenetic trees considered (Smith 1981, Kroh and Smith 2010; Thuy et al. 2011).

The teeth in *Eodiadema* are rhombic in cross section, whereas those described herein are more wedge-shaped in cross section; this type of macrostructure occurs in clypeasterid, fibularid, and astriclypeid lanterns in the Clypeasteroidea (Mani 2004).

4.2 Taxonomic Assignation of the Echinoid Ossicles

The diagnostic features that characterized the families of regular and irregular echinoids, such as the arrangement of the plates in the apical disk, the elements of the perignatic girdle, and the position of the periproct, among others, can only be observed and described in well-preserved complete specimens. Even if the material described in this paper consists of fragments of the test, spines, and several jaw elements, the described ossicles can be at least classified as non-cidaroid, regular echinoids.

Considering that each element (ambulacral and interambulacral plates, spines, hemi-pyramids, rotulae, teeth) is only represented by a unique morphotype and that almost all the material is derived from the same bed, the assumption that all the recovered remnants ultimately belong to a single taxon, could possible be made. In fact, only the ambulacral and interambulacral plates can be related to one group of the "diademataceans", while the rest of the ossicles cannot be assigned with confidence to any single lower order taxonomic group.

Although in some cases, morphological features of a spine enable a designation to a specific taxon, in the described material, the spines cannot be assigned with confidence to any specific taxa. The main features which characterize them are found in primary and secondary spines of non-cidaroid regular echinoids from basal to more derived taxa. The spines described herein are similar to ?*Schizechinus* sp. (Toxopneustidae) illustrated by Kroh (2003), but this family is discarded as it has ambulacral plates following the echinoid style, and imperforate and non-crenulate tubercules. This, coupled with the strong intraspecific variability shown by the primary spines related to the position on the coronae does not allow a closer classification of the material.

The same is true for the hemi-pyramids studied herein, because their morphological features are observed in many regular echinoid taxa. The short pyramidal suture in the studied material allows the cidaroid type to be discarded. Besides, the lack of epiphyses and associated teeth prevents an accurate classification to either an aulodont, stirodont, or camarodont pyramid type.

The rotulae studied herein matches in every detail the rotulae of non-cidaroid echinoids that belongs to a hinge-joint gripping lantern. Our rotulae are similar to those found in ?*Genocidaris catenata* (Family Temnopleuridae) illustrated by Kroh (2003), but this family is discarded because it has imperforate primary tubercles, with or without crenulation, and the compound ambulacral plates in the echinoid style. The studied rotulae are also similar to those of *Paracentrotus lividus* (Family Parechinidae) described by Candia Carnevali et al. (1991), but this family is discarded because it has interambulacral plates with a dense number of subequal tubercles. More accurate assignations are difficult due to the poor fossil record of rotulae in general.

The cross section of the teeth is wedge-shaped and resembles the one found in the irregular Clypeasteroida, but this echinoid taxon appears in the Eocene; this would lead us to consider the possibility that this feature has appeared previously in some taxon during the Cretaceous.

4.3 Micropaleontological Assemblages

The micropaleontological content of the Puerta Curaco and Área 3000 sections show an extremely low richness of calcareous benthic foraminifers. The first section also provided only a single species of the ostracod *Sondagella colchesterensis* Valicenti and Stephens.

The Puerta Curaco section has a lower percentage of barren samples than the Área 3000 section (28% vs. 63%) and also has a greater abundance of specimens (537 specimens vs. 71 specimens, respectively). The productive samples vary from rare to frequent, with the exception of samples PC009b of the Puerta Curaco and AT006 of the Área 3000 with very abundant to abundant foraminiferal tests, respectively.

The assemblages recovered from both sections are quite different. While the lower levels of the Puerta Curaco record an abundant and nearly monotypic fauna composed by more than 99% of *Epistomina loncochensis* Ballent with accessory species belonging to the order Lagenida (Family Vaginulinidae), the lower productive levels of the Área 3000 provide some Spumellaria radiolarians followed by abundant vaginulinids and rare specimens of *Epistomina loncochensis*.

The echinoid remains were recovered in some levels with *Epistomina loncochensis* and/or lagenids, mainly in the samples PC009b, AT006, and AT009.

In the Área 3000 section, a high amount of Lagenida with abundant *Astacolus ambanjabensis* (Espitalié and Sigal) was recovered (sample AT006) representing another monotypic foraminiferal level.

All foraminiferal tests are poorly preserved. Specimens belonging to Lagenida and Robertinida are damaged and have infillings of sediment and framboidal pyrite, but, as was expected by its originally aragonitic composition, the genus *Epistomina* (Order

Robertinida) showed stronger signs of dissolution and recrystallization than those of the lagenids. Some samples contain blackened epistominid tests that, along with the dark claystone rich in organic matter and some traces of pyrite, suggests oxygen-deficient environments.

4.4 Biostratigraphy

Twenty-two levels of the Pilmatué Member of the Agrio Formation from the Puerta Curaco and Área 3000 sections were examined for calcareous nannofossils. Only 20 were fossiliferous. The recovered nannoflora was scarce and poorly preserved, being only frequent in specific levels.

The nannoflora is composed by 23 species of the genera: *Calculites*, *Cretarhabdus*, *Cruciellipsis*, *Cyclagelosphaera*, *Diazomatholithus*, *Eiffellithus*, *Micrantholithus*, *Nannoconus*, *Retecapsa*, *Rhagodiscus*, *Watznaueria*, and *Zeugrhabdotus*. In both sections, the reliable calcareous nannofossils included the presence of *Eiffellithus windii*, *Eiffellithus striatus*, and also *Nannoconus circularis* in Puerta Curaco.

Up to now, the first and last occurrences (FO and LO) of genera *Eiffellithus* and *Nannoconus* were consistent global markers recorded in numerous sections in the Neuquén Basin and have been correlated with ammonite zones (Aguirre-Urreta et al. 2005, 2017; Concheyro et al. 2009; Lescano and Concheyro 2014).

At Puerta Curaco, the presence of *Nannoconus circularis*, *Eiffellithus windii* together with *Eiffellithus striatus* (Table 3) was recognized. All of them could be associated in the CC4A Zone, considering the Mediterranean zonation of the Tethys (Sissingh 1977, modified by Applegate and Bergen 1988) and other studies with nannofossils performed in the Neuquén Basin (Aguirre-Urreta et al. 2017). As nannoflora was identified together with fragments of the ammonite *Decliveites crassicostatus*, an Upper Valanginian age is suggested for this section. At Área 3000, the presence of *Eiffellithus windii* together with *Eiffellithus striatus* (Table 4) was recognized as also recorded at the Puerta Curaco section. For this reason, the nannoflora is included in the CC4A Zone.

The recovered calcareous nannofossils exhibited clear Tethyan affinities, including some cosmopolitan species, and others with a bipolar distribution. Therefore, the nannoflora identified herein would suggest the existence of marine connections with the Tethyan Realm and even part of northwestern Europe during the late Valanginian.

5 Paleoenvironmental Inferences

The echinoid skeletal elements found are generally isolated ossicles and occur scattered in marlstones, mudstones, and claystones. They are always accompanied by the foraminifer *Epistomina loncochensis*, sometimes by lagenids, calcareous nannofossils and sclerites of holothuroidea, and only in one sample by the ostracod *Sondagella colchesterensis* Valicenti and Stephens. Articulated echinoid skeletal elements have not been observed in the field, not even in the sample PC009b, which is the richest fossiliferous horizon.

Among the echinoid remains, the fragments of spines are the most numerous skeletal ossicles, followed by rotulae, hemi-pyramids, teeth, and fragments of test plates. Leaving aside the spines which are very abundant in any echinoid and which may be overestimated since they can be derived from fragmented spines, it is remarkable that the rotulae/spines ratio is very high (0.9) and also the rotulae/hemi-pyramids ratio is greater than 0.5 (1.88). Thus, the echinoid elements recovered could be strongly controlled by the resistance or robustness of the type of skeletal element. Also, the almost total absence of articulated skeletal parts (except for rare fragments formed by few ambulacral or interambulacral plates) would suggest a pre-depositional fragmentation of the coronae. Usually, the signs of fragmentation, dissolution, and abrasion are common among the skeletal elements, and with exception of dissolution, they are interpreted to have resulted from the high energy and transport mechanisms of the shallow water environment in which they occurred.

Regular echinoids which belong to the vagile epifauna usually live on rocky and sand substrates and often hidden among rocks, but also are found in muddy substrates (Mancosu and Nebelsick 2017). Regular echinoid tests are quite resistant, but because of being exposed to scavengers and currents after death, they have a poor fossil record compared to the irregulars that live within the sediment (see discussions in Kier 1977, 1982; Kidwell and Baumiller 1990; Mancosu and Nebelsick 2017). Paleoecological studies on echinoids are often based on complete coronae of echinoids, but also spines can provide clues with respect to the substrate type on which echinoids lived.

The monotypic foraminiferal assemblage of *Epistomina loncochensis* from this basal interval of the Agrio Formation as reported by Sagasti and Ballent (2002) from claystones, marlstones, and mudstones reflects an outer neritic ramp environment. They interpreted *Epistomina loncochensis* as an opportunistic species with the ability to cope with ephemeral dysaerobic to quasi-anaerobic conditions near the sediment–water interface attributed to the transgressive event. Toward higher stratigraphic levels, the echinoid ossicles and the epistominid foraminifera disappear, but the presence of scarce vaginulids are still recorded. According to Koutsoukos and Hart (1990), the lenticular, biconvex, planispirally coiled, smooth, or little-ornamented test and compressed shape of the coiled vaginulinids (*Astacolus*, *Lenticulina*, *Planularia*, *Saracenaria*) indicate they were epifaunal to shallow infaunal active deposit feeders and probably were rapid burrowers adapted to variable bottom conditions.

The fact that almost all echinoid elements were obtained from soft sediments containing *Epistomina loncochensis* (which indicates low oxygen conditions) may reflect their allochthonous nature, because echinoids are unknown in modern environments with depleted oxygen.

6 Concluding Remarks

The present work pointed out the presence of non-cidaroid regular echinoids belonging to the "diademataceans," recovered from Upper Valanginian sediments of the Pilmatué Member, Agrio Formation. They have been found in fine sediments associated with quiet marine environments. The almost total absence of articulated skeletal parts, the ratios between the types of elements found and the signs of fragmentation, dissolution, and abrasion are interpreted as the result of a high energy and transport. Besides, the associated low oxygen tolerant foraminifera fauna would reflect their allochthonous nature.

Findings of the opportunistic benthic foraminifera *Epistomina loncochensis* in the sections could be a response to variations in food and/or light supply, oxygen, salinity, etc. The lagenids could indicate some improvement of environmental conditions at the water/substrate interface. Anyway, the scarcity and the monospecific assemblage of lagenids in our sections would also indicate the persistence of the low oxygen content.

Finally, the consistent presence of nannoflora in both sections allowed to establish an Upper Valanginian age for this noteworthy deposit of regular echinoderm ossicles.

7 Supplementary Information About Nannoflora

Calcareous nannofossils species listed in alphabetical order
Calculites percenis Jeremiah
Cretarhabdus conicus Bramlette and Martini
Cruciellipsis cuvillieri (Manivit) Thierstein
Cyclagelosphaera margerelii Noël
Diazomatholithus lehmannii Nöel
Eiffellithus striatus (Black) Applegate and Bergen
Eiffellithus windii Applegate and Bergen
Micrantholithus hoschulzii (Reinhardt) Thierstein
Micrantholithus obtusus Stradner
Nannoconus cf. *N. bucheri* Brönnimann
Nannoconus circularis Deres and Achéritéguy
Nannoconus cross section
Nannoconus spp.
Nannoconus steinmannii (Kamptner) Deres and Achéritéguy
Retecapsa angustiforata Black
Retacapsa surirella (Deflandre and Fert) Grün in Grün and Allemann
Rhagodiscus asper (Stradner) Reinhardt
Rhagodiscus sp.
Watznaueria barnesiae (Black) Perch-Nielsen
Watznaueria biporta Bukry
Watznaueria fossacincta (Black) Bown
Watznaueria ovata Bukry
Zeugrhabdotus embergeri (Noël) Perch-Nielsen

Acknowledgements. Authors thank G. Cusminsky and E. Bernasconi for inviting us to present the results of our contribution to the Micropaleontological Symposium at the RACAPA 2016. We are very grateful to A. Kroh for his advice on the taxonomy of echinoids and especially for his valuable comments and suggestions; to J. Nebelsick and an anonymous reviewer whose constructive suggestions improved the paper; to Beatriz Aguirre-Urreta for ammonite determination; to M. Barbé and P. Álvarez for their great and patient help with laboratory preparation of samples. We also thank financial support to Agencia Nacional de Promoción Científica y

Tecnológica (ANPCyT) PICT 2013–1413, Universidad Nacional de La Plata, and YPF S.A. This is the contribution R-236 of the Instituto de Estudios Andinos 'Don Pablo Groeber' (IDEAN-UBA-CONICET).

References

Aguirre-Urreta MB, Rawson P (1997) The ammonite sequence in the Agrio formation (Lower Cretaceous). Neuquén Basin, Argentina. Geol Mag 134:449–458

Aguirre-Urreta MB, Rawson PF (2012) Lower Cretaceous ammonites from the Neuquén basin, Argentina: a new heteromorph fauna from the uppermost Agrio formation. Cret Res 35:208–216

Aguirre-Urreta MB, Rawson PF, Concheyro A et al (2005) Lower Cretaceous biostratigraphy of the Neuquén basin. In: Veiga G, Spalletti L, Howell JA, Schwarz E (eds) The Neuquén basin: a case study in sequence stratigraphy and basin dynamics. Geological Society of London, Special Publication, vol 252, pp 57–81

Aguirre-Urreta MB, Morgues FA et al (2007) The Lower Cretaceous Chañarcillo and Neuquén Andean basins: ammonoid biostratigraphy and correlations. Geol J 42:143–173

Aguirre-Urreta MB, Schmitz M et al (2017) A high precision U-Pb radioisotopic age for the Agrio formation, Neuquén basin, Argentina: implications for the chronology of the Hauterivian stage. Cret Res 75:193–204

Applegate J, Bergen J (1988) Cretaceous calcareous nannofossils biostratigraphy of sediments recovered from the Galicia Margin, ODP Leg 103. In: Boillot AC, Winterer EL et al (eds) Initial Reports of the Ocean Drilling Project Program vol 103, pp 293–348

Ballent S, Concheyro A et al (2006) Bioestratigrafía y paleoambiente de la Formación Agrio (Cretácico Inferior), en la provincia de Mendoza, cuenca Neuquina Argentina. Rev Geol Chil 33:47–79

Bown PR, Concheyro A (2004) Lower Cretaceous calcareous nannoplankton from the Neuquén basin, Argentina. Mar Micropaleontol 52:51–84

Bown PR, Young JR (1998) Techniques. In: Bown PR (ed) Calcareous Nannofossils Biostratigraphy. Kluwer Academic Press, Dordrecht, pp 16–28

Candia Carnevali MD, Bonasoro F et al (1991) Microstructure and mechanical design in the lanterno ossicles of the regular sea-urchin *Paracentrotus lividus*. A scanning electron microscope study. Bol Zool 58:1–42

Concheyro A, Lescano M et al (2009) Micropaleontología de la Formación Agrio (Cretácico inferior) en distintos sectores de la cuenca Neuquina, Argentina. Rev Asoc Geol Arg 65:342–361

Edwards A (1963) A preparation technique for calcareous nannoplankton. Micropaleontology 9:103–104

Hanna GD, Church CC (1928) Freezing and thawing to disintegrate shales. J Paleontol 2:131

Kidwell SM, Baumiller T (1990) Experimental disintegration of regular echinoids: roles of temperature, oxygen and decay thresholds. Paleobiology 16:247–271

Kier PM (1977) The poor fossil record of the regular echinoid. Paleobiology 3:168–174

Kier PM (1982) Rapid evolution in echinoids. Palaeontology 25:1–9

Koutsoukos E, Hart MB (1990) Cretaceous foraminiferal morphogroup distribution patterns, palaeocommunities and trophic structures: a case study from the Sergipe basin, Brazil. Earth Environ Sci Trans R Soc Edinb 81:221–246

Kroh A (2003) The Echinodermata of the Langhian (Lower Badenian) of the Molasse zone and the northern Vienna basin (Austria). Ann Natur Mus Wien, Serie A:155–183

Kroh A, Nebelsick JH (2010) Echinoderms and Oligo-Miocene carbonate systems: potential applications in sedimentology and environmental reconstruction. Int Assoc Sedimentol Spec Publ 42:201–228

Kroh A, Smith AB (2010) The phylogeny and classification of post-Palaeozoic echinoids. J Syst Paleontol 8:147–212

Leanza HA, Hugo C (2001) Hoja Geológica Zapala, Hoja 3969–I, 1:250.000. Ins Geol Rec Min Bol 275:1–128

Legarreta L, Gulisano CA (1989) Análisis estratigráfico secuencial de la cuenca Neuquina (Triásico Superior-Terciario Inferior). In: Chebli GA, Spalletti LA (eds) Cuencas sedimentarias argentinas, Ser Corr Geol 6:221–243

Lescano M, Concheyro A (2014) Nanofósiles calcáreos de la Formación Agrio (Cretácico Inferior) en el sector sudoccidental de la Cuenca Neuquina, Argentina. Amegh 46:73–94

Mancosu A, Nebelsick JH (2017) Ecomorphological and taphonomic gradients in clypeasteroid-dominated echinoid assemblages along a mixed siliciclastic-carbonate shelf from the early Miocene of northern Sardinia, Italy. Acta Palaeontol Pol 62:627–646

Mani MS (2004) Progress in invertebrate zoology. Orient Blackswan, pp 1–388

Märkel K (1979) Structure and growth of the cidaroid socket-joint lantern of Aristotle compared to the Hinge-joint lanterns of non-cidaroid regular echinoids (Echinodermata, Echinoidea). Zoomorphologie 94:1–32

Nebelsick JH, Dynowski JF et al (2015) Echinoderms: hierarchically organized light weight skeletons. In: Hamm C (ed) Evolution of lightweight structures: analyses and technical applications, biologically-inspired systems, vol 6. Springer, Germany, pp 141–155

Rodríguez DL (2007) Equinoideos mesozoicos de las cuencas andinas del centro-oeste de Argentina. Unpublished Ph.D. thesis, Facultad de Ciencias Exactas y Naturales, Universidad de Buenos Aires, 1–263 pp

Roth P (1983) Jurassic and Lower Cretaceous calcareous nannofossils in the western North Atlantic (Site 534): biostratigraphy, preservation, and some observations on biogeography and paleoceanography. Init Rep Deep Sea Drilling Proj 76:587–621

Sagasti G, Ballent S (2002) Caracterización microfaunística de una transgresión marina: Formación Agrio (Cretácico inferior), Cuenca Neuquina, Argentina. Geobios 35:721–734

Schwarz E, Spalletti LA et al (2016) First U-Pb SHRIMP age for the Pilmatué Member (Agrio Formation) of the Neuquén Basin, Argentina: implications for the Hauterivian lower boundary. Cret Res 58:223–233

Sissingh W (1977) Biostratigraphy of cretaceous calcareous nannoplankton. Geol Mij 56:37–65

Smith AB (1978) A functional classification of the coronal pores of regular echinoids. Palaeontology 21:759–789

Smith AB (1980) Stereom microstructure of the echinoid test. Special Papers in Palaeontol 25:1–83

Smith AB (1981) Implications of lantern morphology for the phylogeny of post-Palaeozoic echinoids. Palaeontology 24:779–801

Smith AB, Kroh A (2011) The echinoid directory. World wide web electronic publication. http://www.nhm.ac.uk/research-curation/projects/echinoid-directory

Spalletti LA, Veiga GD et al (2011) La Formación Agrio (Cretácico Temprano) en la Cuenca Neuquina. In: Leanza HA, Arregui C, Carbone O, Danielli JC, Vallés JM (eds) Geología y Recursos Naturales del Neuquén, Congreso Geológico Argentino, Relatorio, vol 18, pp 145–160

Thuy B, Gale A et al (2011) A new echinoderm Lagerstätte from the Pliensbachian (Early Jurassic) of the French Ardennes. Swiss J Palaeontol 130:173–185

Lower Cretaceous Benthic Foraminifers and Ostracods of the Middle Magdalena Basin, Colombia: Biostratigraphic and Paleoenvironmental Inferences

Germán D. Patarroyo[1(✉)], Daine Ceolin[2], Gerson Fauth[2], Gustavo A. Torres[1], Claudia P. Cárdenas[3], and Daniel A. Rincón[4]

[1] Universidad Industrial de Santander, Bucaramanga, Colombia
{paleo.patarroyo, gustshao}@gmail.com
[2] ITT Fossil-Instituto Tecnológico de Micropaleontologia, Universidade do Vale do Rio dos Sinos, UNISINOS, São Leopoldo, Brazil
daianeceolin@yahoo.com.br, gersonfauth@gmail.com
[3] Departamento de Geología, Universidad EAFIT, Medellín, Colombia
cpcardenag@eafit.edu.co
[4] Instituto Colombiano del Petróleo (ICP)—ECOPETROL S.A. Piedecuesta, Santander, Colombia
daniel.rincon@ecopetrol.com.co

Abstract. In recent years, there has been an increasing interest in understanding the biostratigraphic framework and depositional environments of the Lower Cretaceous Rosablanca Formation (Middle Magdalena Basin). In order to contribute to this, a micropaleontological study was conducted in two wells that drilled strata of the Rosablanca Formation. Despite the moderate recovery of microfossils, the lower part of the Rosablanca Formation contains several ostracod genera of the families Cytherideidae, Cytheridae, Brachycytheridae, and Cyprididae. This study provides the first evidence of Early Cretaceous ostracods in Colombia. In contrast, the upper part of the Rosablanca Formation contains the benthic foraminifera genera *Lenticulina* and *Choffatella*, along with fragments of the Rotaliina and Textulariina suborders. A transitional to normal marine environment is suggested based on the ecologic preferences of the genera *Fabanella*, *Schuleridea*, and *Perissocytheridea*. The Valanginian to Barremian age for the Rosablanca Formation was corroborated with benthic foraminifera, in concordance with previous studies based on macrofossil assemblages.

Keywords: Benthic foraminifers · Ostracods · Lower Cretaceous
Middle Magdalena Basin · Rosablanca formation · Biostratigraphy
Paleoenvironments

Electronic supplementary material The online version of this chapter (https://doi.org/10.1007/978-3-030-02119-1_2) contains supplementary material, which is available to authorized users.

G. C. Cusminsky et al. (eds.), *Advances in South American Micropaleontology*,
Springer Earth System Sciences, https://doi.org/10.1007/978-3-030-02119-1_2

1 Introduction

Micropaleontological studies play an important role in defining the stratigraphic framework of Cretaceous sedimentary successions. Particularly in Colombia, detailed studies with foraminifers have been conducted on sections and wells of several basins, mostly dealing with Upper Cretaceous units. The Upper Cretaceous of areas such as the Middle Magdalena and the Eastern Cordillera basins has also been studied in specific sections (e.g. Petters 1954; Vergara et al. 1997). In several of the above-mentioned studies, the ostracods have been mentioned as being present but were not analyzed in detail (e.g. Martinez 1989; Vergara 1997).

Despite the broad distribution of Lower Cretaceous successions in the Eastern Cordillera and the Middle Magdalena Basin, micropaleontological studies are scarce. Some factors that have caused this bias are the presence of unsuitable lithologies for fossil preservation (e.g. sandstones, sandy limestones) and the moderate to heavy diagenesis or thermal maturity of the fossiliferous units. For instance, the few foraminiferal and ostracod studies that exist dealt with data from few specific stratigraphic levels that lack detailed taxonomic descriptions (e.g. Petters 1954; Gaona-Narvaez et al. 2013).

Despite the scarcity of micropaleontological data in the Lower Cretaceous successions, there is still a great potential to successfully find microfossils in these strata. In recent years, preliminary surveys have detected stratigraphic levels with a high proportion of benthic foraminifera and ostracods in units such as the Rosablanca and Paja formations (Patarroyo-Camargo et al. 2009; Montenegro 2012). The goal of this contribution is to verify the microfossil potential of the Lower Cretaceous Rosablanca Formation, which has proved to be an additional reservoir in the Middle Magdalena Basin (e.g. Sarmiento 2011). With that purpose, a stratigraphic core that comprised the Rosablanca Formation (well ECO-A) was analyzed for its benthic foraminifera and ostracod content. In addition, we studied cutting samples from a proximal site (well ECO-B). Based on the results, biostratigraphic and paleoecological inferences were proposed for the Rosablanca Formation.

2 Geological Setting

The Middle Magdalena Basin (MMB) is an elongated intermountain basin, surrounded by the Central Cordillera and the Serranía de San Lucas to the West and the Eastern Cordillera to the East (Fig. 1). Their geologic boundaries are the Girardot fold belt, the Bituima-La Salina, the Santa Marta-Bucaramanga, the Espiritu Santo fault systems and the onlap of Neogene rocks over the Serranía de San Lucas and the Central Cordillera (Sarmiento 2011). The lowermost Cretaceous strata of the MMB were deposited in a rift system, probably related to back-arc extension, shifting from Late Jurassic continental to intertidal–shallow marine sedimentation in the Berriasian–Barremian (Ward et al. 1972; Sarmiento 2011). The exposure of Lower Cretaceous rocks is mostly limited to the eastern foothills and the northernmost part of the basin, while Neogene rocks and Quaternary strata crop out in the central part. The analyzed material belongs to two exploratory wells located in the central part of the MMB (Fig. 1).

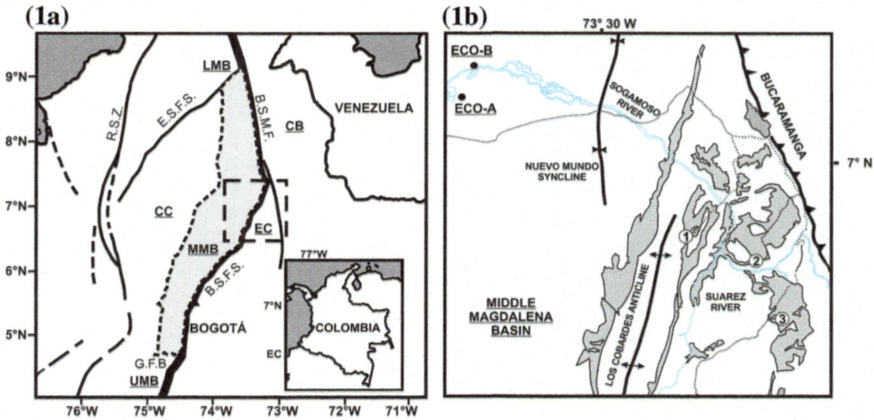

Fig. 1. Study area. 1a Middle Magdalena Basin (MMB; gray area). LMB: Lower Magdalena Basin, EC: Eastern Cordillera Basin, UMB: Upper Magdalena Basin, CB: Catatumbo Basin, CC: Central Cordillera, GFB: Girardot Fold Belt, BSFS: Bituima and La Salina fault systems, BSMF: Bucaramanga-Santa Marta fault system, ESFS: Espíritu Santo fault system, RSZ: Romeral suture zone. The study area is surrounded in the square with dash lines. 1b Location of the two studied wells and distribution of the lowermost Cretaceous units in the area (Los Santos, Cumbre and Rosablanca formations). Lowermost Cretaceous fossiliferous localities: (1) Zapatoca; (2) Los Santos, (3) Curití

The Rosablanca Formation has been stratigraphically defined in the MMB as a succession mostly composed of grainstones, packstones, and evaporites that grade upward into shaly limestones and mudstones (e.g. Morales et al. 1958; Ward et al. 1972; Cardozo and Ramírez 1985). The underlying Los Santos Formation consists of sandstones and sandy siltstones. The overlying Paja Formation consists of black, thinly laminated, locally calcareous shales and biomicritic levels (Fig. 2). Both contacts with the Rosablanca Formation have been proposed as being conformable. The thickness of the Rosablanca Formation in subsurface normally ranges between 90 and 150 m in the central part of the basin, with a progressive width increase toward the Eastern Cordillera (Morales et al. 1958; Ward et al. 1972).

For the Rosablanca Formation, a variety of marine invertebrates and vertebrates have been reported (e.g. Guzmán 1985; Cadena 2011; Gómez-Cruz et al. 2015). Brief studies on microfossils (benthic foraminifers and ostracods) were carried out by Petters (1954) and Montenegro (2012). In general, the invertebrate fossils and the sedimentological features suggest a shallow marine environment with moderate to poorly oxygenated bottom conditions (Guzmán 1985; Gaona-Narvaez et al. 2013). Despite the abundant fossil content, there is not a definite consensus about the age of the Rosablanca Formation in the MMB (Fig. 2). Based on the presence of ammonites, most

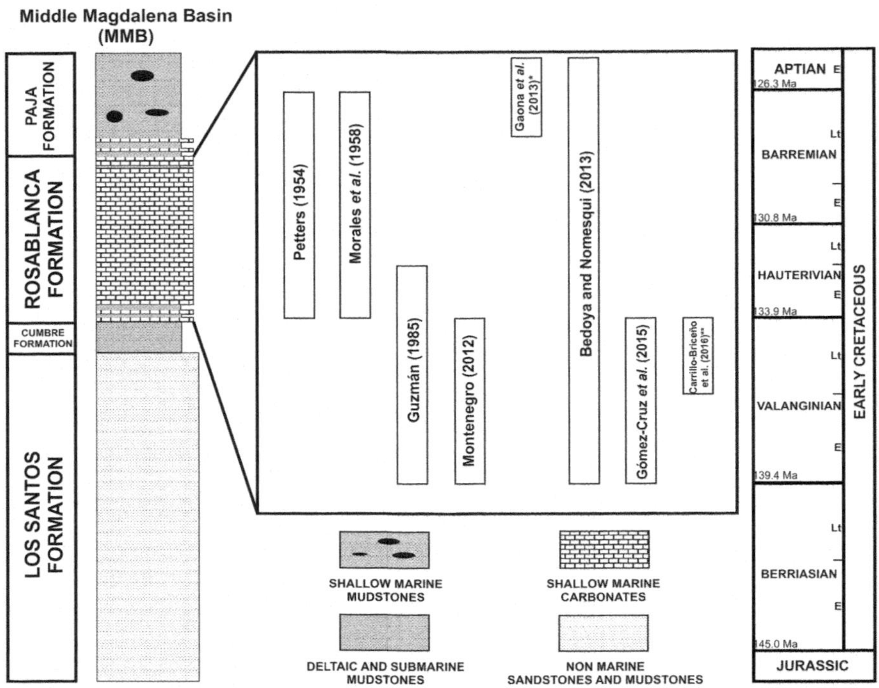

Fig. 2. Generalized stratigraphic column of the Lower Cretaceous succession in the Middle Magdalena Basin. Stratigraphic widths according to Morales et al. (1958) and Sarmiento (2011). Biostratigraphic ranges proposed to the Rosablanca Formation by several authors. *Proposed age for the Rosablanca–Paja stratigraphic contact. **Age assignment for the lower part of Rosablanca Formation

studies agree with a Valanginian–Hauterivian age (e.g. Petters 1954; Guzmán 1985; Carrillo-Briceño et al. 2016; Montenegro 2012) and diachronism along the basin (Ward et al. 1972). According to chemostratigraphic studies, a middle–late Barremian to early Aptian age is proposed for the Rosablanca–Paja limit (Curití area; Gaona-Narvaez et al. 2013) and a Valanginian to early Aptian age for several stratigraphic sections on Los Santos and Zapatoca areas (Figs. 1 and 2; Bedoya and Nomesqui 2013).

3 Materials and Methods

A total of 21 core samples from the well ECO-A and seven cutting samples from well ECO-B were analyzed. Located in the middle part of the MMB, the well ECO-A (7° 11 N, 73° 52 W) scored all the Rosablanca Formation, presenting a thickness of 175 m (3427.5–3593 m). Well ECO-B, located at few kilometers away from ECO-A, had a recovery of 240 m (7° 16 N, 73° 51 W; interval 3669.8–3986.8 m). Table 1 summarizes which samples were analyzed for their benthic foraminifera and ostracod content.

Table 1. Analyzed samples of wells ECO-A and ECO-B

ECO-A						ECO-B	
Sample depth	Type material	Sample depth	Type material	Sample depth	Type material	Sample depth	Type material
[a]3457	Core	3504	Core	3545	Core	3670	Cuttings
3467	Core	3510	Core	3548	Core	[a]3900	Cuttings
3476	Core	3516	Core	3552	Core	3950	Cuttings
[a]3480	Core	3520	Core	3557	Core	[a]3970	Cuttings
3486	Core	3527	Core	[a]3571	Core	[a]3980	Cuttings
3491	Core	3530	Core	[a]3575	Core	[a]3985	Cuttings
3497	Core	3540	Core	[a]3582	Core	3990	Cuttings

[a]Samples where both the foraminiferal and ostracod content were analyzed

The core and cutting samples, comprising limestones, marlstones, and dolomites, were processed at the *Instituto Colombiano del Petróleo* (ICP). They were prepared according to the standard method to extract calcareous microfossils. Approximately, 20–50 g of dried rock were soaked in a 5% H_2O_2 solution for about 6 hours, washed over 63 and 150 μm sieves, and dried at 60 °C. Benthic foraminifers and ostracods were handpicked in the residual material. The total microfossil count is shown in the supplementary material (Appendix 1). Selected specimens were imaged using an EVO MA15 Zeiss scanning electron microscope (SEM) and a SteREO Discovery. V20 at the *Instituto Tecnológico de Micropaleontologia* (*ITT Fossil*) of the *Universidade do Vale do Rio dos Sinos*, Brazil. Since SEM images of some of the foraminifers provide more details than optical images, they were included in the supplementary material (Appendix 2).

4 Results and Discussion

4.1 Systematic Paleontology

The taxonomic identification of the benthic foraminiferal assemblages was based on Bartenstein et al. (1957), Bartenstein and Bolli (1977), Loeblich and Tappan (1987), Bolli et al. (1994), and Neagu and Cîrnaru (2004). The ostracod identification followed Scott (1961) with some modifications alongside other specialized literature. Specifically, for the genus *Reconcavona*, the generic taxonomy of Smith (2000) was adopted. Both the benthic foraminifera and ostracods are stored at the "Regional Collection" of the ICP (Biostratigraphy Team, ICP-ECOPETROL S.A.) in Piedecuesta, Colombia, under their respective catalog numbers (*Archivadores* 10/11, *Bandejas* 46/01).

4.1.1 Benthic Foraminifers

Class Foraminifera D'Orbigny 1826
Subclass Globothalamana Pawlowski et al. 2013
Order Loftusiida Kaminski 2004

Superfamily Loftusioidea Brady 1884
Family Cyclamminidae Marie 1941
Subfamily Choffatellinae Maync 1958
Genus *Choffatella* Schlumberger 1904
Choffatella cruciensis (Pictet and Renevier 1858)
Figure 3, 8a–c
1858 *Operculina cruciensis* Pictet and Renevier: p. 167, pl. 23, figs. 4-c.
1904 *Choffatella decipiens* Schlumberger: p. 763, pl. 18, figs. 1–6.
1977 *Choffatella decipiens* Schlumberger; Bartenstein and Bolli: p. 549, pl. 1, fig. 32.
2004 *Choffatella cruciensis* Pictet and Renevier; Neagu and Cîrnaru: p. 273, pl. 1, figs. 1–20, p. 275, pl. 3, figs. 1–8.
2012 *Choffatella decipiens* Schlumberger; Montenegro: p. 37, pl. 7, figs. 26–27, p. 39, pl. 8, figs. 28–29, p. 43, pl. 10, fig. 36.
Material: 12 specimens from well ECO-A (3457 m).
Remarks: In this study, we followed the proposal of Neagu and Cîrnaru (2004) regarding the synonymy between *Choffatella cruciensis* (Pictet and Renevier 1858) and *Choffatella decipiens* (Schlumberger 1904). Therefore, all previous reports of *Choffatella decipiens* for the Rosablanca Formation and coetaneous units in northern South America will be referred to as *Choffatella cruciensis*.
Stratigraphical Range: Lower Barremian to lower Aptian in Trinidad and Eastern Venezuela (Maync 1950, 1955, 1956; Guillaume et al. 1972; Bartenstein and Bolli 1977; Bolli et al. 1994). Moreover, a Valanginian? to Barremian age is proposed for several *Choffatella* levels in Colombia and Venezuela (Petters 1954; Maync 1956). In the MMB, a Valanginian age is proposed for the analyzed samples in the Zapatoca area (Montenegro 2012). A wider biostratigraphic range, from Upper Jurassic to Cenomanian, is suggested for the genus *Choffatella* in Loeblich and Tappan (1987).
Choffatella? *sogamosae* (Petters 1954)
Figure 3, 1a–c, 9a–b
1858 *Robulina sogamozae* Karsten: p. 114, pl. 6, figs. 5a–b.
1858 *Planulina zapatocensis* Karsten: p. 114, pl. 6, figs. 4a–d
1954 *Choffatella sogamosae* Petters: pl. 24 fig. 1–3, p. 133, figs. 5a–j, p. 134, figs. 6a–d.
2012 *Choffatella sogamosae* Petters; Montenegro: p. 39, pl. 8, figs. 30–31, p. 43, pl. 10, fig. 35.
Material: Two specimens from well ECO-A (3457 and 3480 m).
Remarks: Specimens of *Choffatella sogamosae* were originally described by Karsten (1858) as two different taxa: *Planulina zapatocensis* and *Robulina sogamozae*, both coming from samples of the Rosablanca Formation (Sogamoso Canyon: Zapatoca area). Later, Dietrich (1935) concluded, using samples from the same area, that the specimens described by Karsten belonged to a single genus, assigned as *Haplophragmoides*. The formal designation of this species to *Choffatella* was made by Petters (1954). In its review of some uncertain species of *Choffatella*, Maync (1955) proposed the reassignment of *Choffatella sogamosae* to the genus *Buccicrenata*, based on the absence of a clear labyrinthic wall structure and

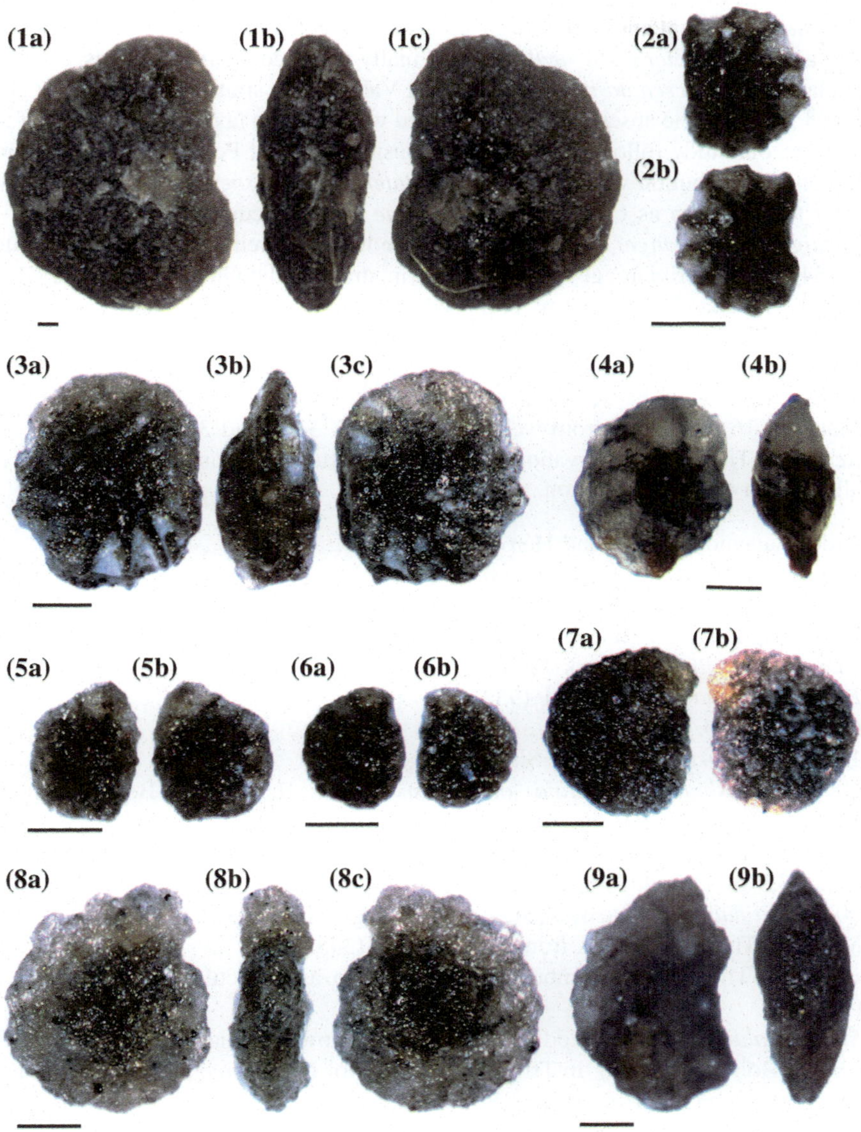

Fig. 3. Benthic foraminifers of the Rosablanca Formation. All scale bars represent 100 μm. 1a–c, 9a–b *Choffatella*? *sogamosae* (ECO-A, samples 3457 and 3480 m); 2a–b, 3a–c *Lenticulina ouachensis* (ECO-A, sample 3457 m); 4a–b *Lenticulina praegultina* (ECO-A, sample 3476 m); 5a–b, 6a–b, 7a–b *Choffatella* sp. (ECO-A, sample 3457 m); 8a–c *Choffatella cruciensis* (ECO-A, sample 3457 m)

alveolus; however, this author could not attribute the analyzed specimens to *Buccicrenata*. A recent study on this genus discussed at least two species for the Lower Cretaceous: *Buccicrenata hedbergi*, originally reported in Aptian beds of Venezuela, and *Buccicrenata italica* reported in Valanginian strata of Italy (BouDagher-Fadel 2001). In the absence of more detailed works on the taxonomic validity of this species, our study followed the original considerations of Petters (1954), since it is the only contribution with images of *Choffatella sogamosae*. Nevertheless, we will refer to this taxon as *Choffatella*? *sogamosae* in the discussion section.
Stratigraphical Range: Upper Hauterivian to lower Barremian according to Petters (1954). A Hauterivian age was suggested in strata of the Guajira Peninsula (Bürgl 1960). A Valanginian age was proposed for a section in the Zapatoca area (Montenegro 2012).
Choffatella sp.
Figure 3, 5a–b; 6a–b; 7a–b
Material: 16 specimens from well ECO-A (3457, 3467 and 3476 m).
Remarks: The poor preservation of both the apertural and umbilical areas did not allow for a specific definition.

Order Lagenida Delage and Hérouard 1896
Superfamily Nodosarioidea Ehrenberg 1838
Family Vaginulinidae Reuss 1860
Subfamily Lenticulininae Chapman et al. 1934
Genus *Lenticulina* Lamarck 1804
Lenticulina ouachensis (Sigal 1952)
Figure 3, 2a–b, 3a–c
1952 *Cristellaria ouachensis* Sigal: S. 16, fig. 10.
1957 *Lenticulina* (*Lenticulina*) *ouachensis* (Sigal); Bartenstein, Bettenstaedt and Bolli: p. 25, pl. 3, figs. 50a–b, pl. 4, figs. 71, 76.
1986 *Lenticulina* (*Lenticulina*) *ouachensis* (Sigal); Bartenstein and Bolli: p. 569, pl. 1, fig. 37; p. 571, pl. 2, figs. 1–2.
2012 *Lenticulina ouachensis* (Sigal); Montenegro: p. 37, pl. 7, figs. 24–25.
Material: Three specimens from well ECO-A (3457 m).
Remarks: The obscured umbilicus is referred to as a circular rim in Bolli et al. (1994).
Stratigraphical Range: Middle Valanginian to lower Aptian is the current consensus (Bolli et al. 1994). In Trinidad, the range of *Lenticulina ouachensis* has been proposed as ranging from middle Barremian to lower Aptian (Bartenstein and Bolli 1986). A Valanginian age was suggested for a section of the Rosablanca Formation in the Zapatoca area (Montenegro 2012).
Lenticulina praegaultina Bartenstein et al. 1957
Figures 3, 4a–b
1957 *Lenticulina* (*Lenticulina*) *praegaultina* Bartenstein, Bettenstaedt and Bolli: p. 24, pl. 3, fig. 48a–b, pl. 4, fig. 63–65.
Material: One specimen from well ECO-A (3476 m).

Remarks: Despite the partial fragmentation of this single specimen, the presence of a slightly elongate outline which suggests a delicate peripheral keel is a distinctive feature among other Lower Cretaceous lenticulinids such as *Lenticulina ouachensis*.
Stratigraphical Range: Barremian (*Lenticulina ouachensis ouachensis* Zone) for the Cuche Formation in Trinidad (Bartenstein et al. 1957). A Barremian to middle Albian age was proposed for analyzed strata in Eastern Venezuela (Serranía del Interior; Guillaume et al. 1972).

4.1.2 Ostracods

Class Ostracoda Latreille 1802
Order Podocopida Sars 1866
Suborder Podocopina Sars 1866
Superfamily Cytheracea Baird 1850
Family Cytherideidae Sars 1925
Subfamily Perissocytherideinae Bold 1963
Genus *Perissocytheridea* Stephenson 1938
Perissocytheridea sp.
Figure 4, a–b
Material: Two carapaces (one male and one female) from well ECO-A (3575 m).
Dimensions: L: 0.622 mm; H: 0.369 mm (female); L: 0.672 mm; H: 0.354 mm (male).
Remarks: This species has the posterior region slightly different from many *Perissocytheridea* species. Despite the poor preservation, it was possible to identify the presence of small tubercles, and the punctation is indicative of more developed ornamentation, as seen in many *Perissocytheridea* species. For these reasons, we choose this generic identification. This species resembles in shape *Perissocytheridea carrenoae* Nicolaidis and Coimbra 2008 and *Perissocytheridea krömmelbeini* Pinto and Ornellas 1970 differing in the posterior region and its ornamentation.
Perissocytheridea? sp.
Figure 4, c
Material: One carapace from well ECO-A (3575 m).
Dimensions: L: 0.405 mm; H: 0.268 mm.
Remarks: This species is left in open nomenclature due to the impossibility to identify tubercles and ornamentation. It differs from *Perissocytheridea* sp. in size and its bluntly pointed posterior region.
Genus *Schuleridea* Swartz and Swain 1946
Schuleridea sp.
Figure 4, d
Material: Eight carapaces from wells ECO-B (3980 and 3985 m) and ECO-A (3571; 3480 and 3457 m).
Dimensions: L: 0.583 mm; H: 0.383 mm (ECO-A, 3480 m).
Remarks: This species differs from *Schuleridea martinensis* Andreu 1983 by being shorter and having a more pronounced valve overlap. Differs from *Schuleridea*

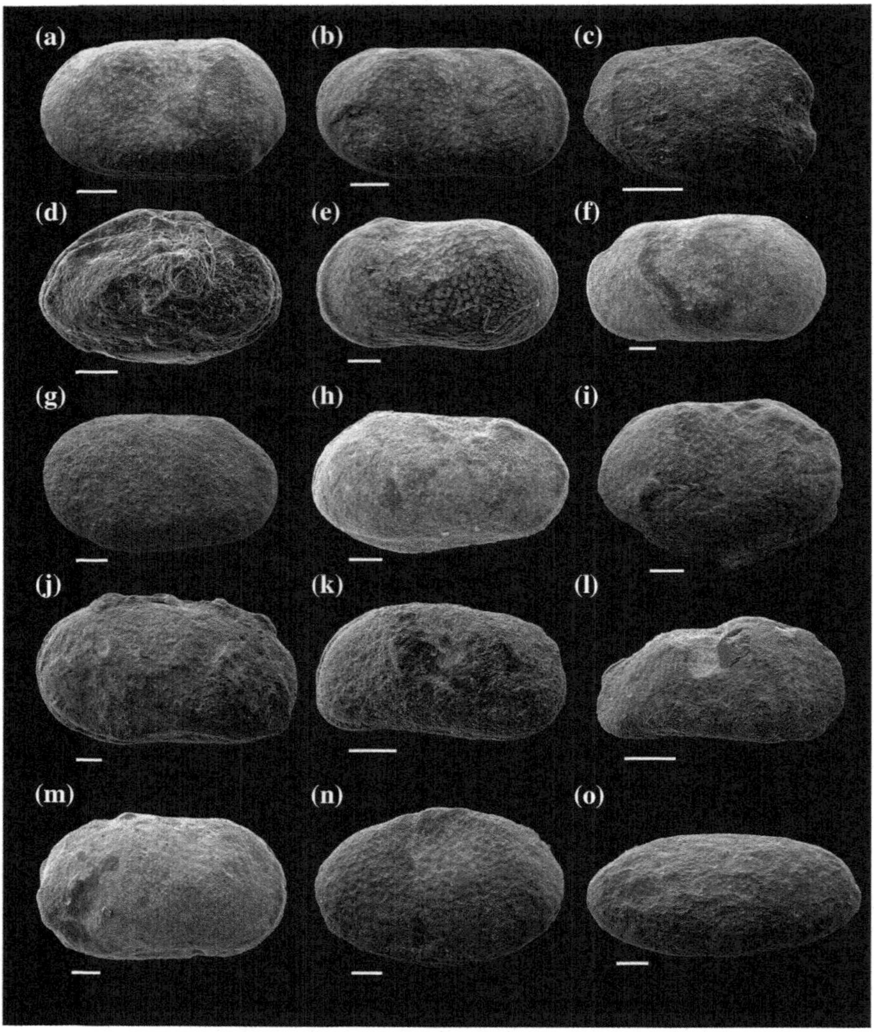

Fig. 4. Ostracods of the Rosablanca Formation. All scale bars represent 100 μm. a–b *Perissocytheridea* sp. A: Female, right lateral view; b male, right lateral view (ECO-A, sample 3.575 m); c *Perissocytheridea*? sp., right lateral view (ECO-A, sample 3571 m); d *Schuleridea* sp., right lateral view (ECO-A, sample 3480 m); e *Fabanella* sp. cf. *Fabanella boloniensis* (Jones), left lateral view (ECO-A, sample 3575 m); f *Fabanella* sp. 1, left lateral view (ECO-A, sample 3575 m); g *Fabanella* sp. 2, right lateral view (ECO-A, sample 3575 m); h *Fabanella* sp. 3, right lateral view (ECO-B, sample 3985 m); i *Fabanella* sp. 4, right lateral view (ECO-B, sample 3985 m); j *Fabanella*? sp., right lateral view (ECO-B, sample 3900 m); k *Reconcavona*? sp., left lateral view (ECO-B, sample 3985 m); l Gen. et sp. Indet., right lateral view (ECO-B, sample 3980 m); m Gen. et sp. indet. 1, right lateral view (ECO-A, sample 3571 m); n Gen. et sp. indet. 2, left lateral view (ECO-B, sample 3980 m); o Gen. et sp. indet. 3, right lateral view (ECO-A, sample 3571 m)

bikfayaensis Bischoff 1990 figured in Colin et al. (2008) (plate 1, figs. 11–14) due to the less developed convexity, its right valve, and its straighter anterior and posterior margins.

Family Cytheridae Baird 1850
Subfamily Cytherideinae Sars 1925
Genus *Fabanella* Martin 1940
Fabanella sp. cf. *Fabanella boloniensis* (Jones 1880)
Figure 4, e
1880 *Cythere boloniensis* Jones, S. 616, Fig. B
1976 *Fabanella boloniensis* (Jones 1880) Brenner, figs. 10, 27
Material: Three carapaces from well ECO-A (3575 m).
Dimensions: L: 0.763 mm; H: 0.428 mm.
Remarks: Differs from *Fabanella boloniensis* (Jones) figured in Brenner 1976 (plate 10, fig. 27) in its more rounded posterior region and a more developed ornamentation. Differs from *Fabanella tuberculata* Brenner 1976 (plate 11, fig. 4) in the absence of dorsomedian tubercles and ornamentations.
Fabanella sp. 1
Figure 4, f
Material: Three carapaces from well ECO-A (3575 m).
Dimensions: L: 0.927 mm; H: 0.510 mm.
Remarks: Differs from *Fabanella leguminoidea* Swain and Brown 1972 (plate 7, fig. 10) in its less concave ventral and dorsal margins and its higher posterior margin.
Fabanella sp. 2
Figure 4, g
Material: One carapace from well ECO-A (3575 m).
Dimensions: L: 0.771 mm; H: 0.475 mm.
Remarks: Differs from *Fabanella* sp. Swain and Brown (1972) (plate 7, fig. 8) in its less obliquely rounded anterior margin without a marginal rim, suboval concavity, and ornamentation.
Fabanella sp. 3
Figure 4, h
Material: One carapace from well ECO-B (3985 m).
Dimensions: L: 0.800 mm; H: 0.452 mm.
Remarks: Differs from *Fabanella boloniensis* (Jones) figured in Brenner (1976) (plate 10, figs. 26–29), *Fabanella* sp. 1 and *Fabanella* sp. 2 in its straight dorsal margin, the narrower and symmetric anterior margin, and its smooth surface.
Fabanella sp. 4
Figure 4, i
Material: One carapace from well ECO-B (3985 m).
Dimensions: L: 0.741 mm; H: 0.449 mm.
Remarks: Differs from *Fabanella boloniensis* (Jones), *Fabanella* sp. 1, *Fabanella* sp. 2, and *Fabanella* sp. 3 in its bluntly pointed posterior margin, its dorsal concavity, and its more symmetric anterior margin.
Fabanella? sp.

Figure 4, j

Material: Six carapaces from well ECO-B (3900 and 3985 m).

Dimensions: L: 1.075 mm; H: 0.654 mm (3900 m).

Remarks: This species is left in open nomenclature due to the poor preservation of the specimens that did not allow for a proper analysis of its eye tubercle and its dorsal and anterior margins. Among other characteristics, it differs from other *Fabanella* species in having a more robust carapace.

Superfamily Cypridacea Baird 1845

Family Cyprididae Baird 1845

Subfamily Candoniane Kaufmann 1900

Genus *Reconcavona* Krömmelbein 1962

Reconcavona? sp.

Figure 4, k

Material: Three carapaces from well ECO-B (3980 and 3985 m).

Dimensions: L: 0.544 mm; H: 0.299 mm (3985 m).

Remarks: This species is left in open nomenclature due to the impossibility of properly observing its anterodorsal node. In addition, its valve overlap is also opposite to that of the generic diagnosis, though according to Sames (2011) this is not a generic feature.

Suborder Cytherocopina Baird 1850

Superfamily Cytheroidea Baird 1850

Family Brachycytheridae Puri 1954

Gen. *et* sp. indet.

Figure 4, l

Material: One carapace from well ECO-B (3980 m).

Dimensions: L: 0.496 mm; H: 0.263 mm.

Remarks: This species shares some external features such as the outline and the inflated ventral region that allow us to classify it in the family Brachycytheridae (e.g. *Sapucariella*). According to Puckett (2002) and Puckett et al. (2016), the evolution of this family began in the African realm with species at the end of the Early Cretaceous (late Albian) having already spread across the margins of Gondwana from Madagascar, India, and North Africa to Venezuela. This specimen could indicate the presence of this family in Colombia earlier than previously recorded.

Indeterminate taxa

Gen. *et* sp. indet. 1

Figure 4, m

Material: 13 carapaces from well ECO-A (3571; 3575 and 3582 m)

Dimensions: L: 0.898 mm; H: 0.547 mm (3571 m).

Remarks: These specimens resemble the genus *Cytherella* Jones, but its valve overlap is the inverse of the generic description.

Gen. *et* sp. indet. 2

Figure 4, n

Material: Two carapaces from well ECO-B (3970 and 3980 m).

Dimensions: L: 0.851 mm; H: 0.564 mm (3980 m).
Remarks: The specimens are smashed making it difficult to discern the real shape of their carapaces.
Gen. *et* sp. indet. 3
Figure 4, o
Material: One carapace from well ECO-A (3571 m).
Dimensions: L: 0.893 mm; H: 0.418 mm.
Remarks: The recovered specimen is covered with sediments, making it impossible to assess to more details.

4.2 Faunal Features

In general, from the 28 analyzed samples in both wells, most of them presented a low to moderate recovery of benthic foraminifera and ostracods (<50 specimens/sample). In ten samples where foraminifers were recovered in both wells, three genera were identified but left in open nomenclature, comprising calcareous (lenticulinids, buliminids) and agglutinated tests (choffatellids). In the well ECO-A, *Choffatella* was the most common genus in the identified assemblages with two species, while the two species of *Lenticulina* observed were present in a smaller amount. The benthic foraminifera in well ECO-B were scarce, being comprised mostly of poorly preserved agglutinated and calcareous forms.

From the nine samples studied for ostracods in both wells, it was possible to identify ten species, with nine in open nomenclature and one species left as having affinity with *Fabanella boloniensis*. All of them belong to the families Cytherideidae, Cytheridae, Brachycytheridae, and Cyprididae, while three species could not be determined. The most representative genus was *Fabanella* with six species, followed by *Perissocytheridea* with two species, and *Schuleridea* and *Reconcavona* with one. The specific abundance and the stratigraphic distribution of the fauna are illustrated in Figs. 5 and 6.

4.2.1 Faunal Composition of Well ECO-A

In well ECO-A, (3457–3582 m), from 21 samples, 12 were barren for foraminifers and ostracods. The benthic foraminiferal abundance was low (<30 specimens/sample), with moderate to poor preservation, hampering the specific taxonomic description in most of the specimens. Evidence of recrystallization or fragmentation of the delicate ornamentation was observed in most of the calcareous tests (e.g. the peripheral keel in the single specimen of *Lenticulina praegaultina*). No sign of dissolution or mechanic abrasion was observed in the collected specimens. Nevertheless, two assemblages were identified: an agglutinated one, composed by the taxa *Choffatella cruciensis* (Pictet and Renevier 1858) and *Choffatella*? *sogamosae* (Petters 1954), and a calcareous one, represented by *Lenticulina ouachensis* (Sigal 1952) and *Lenticulina praegaultina* (Bartenstein et al. 1957). The stratigraphic distribution of the foraminifers was mainly restricted to the upper portion of the well. This interval comprises mostly marlstones and limestones (Fig. 5).

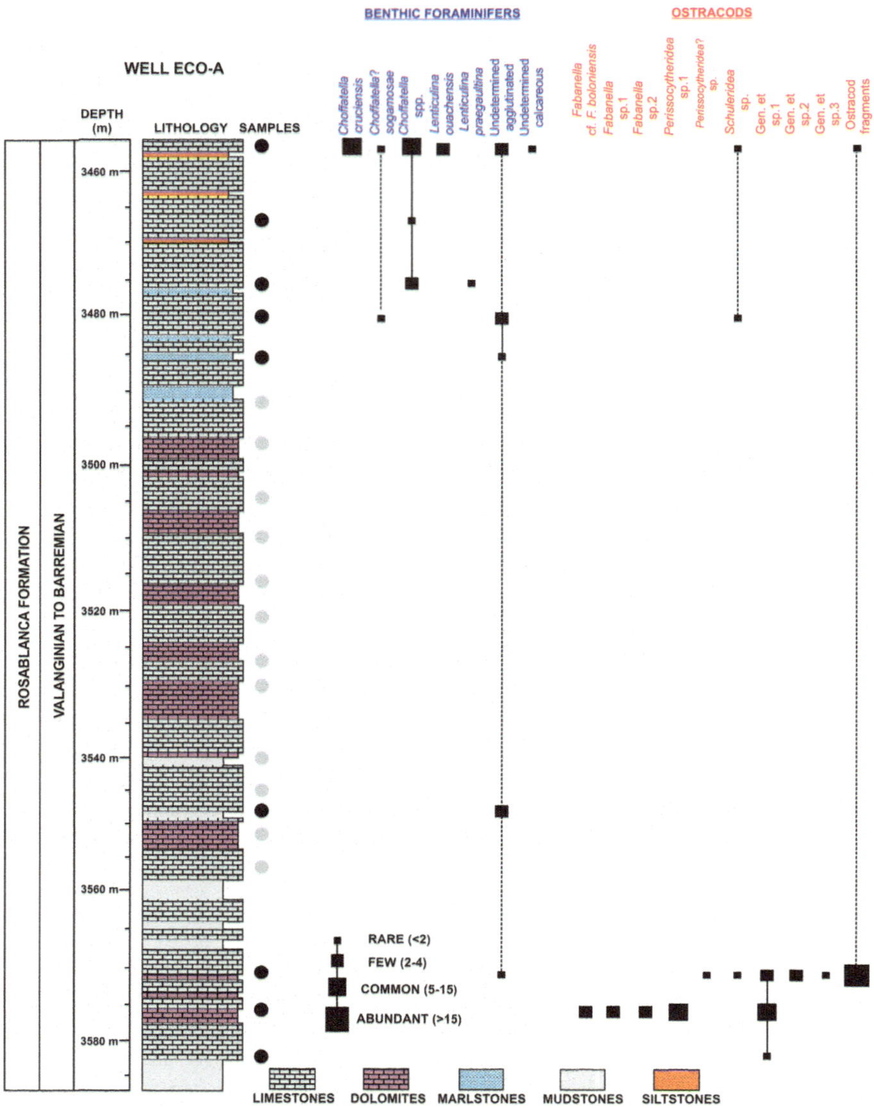

Fig. 5. Stratigraphic distribution (total counts) of the analyzed microfauna from well ECO-A. In blue: benthic foraminifers, in pink: ostracods. The sterile samples are represented by gray dots

The ostracod assemblages were identified in two intervals (3457–3476 and 3571–3582 m), with the deeper one having presented a higher abundance (>10 specimens/sample). Despite their poor preservation and the presence of only carapaces, the genera *Fabanella*, *Perissocytheridea,* and *Schuleridea* were identified. *Fabanella* was the most abundant and diversified genus, with three species, such as *Fabanella*

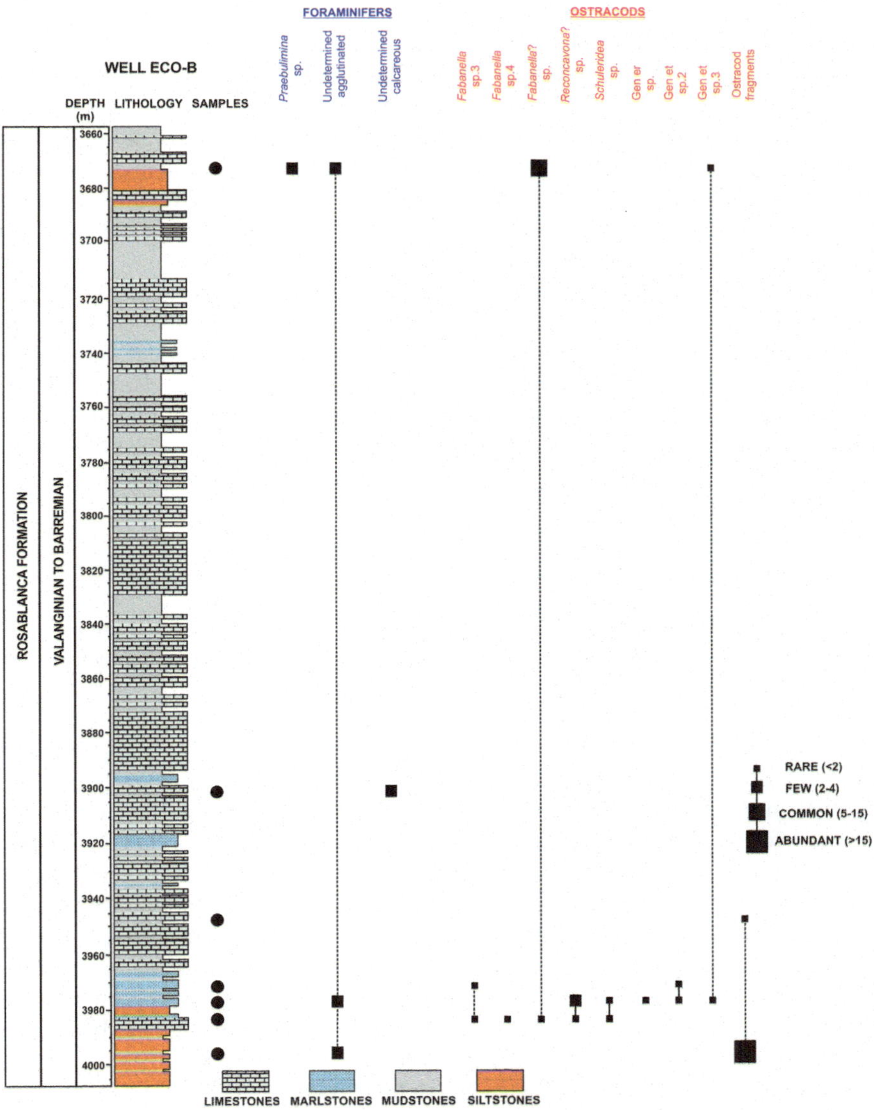

Fig. 6. Stratigraphic distribution (total counts) of the analyzed microfauna from well ECO-B. In blue: benthic foraminifers, in pink: ostracods. The sterile sample is represented by a gray dot

sp. cf. *F. boloniensis* (Jones 1880), having been recovered at the depth of 3575 m. In addition, sexual dimorphism in *Perissocytheridea* and juvenile specimens of the genus *Schuleridea* were identified. The lithology of the samples in which ostracods were recovered corresponded to limestones (Fig. 5).

4.2.2 Faunal Composition of Well ECO-B

The analysis of this well involved seven samples taken from the interval 3670–3990 m, where only sample 3950 m was barren. The benthic foraminifera assemblage identified was characterized by a low abundance (<5 specimens/sample) and poor preservation contrasting with the ones from well ECO-A, which were more abundant and diversified. A heavily recrystallized specimen of *Praebulimina* (not described in the taxonomy section) and calcareous and agglutinated tests belonging to the Rotaliida and Loftusiida, respectively, were identified. The samples with foraminifers correspond to marlstones, mudstones, and siltstones (Fig. 6).

All the four samples analyzed for ostracods contained carapaces. The genera *Fabanella, Reconcavona,* and *Schuleridea* (juvenile forms) were identified. As in well ECO-A, *Fabanella* was the most common genus in the examined samples, allowing for the identification of at least three species that were left in open nomenclature, followed by *Schuleridea* and *Reconcavona* with one species each. In general, most of the collected carapaces show evidence of recrystallization or fragmentation. The lithologies of the samples where ostracods were recovered from correspond to siltstones and limestones (Fig. 6).

4.3 Biostratigraphic Inferences

The presence of benthic foraminifera and ostracods in wells ECO-A and ECO-B provides new insights not only on the biostratigraphy of the Lower Cretaceous successions of Colombia, but also of northern South America.

4.3.1 Benthic Foraminifers

In northern South America, the age of the described foraminiferal levels has been defined by macrofossils. For instance, Maync (1956) proposed a Hauterivian–Barremian age for the *Choffatella cruciensis* levels in Trinidad, based on previous reports of ammonites (genera *Pulchellia* and *Karstenia* among others). In a later study in Eastern Venezuela, Guillaume et al. (1972) integrated both foraminiferal (benthic and planktonic) and ammonite data, proposing biozones with a lower Barremian to Cenomanian range (Fig. 7). According to their proposal, the *Choffatella cruciensis* levels would range between lower Barremian and lower Aptian. Upper Aptian ammonite genera such as *Cheloniceras*, *Colombiceras* were found in underlying levels alongside *Choffatella cruciensis.* On the other hand, a revision of the detailed micropaleontological studies conducted by Helmut Bartenstein in Trinidad illustrates that *Choffatella cruciensis* has only ever been mentioned in Barremian strata (Bartenstein and Bolli 1977).

The presence of the *Choffatella* levels in Colombia has been only discussed by Petters (1954), using data from restricted outcrops and core material around the MMB and the UMB (Fig. 7). To this author, the presence of different ammonite assemblages suggests distinct ages for the *Choffatella* levels. For instance, in the Villeta region (~400 km south of the studied area), the presence of the ammonite *Astieria* suggested a Hauterivian age to this author. In contrast, a late Hauterivian to lowermost Barremian (?) age was proposed in the northern localities of the MMB, based on the presence of the genera *Pseudohaploceras, Nicklesia,* and *Karsteniceras.* In strata belonging to the

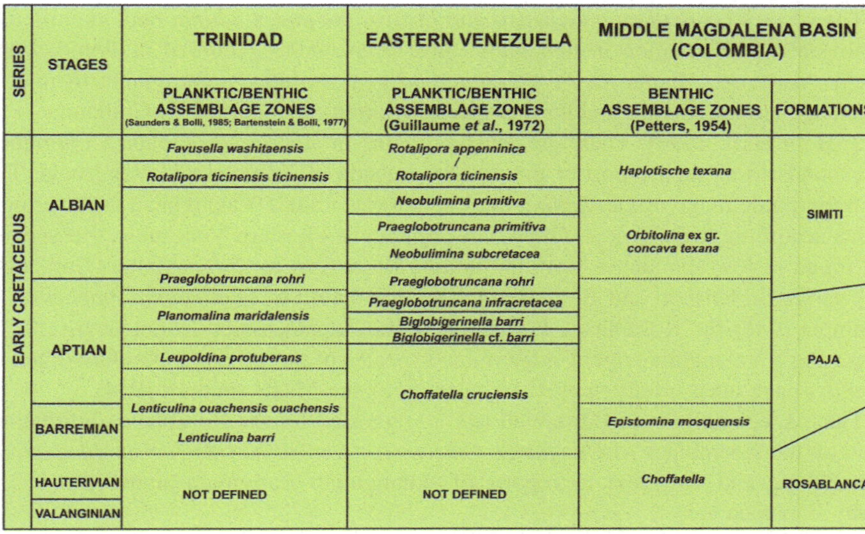

Fig. 7. Correlation of the proposed foraminiferal biozones from northern South America (Modified from Bolli et al. 1994). The lithostratigraphy of the Middle Magdalena Basin was based on Ward et al. (1972) and Sarmiento (2011). The age of the *Choffatella* levels in Colombia was extended to Valanginian based on the results of this study

Rosablanca Formation in Los Santos area, the occurrence of the Barremian ammonite *Pulchellia galeata* (wrongly assigned as *P. galatea*) in the overlying Paja Formation indicated a late Hauterivian age for the *Choffatella* levels (Petters 1954). Two aspects difficult a deeper discussion about these classic localities: (1) the lack of detailed stratigraphic columns where *Choffatella* and other foraminiferal levels were found and (2) the absence of detailed taxonomy descriptions of the associated macrofauna.

A recent study provided new insights into the biostratigraphic distribution of the *Choffatella* levels of Colombia. In a stratigraphic section of the Rosablanca Formation (Zapatoca), the ammonite genus *Karakaschiceras* was found in association with *Choffatella*, suggesting at least a Valanginian age for this section (Pedro Patarroyo, personal communication). Despite the lack of detailed micropaleontological studies, the Zapatoca area comprises the most promising sites for finding microfossils in the Rosablanca Formation. For instance, several authors observed foraminifers and ostracods in their analyzed sections, (e.g. Guzmán 1985; Bedoya and Nomesqui 2013; Gómez-Cruz et al. 2015).

Conversely, many species of the genus *Lenticulina* have been used as biostratigraphical markers in the lowermost Cretaceous strata of the Tethyan and Boreal realms (e.g. Bartenstein 1985; Bolli et al. 1994). For instance, the *Lenticulina barri* and *Lenticulina ouachensis ouachensis* biozones were proposed to define the Barremian and lower Aptian in Trinidad (Fig. 7; Bartenstein and Bolli 1977; Bartenstein 1985).

The species *Lenticulina ouachensis* and *Lenticulina praegaultina*, both identified in the Rosablanca Formation in this study, have been mainly recorded in Trinidad and Eastern Venezuela. In the case of *Lenticulina praegaultina*, a Barremian to middle Aptian age has been suggested for this species (Bartenstein et al. 1957; Guillaume et al. 1972; Bolli et al. 1994). Therefore, their presence in strata of Rosablanca Formation corresponds to the first report of this species outside of Trinidad and Venezuela. The biostratigraphic range of *Lenticulina ouachensis* was middle Valanginian to lowermost Aptian according to Bolli et al. (1994). In Trinidad and Eastern Venezuela, this species was found in low abundance from the middle Barremian to lower Aptian (Guillaume et al. 1972; Bartenstein and Bolli 1986). The first record of *Lenticulina ouachensis* in Colombia is for the Rosablanca Formation in the Zapatoca area (Montenegro 2012), suggesting a Valanginian age. Undetermined species of the genus *Lenticulina* (cited as *Robulus*) have been briefly spotted in outcrops of the MMB (Petters 1954).

Thus, a Valanginian to Barremian age is suggested for the Rosablanca Formation, based on the assemblage of *Choffatella cruciensis–Lenticulina ouachensis–Lenticulina praegaultina* and the previous reports of Valanginian–Barremian ammonites in the nearby Zapatoca area.

4.3.2 Ostracods

Compared to the scarce foraminiferal data of the Mesozoic of northern South America, reports of ostracod levels are almost nonexistent. There are brief mentions of *Pontocyprella* sp. aff. *P. gracilis* and possibly new species of *Bairdia*, *Cytherella,* and *Cytherura*? in the late Aptian of Trinidad (Bartenslein et al. 1966). An unpublished Ph. D. thesis studied additional surface and subsurface samples from Trinidad and Eastern Venezuela, describing 77 ostracod species where 37 might correspond to new forms (Stacy 1966; Bolli et al. 1994).

In general, studies about fossil ostracods in Colombia are rare, dealing with assemblages from the late Miocene of the Colombian Caribbean and the early to middle Miocene from the Amazon region (e.g. Bold 1966; Muñoz-Torres et al. 2006). For the MMB, the genus *Darwinula* has been briefly reported in continental Jurassic strata (Bruckner 1954; Ward et al. 1972). Recently, in samples of the Rosablanca Formation (Zapatoca area), the ostracod genera *Paracypris* and *Cytherella* were observed (Montenegro 2012).

The identified ostracod fauna for the Rosablanca Formation was composed by the genera *Fabanella*, *Perissocytheridea*, *Schuleridea,* and *Reconcavona*, all of them with a relatively wide stratigraphic range. For example, *Fabanella* is a well-documented genus from the Jurassic to the Upper Cretaceous of England, North Atlantic Coast, and Africa (e.g. Brenner 1976; Andreu 2002; Tibert et al. 2003). The genus *Schuleridea* also has a broad stratigraphical range, being found from Upper Jurassic to Miocene strata in the Atlantic coastal region (Swain and Brown 1972). In many other localities of Europe, this genus was recorded from the Jurassic to the Upper Cretaceous (e.g. Brenner 1976; Horne 1995). In South America, unpublished data from Venezuela identifies this genus in the Aptian (Babinot and Colin 1988).

Reports of the genus *Perissocytheridea* show their presence in the Late Cretaceous of Africa, South America, Middle East, and southern Europe (Andreu 1996; Viviers et al. 2011; Piovesan et al. 2016). The genus *Reconcavona* is frequently recorded in

Jurassic–Lower Cretaceous strata of northeastern Brazil (e.g. Krömmelbein 1962; Smith 2000; Coimbra et al. 2002; Guzmán-González et al. 2016). The presence of this non-marine genus in the Rosablanca Formation will be discussed in the next section.

4.4 Paleoenvironmental Inferences

The identified foraminiferal assemblages might suggest a very shallow marine environment for the analyzed samples of the Rosablanca Formation. For instance, the presence of *Choffatella* species has been described in Trinidad as being diagnostic of restricted environments, possibly lagoon settings (Koutsoukos and Merrick 1986). Paleoenvironmental inferences, based on macrofossil assemblages, sedimentological and petrographical studies, support a shallow marine to marginal environment with evaporitic episodes for the Rosablanca Formation (Ward et al. 1972; Cardozo and Ramírez 1985; Guzmán 1985; Gaona-Narvaez et al. 2013; Bedoya and Nomesqui 2013). Moreover, the abundance of *Lenticulina* for the Barremian of Trinidad and Eastern Venezuela has been associated with low oxygen bottom conditions (e.g. Bartenstein et al. 1957; Guillaume et al. 1972; Koutsoukos and Merrick 1986). Even though there is a low proportion of *Lenticulina* in our samples, its presence might be in agreement with previous reports of low oxygen conditions for some parts of Rosablanca Formation (e.g. Guzmán 1985; Gaona-Narvaez et al. 2013; Bedoya and Nomesqui 2013).

Ostracod assemblages had a better recovery at the base of the studied sections, especially in well ECO-A (Fig. 5). *Fabanella* was the most representative genus, and it was documented in marginal marine and brackish water environments of several Jurassic–Lower Cretaceous localities in Europe, Africa, and North America (e.g. Andreu 2002; Tibert et al. 2003). In well ECO-A, *Fabanella* cf. *Fabanella boloniensis* and *Fabanella* sp. 1 were associated with *Perissocytheridea* sp. 1, which is characteristic of brackish water environments (e.g. Babinot 1988; Nicolaidis and Coimbra 2008). In the Late Cretaceous, this genus has been reported being tolerant of different gradients of salinity, having been found mainly associated with both freshwater and marine species (Piovesan et al. 2016).

In well ECO-B, the ostracod fauna displayed less abundance and diversity than in well ECO-A (Fig. 6). Nonetheless, *Fabanella* was also the most abundant genus, occurring alongside *Schuleridea* and *Reconcavona.* According to Neale (1973), the presence of *Schuleridea* with other salinity-tolerant forms and mesohaline to low polyhaline genera such as *Fabanella* suggests an environment that is not truly euhaline in the well ECO-B. Conversely, the presence of the non-marine genus *Reconcavona* with the referred marine taxa might suggest a transitional environment in which the transport of carapaces was possible. According to previous authors, the depositional environment of the Rosablanca Formation must have been influenced by tides and shallow currents (e.g. Cardozo and Ramírez 1985). Therefore, it is possible that this genus was transported from a non-marine setting as a product of hydrodynamic processes (e.g. tidal, currents, sea-level variations).

The low oxygenation episode suggested by benthic foraminifers and others fossil groups in the Rosablanca Formation is corroborated by the presence of *Schuleridea* in the same level of *Lenticulina* at the top of well ECO-A (Sample 3457 m). The presence of this genus in a low diversified fauna suggested low oxygenation in sediments of England (Slipper 2009).

A final aspect of our study is the stratigraphic distribution of the microfossils along the Rosablanca Formation, with ostracods being common and more diverse in its lower sections, in contrast to the presence of the lenticulinids and choffatellids in its upper part (Figs. 5 and 6). Since only a few samples from well ECO-B were analyzed, we did not propose a correlation between the two studied wells. This aspect has been observed in less detail for other wells in the MMB that drilled the Rosablanca Formation (Duque-Caro†, personal communication). In general terms, this transition in the microfossil assemblages reflected how the restricted environments with shallow marine water at the base of the Rosablanca Formation deepened, revealing as in previous studies the initial stages of the Cretaceous marine incursion in the MMB and the Eastern Cordillera (e.g. Sarmiento 2011).

5 Conclusions

The micropaleontological study of two wells (ECO-A and ECO-B) that drilled strata of the Lower Cretaceous Rosablanca Formation resulted in the identification of a calcareous and agglutinated benthic foraminiferal assemblage. It is composed of four species belonging to the genera *Choffatella* and *Lenticulina* (with the first description in Colombia of *Lenticulina praegaultina*) and the first mention of ten ostracod species belonging to the genera *Fabanella*, *Perissocytheridea*, *Schuleridea,* and *Reconcavona* for the Early Cretaceous of Colombia.

In general, the benthic foraminiferal assemblage indicates a Valanginian to Barremian age for the Rosablanca Formation, in concordance with previous studies based on macrofossil assemblages, mainly ammonites. The ostracod and foraminiferal assemblages suggest a transitional to marginal marine setting, probably related to the early stages of the Cretaceous marine ingression in northern South America. The low oxygenation might be corroborated by the benthic foraminifera *Lenticulina* and the ostracod *Schuleridea* recovered at the top of well ECO-A.

Further studies with microfossils in other sections of the Middle Magdalena Basin will help to improve the paleoenvironmental and biostratigraphic understanding of this region.

Acknowledgements. This contribution wishes to pay tribute to Colombian micropaleontologist Hermann Duque-Caro (1927–2015), whose contagious curiosity and initial discussions about the Cretaceous foraminifers in Colombia were determinant for the first author. This contribution was possible thanks to the support of ICP-ECOPETROL S.A. (Biostratigraphy Team). The authors thank Michele Goulart da Silva for the SEM work and Guilherme Krahl for the optical images in the Stereo Microscope V20. Daiane Ceolin thanks CAPES from the grant (proc. 88,887.111026/2015-00). Two referees contributed with valuable comments on an earlier version of this manuscript. This manuscript was made in collaboration with the CNPq project (309918/2015-4).

References

Andreu B (1983) New species of ostracodes from the South-Pyrenean Albian and Cenomanian (Sierra D'Aulet, Spain). Bull Centres Rech Explor Prod Elf-Aquitaine 7:1–43

Andreu B. (1996) New species of ostracodes in the upper Turonian (?)-Coniacian (?)-Santonian from the Boulmane area, Middle Atlas, Morocco. Systematics, Biostratigraphy and Paleoecology, Paleobiogeography of the assemblages. In: Jardiné S (ed) 12e colloque de micropaléontologie africaine—2e Colloque de stratigraphie et paléogéographie de l'Atlantique Sud, Angers, Juille 1994. Bull Centres Rech Explor Prod Elf-Aquitaine Pau. Memoire, vol 16, pp 483–509

Andreu B (2002) Cretaceous ostracod biochronology of Morocco. Ecl Geol Helv 95:133–152

Babinot JF (1988) Première donnée sur les ostracodes du Cénomanien de Yougoslavie (Istrie du Sud). Geobios 21:5–15

Babinot JF, Colin JP (1988) Paleobiogeography of tethyan cretaceous marine ostracods. In: Hanai T, Ikeya N et al (eds) Evolutionary biology of ostracoda, its fundamentals and applications, 1st edn. Elsevier, Amsterdam, pp 823–839

Baird W (1845) Arrangement of British entomostraca. Zoologist 1:193–197

Baird W (1850) The natural history of the British Entomostraca. The Ray Society, London

Bartenstein H (1985) Stratigraphic pattern of index foraminifera in the lower cretaceous of trinidad. Newsl Strat 14(2):110–117

Bartenstein H, Bettenstaedt F et al (1957) Die Foraminiferen der Unterkreide von Trinidad, B. W. I. Erster Teil: Cuche und Toco Formation. Ecl Geol Helv 50(1):5–56

Bartenslein H, Bettenslaedt F et al (1966) Die Foraminiferen der Unterkreide von Trinidad, WI Zweiter Tei\: Maridale-Formation (Typlokalitlil). Ecl Geol Helv 59:129–177

Bartenstein H, Bolli HM (1977) The Foraminifera in the Lower Cretaceous of Trinidad, W.I. Part 4: Cuche Formation, upper part; *Leupoldina protuberans* Zone. Ecl Geol Helv 70(2):543–573

Bartenstein H, Bolli HM (1986) The foraminifera in the lower cretaceous of trinidad. W.I. Part 5: Maridale Formation, upper part: *Hedbergella rohri* Zone. Ecl Geol Helv 79:945–999

Bedoya C, Nomesqui HE (2013) Estratigrafía, quimioestratigrafía y petrografía de la Formación Rosablanca: Implicaciones para la identificación de eventos anóxicos oceánicos en la Cordillera Oriental Colombiana. Universidad de Caldas, Graduate Project, pp 1–84

Bischoff G (1990) Ostracoden-Studien im Libanon, 5: the genus *Schuleridea* in the Late Jurassic and Early Cretaceous. Senck. Lethaea 71:85–137

Bolli H, Beckmann J et al (1994) Benthic foraminiferal biostratigraphy of the South Caribbean region. Cambridge University Press, Cambridge, pp 1–408

BouDagher-Fadel MK (2001) The taxonony and evolution of the foraminiferal genus *Buccicrenata* Loeblich and Tappan. Micropal 47(2):168–172

Brady HB (1884) Report on the foraminifera dredged by H.M.S. Challenger during the years 1873–1876. Report of the scientific results of the voyage of H.M.S. Challenger, 1873–1876. Zoology 9:1–814

Brenner P (1976) Ostracoda and charophyta from the Spanish Wealden (Systematics, ecology, stratigraphy, paleogeography). Palaeont Abt A 152:113–201

Bruckner WD (1954) Note on some fossils from the Girón Group in the Rio Lebrija Valley, Department of Santander, Colombia. J Paleont 28:112–113

Bürgl H (1960) Geología de la Península de la Guajira. Bol Geol 6(1–3):129–168

Cadena EA (2011) Potential earliest record of podocnemidoid turtles from the early Cretaceous (Valanginian) of Colombia. J Paleont 85(5):877–881

Cardozo E, Ramírez C (1985) Ambientes de depósito de la Formación Rosablanca: Área de Villa de Leiva. In: Etayo-Serna F, Laverde F (eds) Proyecto Cretácico. Publicación Geológica Especial 16 Ingeominas, Bogotá, Chapter 13, pp 1–13

Carrillo-Briceño JD, Cadena EA et al (2016) First record of a hybodont shark (Chondrichthyes: Hybodontiformes) from the lower Cretaceous of Colombia. Neot Biodiv 2(1):81–86

Chapman F, Parr WJ et al (1934) Tertiary foraminifera of Victoria, Australia. The Balcombian deposits of Port Phillip, Part lll. Linnean Society. J Zool 38(262):553–577

Colin JP, Dejaz J et al (2008) Ostracodes of the Bhanness Formation, Kimmeridgian (Upper Jurassic) of Central Lebanon: Taxonomy, palaeoecology and palaeobiogeography Ann Paléont 94:13–27

Coimbra JC, Arai M et al (2002) Biostratigraphy of Lower Cretaceous microfossils from the Araripe basin, northeastern Brazil. Geobios 35:687–698

D'Orbigny A (1826) Tableau méthodique de la classe des Céphalopodes, 3-me ordre, Foraminifères. An Sci Nat 7:245–314

Delage Y, Herouard E (1896) Traite de Zoologie Concrète. Tome I. La Cellule et Ies Protozoaires. Paris, 1–584

Dietrich W (1935) Zur stratigraphie der kolumbianischen Ostkordillere. Zentralblatt für Mineralogie, Geol und Palänt B(3):74–82

Ehrenberg CG (1838) Die Infusionsthierchen als vollkommene Organismen. Leipzig, 1–547

Gaona-Narvaez T, Maurrasse FJ et al (2013) Geochemistry, palaeoenvironments and timing of Aptian organic-rich beds of the Paja Formation (Curití, Eastern Cordillera, Colombia). In: Bojar AV, Melinte-Dobrinescu MC et al (eds) Isotopic studies in Cretaceous research. Geological Society Special Publications, vol 382. London. https://doi.org/10.1144/sp382.6

Gómez-Cruz A, Bermúdez HD et al (2015) A new species of *Diaulax* Bell, 1863 (Brachyura: Dialucidae) in the early Cretaceous of the Rosablanca Formation. Colombia. Bol Soc Geol Mex 67(1):103–112

Guillaume HA, Bolli HM et al (1972) Estratigrafía del Cretáceo Inferior en la Serranía del Interior, oriente de Venezuela. Memorias del IV Congreso Geológico Venezolano 3:1619–1655

Guzmán G (1985) Los Griferidos Infracretácicos *Aetostreon couloni* y *Ceratostreon boussingaulti*, de la Formación Rosablanca, como Indicadores de oscilaciones marinas. In: Etayo-Serna F, Laverde F (eds) Proyecto Cretácico. Publicación Geológica Especial 16 Ingeominas, Bogotá, Chapter 11, p 1–16

Guzmán-González J, Piovesan EK et al (2016) Non-marine ostracoda from the Aliança Formation in the north-central portion of the Jatobá Basin, Northeastern Brazil. Rev Bras Paleont 19(1):15–24

Horne DJ (1995) A revised ostracod biostratigraphy for the Purbeck-Wealden of England. Cret Res 16:639–663

Jones TR (1880) Sur le "Calcaire a Cypris" du Boulonnais. Bull Soc Geol France (3ème série) 8:615–616

Kaminski MA (2004) The year 2000 classification of the agglutinated foraminifera. In: Bubik M, Kaminski MA (eds) Proceedings of the sixth international workshop on agglutinated foraminifera, vol 8, pp 237–256. Grzybowski Foundation Special Publication

Karsten H (1858) Über die geognostischen Verhältnisse des westlichen Columbien. Der heutigen Republiken Neu-Granada und Equador. Amtlicher Bericht über die zweiunddreissigste Versammlung deutscher Naturforscher und Ärzte zu Wien 32:20–117

Kaufmann A (1900) Cypriden und Darwinuliden der Schweiz. Rev Suis Zool 8:209–423

Koutsoukos EAM, Merrick KA (1986) Foraminiferal paleoenvironments from the Barremian to Maastrichtian of Trinidad, West Indies. In: Transactions of the First Geological Conference of the Geological Society of Trinidad, Port of Spain, pp 85–101, 10–12 July 1985

Krömmelbein K (1962) Zur Taxionomie und Biochronologie stratigraphisch wichtiger Ostracoden-Arten aus der oberjurassisch?-unterkretazischen Bahia-Serie (Wealden-Fazies) NE-Brasiliens. Senck Lethaea 43(6):437–527

Lamarck JB (1804) Suite des mémoires sur les fossiles des environs de Paris: Museum Natural de Histoire Paris. Ann vol 5:(a) 179–188, (b):237–245, (c):349-357

Latreille PA (1802) Histoire naturelle, générale et particulière des Crustacés et des Insectes. Histoires des Cypris et des Cytherées 8(4):232–254

Loeblich A, Tappan H (1987) Foraminiferal genera and their classification. Van Nostrand Reinhold, New York, pp 1–210

Marie P (1941) Les foraminifères de la craie à *Belemnitella mucronata* du Bassin de Paris. Museum Natural de Histoire. Mémoire 12(1):1–296

Martin GPR (1940) Ostracoden des norddeutschen Purbeck und Wealden. Senckenbergiana 22 (5–6):275–361

Martinez JI (1989) Foraminiferal biostratigraphy and paleoenvironments of the Maastrichtian Colon mudstones of northern South America. Micropal 35:97–113

Maync W (1950) The foraminiferal genus *Choffatella* (Schlumberger) in the lower Cretaceous (Urgonian) of the Caribbean region (Venezuela, Cuba, Mexico and Florida). Eclog Geol Helv 42(2):529–547

Maync W (1955) On some erroneous or questionable determinations of *Choffatella*. Micropal 1 (3):269–272

Maync W (1956) On the age of *Choffatella*-bearing beds in Venezuela. Micropal 2:92

Maync W (1958) *Ammocycloloculina*, n. gen., an unknown foraminiferal genus. Contrib Cush Found Foram Res 9(3):53–57

Montenegro LC (2012) Análisis del componente micropaleontológico del segmento superior de la Formación Rosablanca, en el área de Zapatoca. Universidad Nacional de Colombia, Santander. Graduate Project, pp 1–43

Morales LG, Podesta DJ et al (1958) General geology and oil occurrences of Middle Magdalena Valley, Colombia. In: Weeks LG (ed) Habitat of Oil. An AAPG symposium. American Association of Petroleum Geologists, Tulsa, pp 641–695

Muñoz-Torres F, Whatley RC et al (2006) Miocene ostracod (Crustacea) biostratigraphy of the upper Amazon Basin and evolution of the genus *Cyprideis*. J S Am Ear Sci 21:75–86

Neagu T, Cîrnaru P (2004) Genus *Choffatella* (Schlumberger), 1904 (Foraminifera) in the lower Aptian (Bedoulian) from southern Dobrogea and SE part of the Romanian Plain. Ac Palaeont Rom 4:269–275

Neale JW (1973) Ostracoda as means of correlation in the Boreal Lower Cretaceous, with special reference to the British marine ostracoda. In: Casey R, Rawson PF (eds) The Boreal Lower Cretaceous. Geological Journal Special Issue, London, pp 169–184

Nicolaidis DD, Coimbra JC. (2008) *Perissocytheridea carrenoae* sp. nov. (Crustacea, Ostracoda) and associated calcareous microfauna from Yecua Formation (Miocene) Bolivia. Rev. Bras. Paleont. 11(3):179–186

Patarroyo-Camargo GD, Patarroyo P et al (2009) Foraminíferos bentónicos en el Barremiano inferior de la Formación Paja (Boyacá-Santander, Colombia): Evidencias preliminares de un posible bioevento. Geol Col 34:97–108

Pawlowski J, Holzmann et al (2013) New supraordinal classification of Foraminifera: molecules meet morphology. Mar Micropal 100:1–10

Petters V (1954) Typical foraminiferal horizons in the lower Cretaceous of Colombia, S.A. Contrib. Cush Found Foram Res 112(5):128–137

Pictet F, Renevier J (1858) Description des fossiles du terrain aptien de la Pert e du Rhône et des environs de Ste Croix Genève

Pinto ID, Ornellas LP (1970) A new brackishwater ostracode *Perissocytheridea krömmelbeini* Pinto and Ornellas, sp. nov., from Southern Brazil. Escola de Geologia de Porto Alegre, Sp Pub 20:1–19

Piovesan EK, Cabral MC et al (2016) *Fossocytheridea* Swain and Brown and *Perissocytheridea* Stephenson (Ostracoda): insights into paleosalinity gradients of Late Cretaceous deposits from Brazil and Portugal. Rev Bras Paleont 18(1):21–30

Puckett MT (2002) Systematics and paleobiogeography of brachycytherine Ostracoda. Micropal 48(2):1–87

Puckett MT, Andreu B et al (2016) The evolution of the Brachycytheride ostracoda in the context of the breakup of Pangea. Rev Micropal 59:97–167

Puri HS (1954) Contribution to the study of the Miocene of the Florida Panhandle. Part 3: ostracoda. Florida Geological Survey, Geol Bull 36:215–345

Reuss AE (1860) Uber *Lingulinopsis*, eine neue Foraminiferen-Gattung aus dem bohmischen Planer: K. Biihm. Gesell. Wiss. Prag, math-naturw. CI, Sitzungsber pp 23–24

Sames B (2011) Early Cretaceous *Cypridea* Bosquet 1852 in North America and Europe. In: Sames, B., Taxonomic studies in Early Cretaceous nonmarine Ostracoda of North America. Micropal 57(4–5):345–431

Sarmiento LF (2011) Middle Magdalena Basin. In: Cediel F, Colmenares F (eds) Petroleum geology of Colombia, vol 11. ANH. Fondo Editorial Universidad EAFIT, Medellín, pp 1–193

Sars GO (1866) Oversigt af Norges marine Ostracoder. Forhandlinger i Videnskabs-Selskabet i Christiania 1:1–130

Sars GO (1922–1925) An account of the crustacea of Norway with short descriptions and figures of all species. Crustacea of Norway: Cypridinidae, Conchoeciidae, Polycopidae (1922) Polycopidae, Cytherellidae, Cypridae (1923) Cypridae (1925) Cypridae, Cytheridae (1925). Cyther, 9th edn. Bergen, 1–277

Schlumberger C (1904) Note sur le genre *Choffatella* n. gen Bull Soc Géol France 4(4):763

Scott HW (1961) Classification of Ostracoda. Q74–Q91. In: Moore RC, Pitrat CW (eds) Treatise on invertebrate paleontology, part Q, Arthropoda 3: Crustacea, Ostracoda. Geological Society of America, Boulder and University of Kansas Press, Lawrence, pp 1–465

Sigal J (1952) Aperçu stratigraphique sur la Micropaléontologie du Crétacé. XIXe Congrès Géol. International Monographie Régional, série 1. Algérie 26:1–45

Slipper IJ (2009) Marine Lower Cretaceous. In: Whittaker JE, Hart MB (eds) Ostracods in British stratigraphy. The Micropalaeontological Society Special Publications, London, pp 309–343

Smith RJ (2000) Morphology and ontogeny of Cretaceous ostracods with preserved appendages from Brazil. Palaeont 43(1):63–98

Stacy HE (1966) The lower Cretaceous microfauna from Trinidad and adjacent areas, pp 1–228. Dissertation, University of Michigan

Stephenson MB (1938) Miocene and Pliocene ostracoda of the genus *Cytheridea* from Florida. Jour Paleont 12:127–148

Swain F, Brown PM (1972) Lower Cretaceous, Jurassic (?), and Triassic Ostracoda from the Atlantic Coastal Region. Geol Surv Prof Paper, Washington, p 795

Swartz FM, Swain FM (1946) Ostracoda from the Upper Jurassic Cotton Valley Group of Lousiana and Arkansas. J Paleont 20:362–373

Tibert NE, Colin JP et al (2003) Revision of the ostracode genus *Fossocytheridea* Swain and Brown, 1964: Mesozoic ancestral root for the modern eurytopic *Cyprideis* Jones. Micropal 49:205–230

van den Bold WA (1963) Upper miocene and pliocene ostracoda of trinidad. Micropal 9:361–424

van den Bold WA (1966) Upper miocene ostracoda from the tubará formation (Northern Colombia). Micropal 12(3):360–364

Vergara L (1997) Stratigraphy, foraminiferal assemblages and paleoenvironments in the Late Cretaceous of the Upper Magdalena Valley, Colombia (Part I). J S Am Ear Sc 10:111–132

Vergara L, Rodríguez G et al (1997) Agglutinated foraminifera and sequence stratigraphy from the Chipaque Formation (Upper Cretaceous) of El Crucero section. Colombia, South America Micropal 43(2):185–201

Viviers MC, Koutsoukos EAM et al (2011) Stratigraphy and biogeographic affinities of the late Aptian-Campanian ostracods of the Potiguar and Sergipe basins in northeastern Brazil. Cret Res 21:407–455

Ward DE, Goldsmith R et al (1972) Geología de los cuadrángulos H-12 Bucaramanga y H-13 Pamplona, Departamento de Santander. Bol Geol 21(1–3):1–132

Turonian–Santonian Calcareous Nannofossil Biozonation for La Luna Formation, Middle Magdalena Valley Basin, Northern Colombia

Juan P. Pérez Panera[1,2(✉)], Francisco J. Parra[3,4,5], Rosa E. Navarrete[3], Christian Sánchez[6], Diana Daza[6], Manuel Rodriguez[6], and Mario Prince[6]

[1] CONICET—Laboratorio de Bioestratigrafía, Gerencia de I+D Geociencias, YPF Tecnología S.A., Avenida del Petróleo Argentino s/n entre 129 y 143, Buenos Aires 1923 Berisso, Argentina
juan.p.panera@ypftecnologia.com

[2] División Paleozoología Invertebrados, Universidad Nacional de La Plata, Buenos Aires Museo de La Plata. Paseo del Bosque s/n, 1900 La Plata, Buenos Aires, Argentina

[3] PALEOSEDES E. U. Departamento de Bioestratigrafía, Transversal 27 #57-49., 111311 Bogotá, Colombia
paleosedes@gmail.com
https://paleosedes.tk

[4] Géosciences-Environnement Toulouse, Université de Toulouse. UPS (SVT-OMP), CNRS, IRD, 14 Avenue Édouard Belin, 31400 Toulouse, France

[5] Laboratorio de Bioquímico-Estratigrafía, Facultad de Geociencias, Universidad Nacional de Colombia, Ciudad Universitaria, Edificio Manuel Ancizar (425), Bogotá, D.C, Colombia

[6] Gran Tierra Energy Colombia Ltd. Calle, 113 No. 7-80 Piso 17, Bogotá, Colombia

Abstract. A well site biostratigraphic study was carried out in the Chuira-2 ST-1 well, through the Umir and La Luna Formations in the Middle Magdalena Valley Basin (MMVB), north of Colombia. Results of the calcareous nannofossil analysis are presented and compared to local biozonations for planktonic foraminifera and palinomorphs for La Luna Formation in the MMVB and calcareous nannofossils for the Maracaibo Basin. Biostratigraphic analyses result in the identification of different time intervals and sediment thickness of an inverse fault traversed by the well. Six local calcareous nannofossil biozones, based on eleven dissolution-resistant taxa bioevents, for the lower to the upper part of the La Luna Formation in the MMVB are proposed—*Micula adumbrata* Interval Biozone, comprising the early Turonian; *Eprolithus moratus* Interval Biozone, for the middle Turonian to the latest Turonian; *Micula staurophora* Interval Biozone, for the early Coniacian to the early late Coniacian; *Micula adumbrata–Micula concava* Interval Biozone, of late Coniacian age; *Micula concava* Interval Biozone, of early Santonian age; and *Zeugrhabdotus scutula* Interval Biozone, which corresponds to the early Santonian to late Santonian.

Keywords: Biostratigraphy · Late cretaceous · La Luna Formation
Calcareous nannofossil

G. C. Cusminsky et al. (eds.), *Advances in South American Micropaleontology*, Springer Earth System Sciences, https://doi.org/10.1007/978-3-030-02119-1_3

1 Introduction

Well site biostratigraphy was performed at Chuira-2 and Chuira-2 ST-1 wells in the Middle Magdalena Valley (MMVB) through Umir and La Luna Formations (Fig. 1). Biostratigraphy was carried out by studying the foraminifera, palinomorphs and calcareous nannofossil content. These analyses allowed estimation of sediment age, involved time of the unconformities, thickness of fault repetition and estimation of reservoir depth in real time during the perforation. Although calcareous nannofossils in La Luna Formation were much affected by dissolution and recovered taxa were mostly dissolution-resistant species, they showed to be biostratigraphically useful and correlate very well with foraminifera and palinomorph local biozonations (Navarrete et al., this volume). Although nannofossils are very useful for biostratigraphical analysis for the

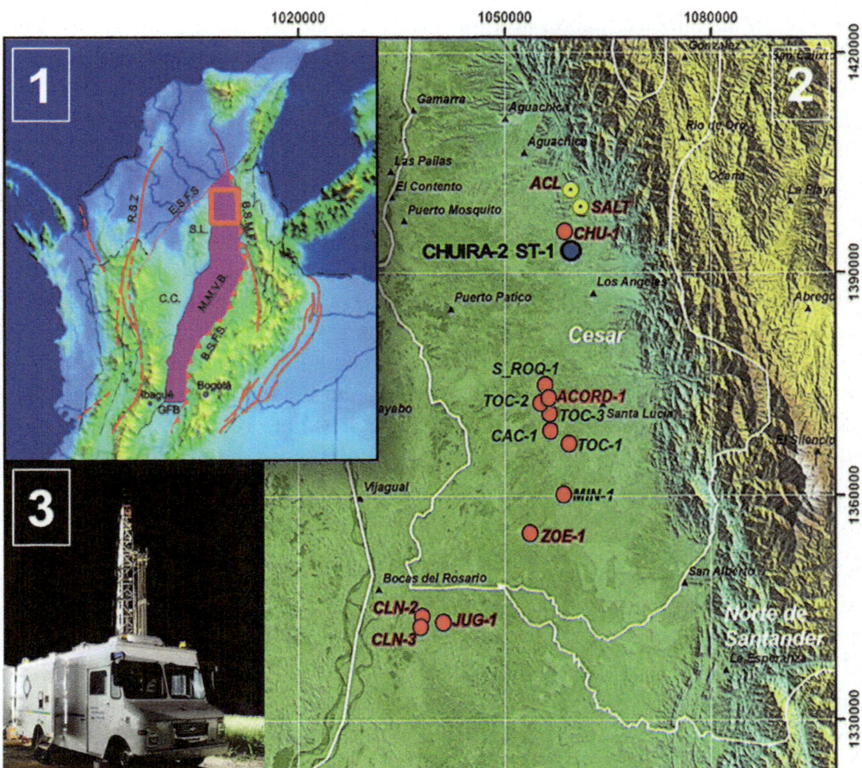

Fig. 1 Location map of the Chuira-2 ST-1 well. **1.1** Middle Magdalena Valley Basin in north of Colombia. BSMF: Bucaramanga-Santa Marta Fault system; BSFS: Bituima and La Salina Fault system; GFB: Girardot fold belt; CC: central cordillera basement; SL: Serranía de San Lucas; ESFS: Espíritu Santo Fault system. Red dots: wells, yellow dots: outcrops, blue dot: Chuira-2 ST-1. **1.2** Detail of the Chuira-2 ST-1 location at the MMVB. **1.3** Paleosedes EU Motile Unite at Chuira-2 ST-1 well site

oil and gas industry, there are only few public studies for such an important formation as La Luna and only for the western Venezuelan area. In this contribution, we present the first results of calcareous nannofossils from the La Luna Formation in the Colombian MMVB and a biozonation scheme correlated with foraminifera and palinomorph biozones. Data presented correspond to the Chuira-2 ST-1 well, as communicated during the "Mesozoic and Cenozoic of South America and Antarctica Micropaleontology Symposium" of the "*11 Congreso de la Asociación Paleontológica Argentina*" in November 2016 at General Roca, Argentina (Navarrete et al. 2016), but it has been also compared with other studies developed by the Biostratigraphic Consultant Paleosedes EU, in other wells and outcrops in the area.

2 Geological Setting and Previous Works

The Middle Magdalena Valley Basin (MMVB) is located at the north of Colombia and consists of an asymmetric depression between the Central and Eastern Cordilleras of the Colombian Andes, which spans 30,000 km^2 with south–north orientation (Fig. 1). At the north, it is limited with the Espíritu Santo Fault system, at northeast with the Bucaramanga-Santa Marta Fault system, at the southeast with the Bituima and La Salina Fault system, at the south with the Girardot fold belt and at the west it onlaps with Neogene sediments of the Serranía de San Lucas and Central Cordillera basement (Rolón 2004; Barrero et al. 2007; Prince et al. 2008). The infill of the basin started in the Triassic with an epicontinental transgression that lasts up to the earliest Cretaceous in relation to the opening of the proto-Caribbean Sea during the separation of North and South America (Cooper et al. 1995). During the Early Cretaceous, the basin became continental to marginal marine and shallow marine. From the Cenomanian to Santonian, the basin evolved to a back-arc and sedimentation was mainly epicontinental shallow marine. La Luna Formation was deposited during the Turonian–Santonian, and it is one of the best liquid-hydrocarbon generation source rocks in the world (West 1994; Konyukhov 2012, see Navarrete et al. this volume for details). Sedimentation of the basin goes from marginal marine to continental from the Campanian to the Neogene. For a generalized lithostratigraphic column of the basin, see Navarrete et al. (this volume).

La Luna Formation was defined by Garner (1926) at the La Luna Creek in the Perijá area, northwestern Venezuela as "La Luna limestone". Later, Hedberg and Sass (1937) formally describe it as a formation. Lithology corresponds to black calcareous shale and limestones characterized by alternating beds of laminated gray bituminous limestone and black calcareous shale with large discoidal limestone concretions. It has been described and nominated in the Maracaibo Basin, Venezuela, but later introduced in the Colombian basins Catatumbo (Notestein et al. 1944) and MMVB (Morales 1958). In the MMVB, La Luna Formation has less than 400 m thickness (Galvis-Portilla et al. 2014; Rodríguez et al. 2016) and starts with a sharp contact with the underlying Albian limestones of Tablazo Formation. At the top, it ends with a mature glauconitic layer, whereas the Umir Formation rests by a paraconformity. Morales (1958) divided it into three members: the lower Salada Member, the middle Pujamana Member, and the upper Galembo Member. Albeit this subdivision in members is widely used in the oil and gas industry in the Barco Concession and Catatumbo Basin, sometimes they are not easy to recognize, and in this contribution we do not distinguish them. It is worthwhile to notice that although along the Catatumbo and MMVB Colombian basins and the Maracaibo Basin in Venezuela, La Luna Formation could be easily correlated and its boundaries had similar characteristics, they are strongly diachronic. In the Maracaibo Basin it could be Cenomanian to Campanian (de Romero et al. 2003) and in the MMVB is mostly Turonian to Santonian. In Fig. 2, La Luna Formation stratigraphic section in Chuira-2 ST-1 is summarized. For further information and characteristics of this formation, see Navarrete et al. (this volume). First and most biostratigraphic works in this formation are based on ammonites (Liddle 1928; Sutton 1946; Renz 1959, 1977, 1982), but correlation and resolution are compromised because ammonites are not very frequent and global correlation could be difficult (de Romero et al. 2003). Inoceramids have the same problem and furthermore, macrofossils are not very useful for subsurface biostratigraphy. Most recent biostratigraphic works are based on foraminifera, calcareous nannofossils, and palinomorphs, but most of them are studies for oil companies and there is little public literature. Most public results are abstracts (Farias et al. 1995, 1996, 2000; Truskowski et al. 1995, 1996; de Romero and Odreman 1996; Lorente et al. 1996; de Romero et al. 2000; Navarrete et al. 2016). Only few detailed papers based on microfossil biostratigraphy have been published. Petters (1955) introduced a benthonic foraminifera biozonation for La Luna Formation in the MMVB, in which he also described an abundant assemblage of planktonic foraminifers, radiolarians, and ammonites. Later, for the same basin, Tchegliakova and Pérez (1995) recognized one planktonic foraminifera biozone (*Dicarinella concavata*) for the Coniacian of La Luna Formation in central Eastern Cordillera (Colombia). In the Maracaibo Basin, northwestern Venezuela, Boesi et al. (1988) studied the foraminifera and calcareous nannofossils in surface sections of the southwestern part of the basin. De Romero and Galea-Álvarez (1995) studied the foraminifera of the phosphate horizons and Tres Esquinas Member in the upper part of the formation at the southern part of the basin. Davies et al. (1999) presented the biostratigraphic results of planktonic foraminifera in the Alpuf-6 well core at the northwestern part of Maracaibo Basin and compare their results with $\delta^{13}C_{org}$ analysis. Finally, the study of de Romero et al. (2003) is the only comprehensive contribution about calcareous nannofossils from La Luna Formation. In that study, authors analyzed one well and four sections in different areas of the Maracaibo Basin. These are the Totumo-3 well in Perijá, the San Pedro del Río Section in Táchira, the Río Loro Section in Táchira at the border of Mérida, and the Carora and La Peña/San Felipe Sections in Trujillo–Lara. In their contribution, authors compare calcareous nannofossils events with foraminifera and $\delta^{13}C_{org}$ curves developing a well-constrained age and sedimentation rate model for La Luna Formation in the Maracaibo Basin.

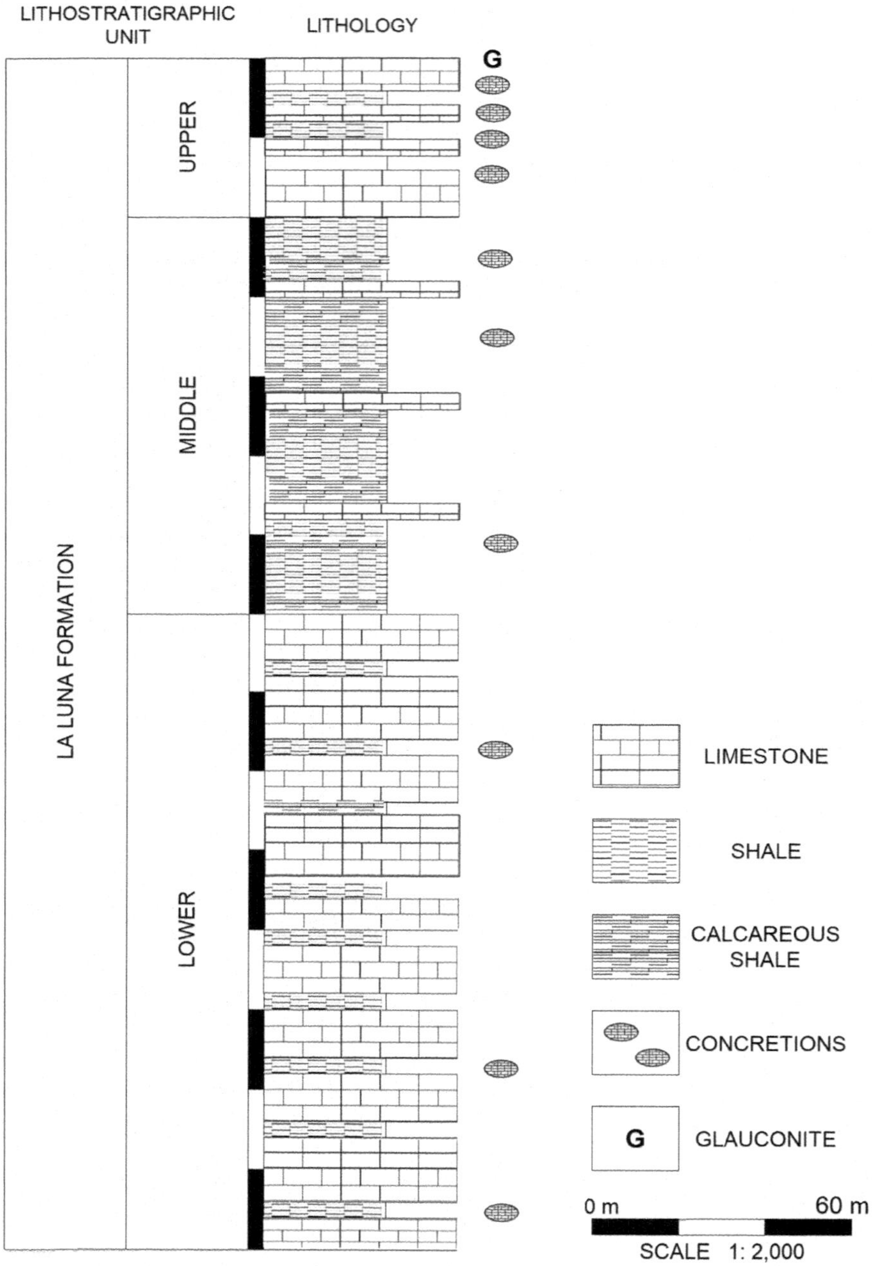

Fig. 2 Stratigraphic section of La Luna Formation in the Chuira-2 ST-1 well, MMVB

3 Materials and Methods

Thirty-six drill samples from Umir and La Luna Formations (2 and 34, respectively) from the Chuira-2 ST-1 well were provided by Petronorte-UT Midas (now Gran Tierra Energy Colombia, Ltd.) during a well site biostratigraphic analysis. The samples were prepared for study in polarized light using a simplification of the gravity settling technique (Gardet 1955; Bramlette and Sullivan 1961) at the motile laboratory of Paleosedes EU. In order to get semi-quantitative data, a 500 µm volume of sediment was diluted in 10 ml of water and then washed with an ultrasonic for 30 s at 40 kHz frequency. After that treatment, 200 µl of the solution was used for the slide preparation using a micropipette. The effort of the analysis per sample was standardized at 400 fields of view, and when less fields were observed, data was pondered. In each case, two or three transects were additionally studied after data collecting, in order to search for rare species with biostratigraphic or paleoenvironmental significance. Data was plotted on a distribution chart along with preservation, abundance, and species richness (Fig. 3). The preservation criteria is as follows: G: Good, most specimens exhibit little or no secondary alteration; M: Moderate, specimens exhibit the effects of secondary alteration from etching and/or overgrowth (identification of species not impaired); P: Poor, Specimens exhibit intense effects of secondary alteration from etching and/or overgrowth (identification of species impaired but possible in some cases). Total abundance was calculated by dividing the total specimen count by the total fields of view and then assigned to a letter category. Letters used in the distribution chart (Fig. 3) to denote nannofossil species abundances are keyed to the 10 log of the number of specimens of a particular taxon likely to be observed in any one field of view of the microscope. This and the corresponding logs are determined as follows: V: Very abundant, more than ten specimens per field of view; A: Abundant, 1–10 specimens per field of view; C: Common, one specimen per 2–10 fields of view; F: Few, one specimen per 11–100 fields of view; R: Rare, one specimen per 101–1000 fields of view. A petrographic microscope with polarized light at a magnification of 1000X was used for nannofossil analysis and counting. For biostratigraphic interpretation, nannofossil events were evaluated according to the standard biozonation of Burnett (1998) and the local for Venezuelan La Luna Formation of de Romero et al. (2003). Biozone age calibration is according to Ogg and Hinnov (2012). Photographs were taken during well site analysis and in office laboratory. All slides are stored at the Paleosedes EU repository.

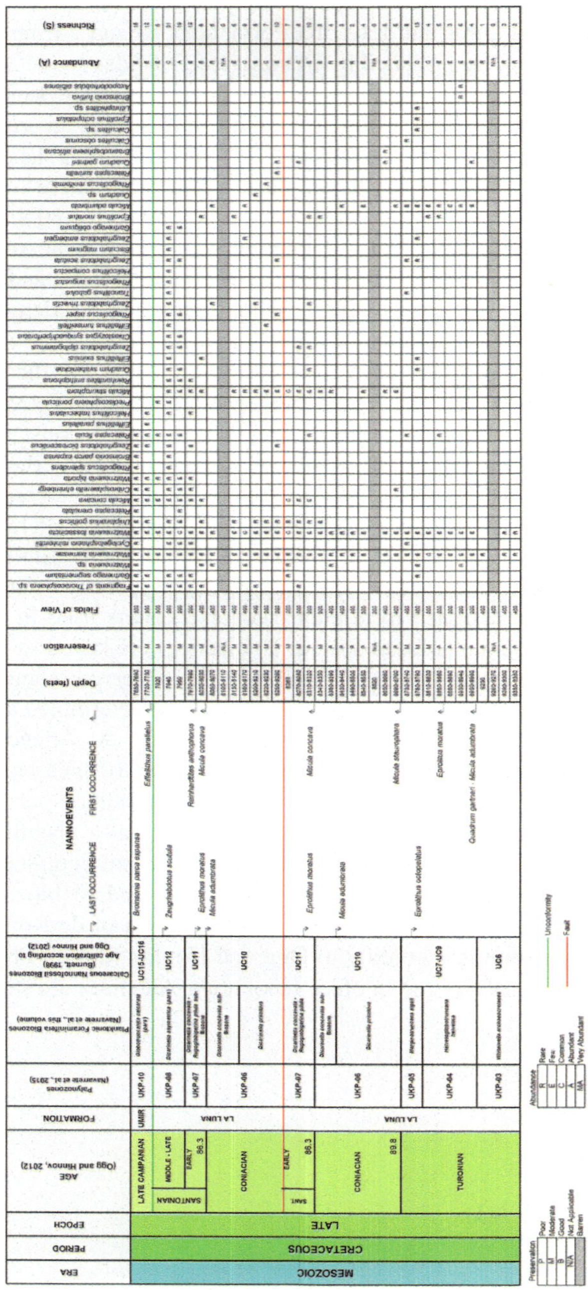

Fig. 3 Distribution chart, nannoevents, and biostratigraphic interpretation of calcareous nannofossils from the Chuira-2 ST-1 well according to Burnett (1998) biozones, calibrated with Ogg and Hinnov (2012). Correlation with local palinomorphs and planktonic foraminifera biozones for the MMVB (Navarrete et al. 2015, this volume)

4 Results and Discussion

As pointed out by de Romero et al. (2003), calcareous nannofossils from La Luna Formation are highly affected by dissolution. Sometimes, also overimposed overgrowth is observed. This is of significant importance since assemblage composition and abundance are strongly affected. Even, taxonomic identification is hard and time-consuming. Dissolution is mostly present in all stratigraphic levels but could vary in intensity affecting species richness as much as from 1 to 31 (Fig. 3). As also noticed by de Romero et al. (2003), this particular and rare levels where samples reach high species richness lead to the conclusion that the nannofossil assemblages in La Luna Formation were originally abundant and diverse, but due to taphonomic processes, only high dissolution-resistant species are consistent and well represented. These are species of the genus *Watznaueria* and of the family Polycyclolithaceae. Species of this family are the most useful for biostratigraphic interpretation.

Calcareous nannofossil biostratigraphy in Chuira-2 ST-1 well. Calcareous nannofossils were investigated from the late Cretaceous–Danian Umir Formation up to the end of the perforation in the lower part of La Luna Formation. In this contribution, only the lowermost levels of the Umir Formation and the complete section of the perforated La Luna Formation are presented (Fig. 3). Most important species are illustrated in Plates 1, 2, and 3. Biostratigraphy results are presented in eight intervals, according to the normal development of the perforation, from the youngest to oldest levels.

First Interval. Late Campanian (7680–7690 to 7720–7730 ft).

Assemblages in this interval are the most abundant and diverse of the Umir Formation although they are dominated by fragments of calcareous dinoflagellates. Other well-represented species are *Gartnerago segmentatum, Watznaueria barnesae, W. fossacincta, Zeugrhabdotus bicrescenticus,* and the Polycyclolithaceae *Micula concava* and *Uniplanarius gothicus*. The key species *Broinsonia parca expansa* and *Eiffellithus parallelus* allow interpretation of late Campanian age for this interval, biozones UC15 to UC16 of Brunett (1998). This interval corresponds to the *Globotruncanita calcarata* planktonic foraminifera biozone. Between the 7720–7730 and 7920 ft of the below interval, sediments yield late early and early late Campanian foraminifers of the *Globotruncanita elevata* and *Globotruncana ventricosa* biozones, but those 190 ft sediments were not studied by calcareous nannofossils (see Navarrete et al., this volume).

Second Interval. Middle to late Santonian (7920–7950 ft).

A mature glauconitic level has been observed at the top of this interval. This corresponds to the top of La Luna Formation, which overlies in paraconformity with the upper Umir Formation. This paraconformity involves the latest Santonian to the early late Campanian ($\sim$4.4 Myr; see Navarrete et al. this volume). Calcareous nannofossils in this interval have moderate preservation, and the most diverse and rich assemblages in Churia-2 ST-1 occur here (7940 ft, Richness: 31, see Fig. 3). Most abundant species are *Chiastozygus synquadriperforatus, Cribrosphaerella ehrenbergii, Cyclagelosphaera reinhardtii, Gartnerago obliquum, G. segmentatum, Eiffellithus eximius, Prediscosphaera ponticula, Reinhardtites anthophorus, Retecapsa ficula, Rhagodiscus asper, Thoracosphaera* sp. (fragments), *Watznaueria barnesae,*

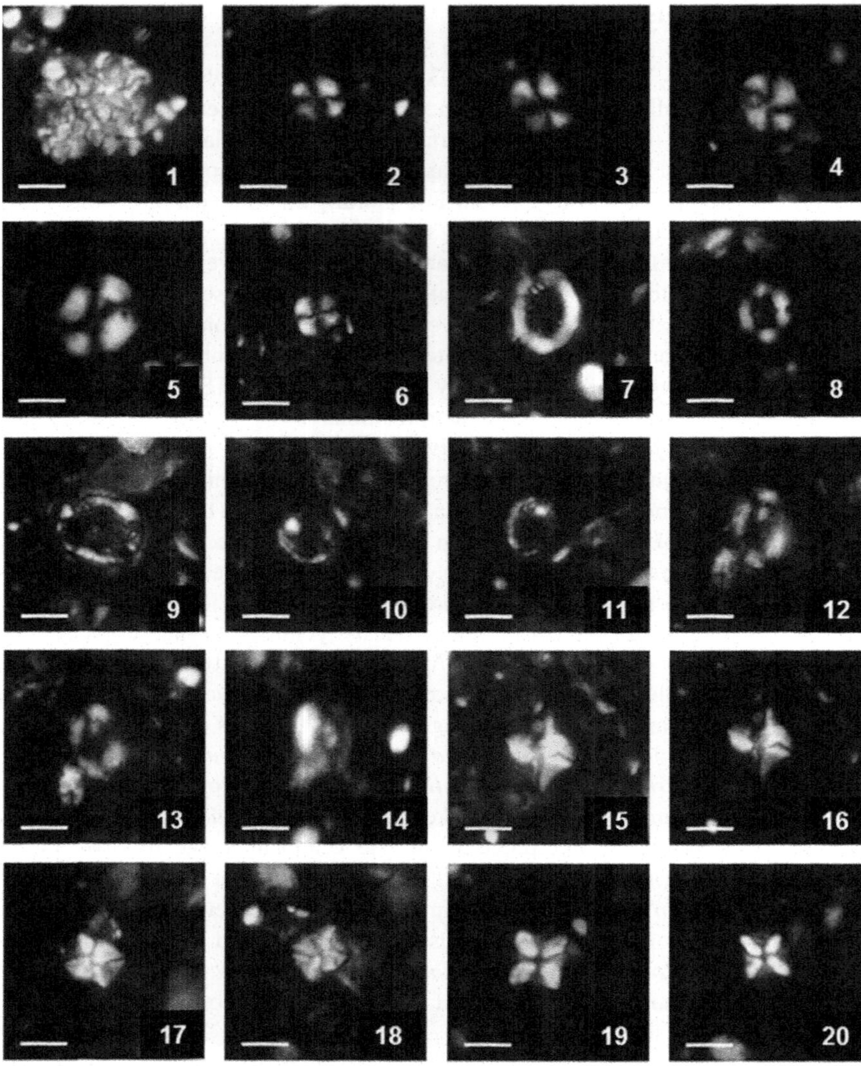

Plate 1 Calcareous nannofossils recovered at Chuira-2 ST-1 well. Scale bar: 5 μm **1** fragments of *Thoracosphaera* sp., sample 7680–7690 ft; **2–4** *Watznaueria barnesae*, sample 7680–7690 ft; **5** *Watznaueria fossacincta*, sample 7680–7690 ft; **6**. *Cyclagelosphaera reinhardtii*, sample 7680–7690 ft; **7**. *Retecapsa crenulata*, sample 7680–7690 ft; **8** *Retecapsa ficula*, sample 7680–7690 ft; **9–11** *Cribrosphaerella ehrenbergi*, sample 7680–7690 ft; **12–13** *Broinsonia parca expansa*, sample 7680–7690 ft; **14** *Zeugrhabdotus bicrescenticus*, sample 7680–7690 ft; **15–18** *Micula concava*, sample 7680–7690 ft; **19–20** *Micula staurophora*, sample 7680–7690 ft

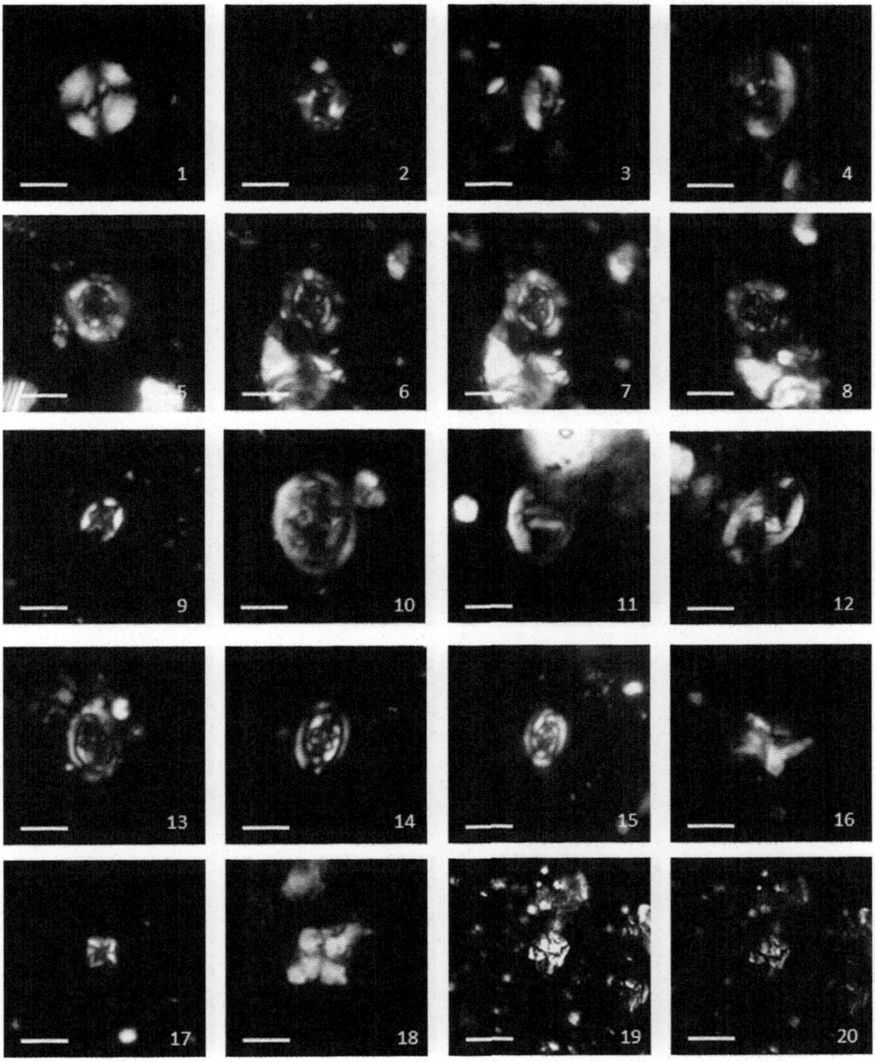

Plate 2 Calcareous nannofossils recovered at Chuira-2 ST-1 well. Scale bar: 5 μm. **1** *Watznaueria barnesae*, sample 7940 ft; **2** *Biscutum magnum*, sample 7940 ft; **3** *Rhagodiscus angustus*, sample 7940 ft; **4** *Rhagodiscus asper*, sample 7940 ft; **5–8** *Prediscosphaera ponticula*, sample 7940 ft; **9**. *Helicolithus compactus*, sample 7940 ft; **10** *Reinhardtites anthophorus*, sample 7940 ft; **11**. *Zeugrhabdotus diplogrammus*, sample 7940 ft; **12** *Zeugrhabdotus embergerii*, sample 7940 ft; **13**. *Zeugrhabdotus scutula*, sample 7940 ft; **14–15** *Zeugrhabdotus trivectis*, sample 7940 ft; **16** *Micula concava*, sample 7940 ft; **17** *Quadrum svabenickae*, sample 7940 ft; **18** *Uniplanarius gothicus*, sample 7940 ft; **19–20** *Eprolithus moratus*, sample 8340–8350 ft

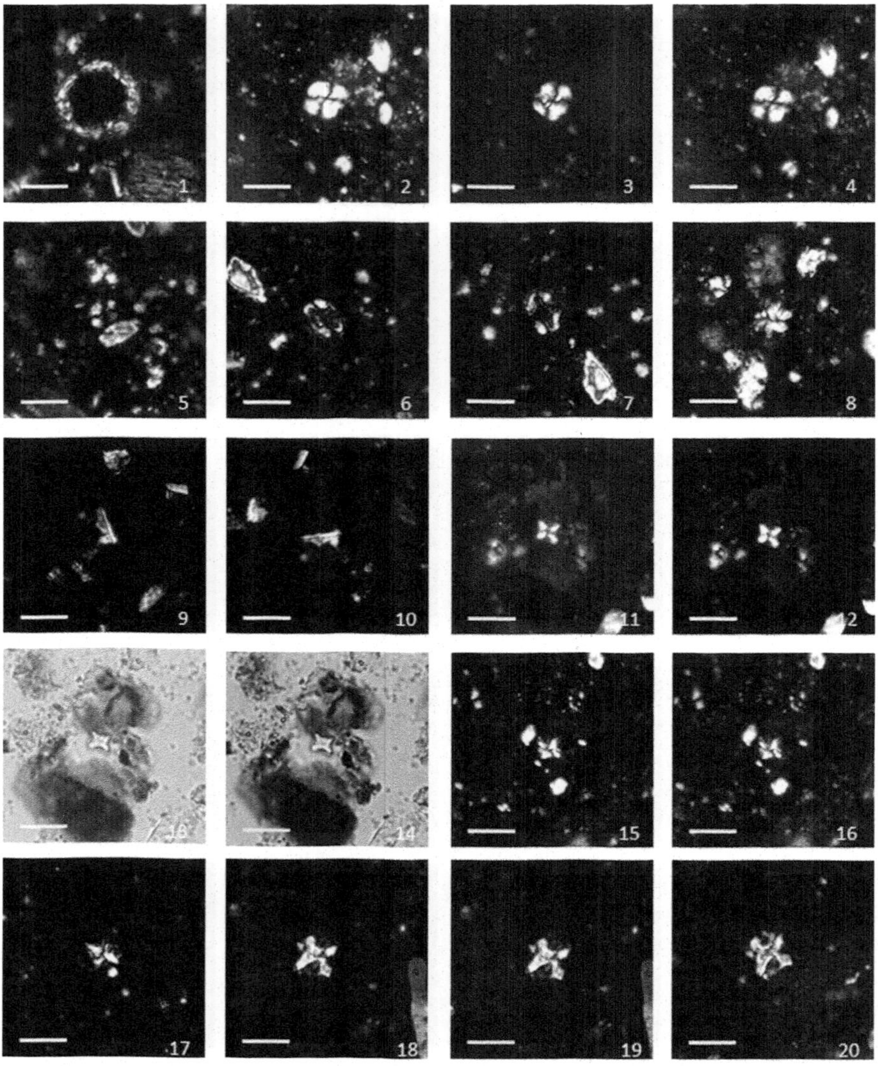

Plate 3 Calcareous nannofossils recovered at Chuira-2 ST-1 well. Scale bar: 8 μm **1** *Watznaueria* sp. coccosphere, sample 8780–8790 ft; **2–4** *Watznaueria barnesae*, sample 8780–8790 ft; **5** *Watznaueria fossacincta*, sample 8780–8790 ft; **6–7** *Eiffellithus eximius*, sample 8780–8790 ft; **8**. *Eprolithus octopetalus*, sample 8780–8790 ft; **9–10** *Lithraphidites* sp., sample 8780–8790 ft; **11–20** *Micula adumbrata*, sample 8780–8790 ft

W. biporta, *W. fossacincta*, *Zeugrhabdotus bicrescenticus*, *Z. diplogrammus*, *Z. trivectis*, *Z. scutula,* and the Polycyclolithaceae *Micula concava*, *M. staurophora*, *Quadrum svabenickae,* and *Uniplanarius gothicus*. The last occurrence (LO) of *Zeugrhabdotus scutula* was observed within this interval and is here highlighted as an important nannoevent for local correlation. This interval corresponds to the UC12 biozone and *Dicarinella asymetrica* planktonic foraminifera biozone.

Third Interval. Early Santonian (7970–7980 to 8020–8030 ft).

Preservation of this interval is moderate, and species richness is moderately high for the La Luna Formation. Most abundant species are *Cyclagelosphaera reinhardtii* and Watznauerias. The Polycyclolithaceae are rare but constant in the assemblages. Three important nannoevents occur in this interval which are the LO of *Eprolithus moratus* and the first occurrence (FO) of *Micula concava* and *Reinhardtites anthophorus.* Those FOs are widely used as markers for the Coniacian/Santonian boundary (Sissingh 1977; Crux 1982; Perch-Nielsen 1979, 1985; Burnett 1998; Melinte and Lamolda 2002). The latter has been questioned by some authors (Burnett 1998), but seems to be consistent in all sites investigated at the MMVB. This interval is correlated with UC11 biozone and *Dicarinella concavata–Rugoglobigerina pilula* planktonic foraminifera sub-biozone of Navarrete et al. (this volume).

Fourth Interval. Coniacian (8060–8070 to 8250–8260 ft).

Assemblages of this interval have moderate preservation and low richness. Most abundant and consistent species are Watznauerias and *Micula staurophora. Micula adumbrata* has its LO at the top of this interval, and in this study, this species is considered to be biostratigraphically important. A fault has been observed at the base of this interval, and it is most possible that the fault is placed until the base of the Coniacian. This interval is correlated with the UC10 biozone, which according to Ogg and Hinnov (2012), comprises all the Coniacian and the planktonic foraminifera biozones of *Dicarinella concavata* and *Dicarinella primitiva* (Navarrete et al., this volume).

Fifth Interval. Early Santonian (8263 to 8310–8320 ft).

Below the fault, the early Santonian assemblage is repeated. It showed little differences in abundance and species composition than in the upper early Santonian interval, being the most abundant the Watznauerias, *Micula concava*, *M. staurophora*, and *Uniplanarius gothicus.* Also, the upper early Santonian interval is more or less 110 ft thick and this one, below the fault, is 80 ft. This could be due to two reasons. The first one is that in the repeated interval, the fault is placed somewhere under the top of the early Santonian in such a way that the uppermost part is missing, or because the fault differentially affected the thickness of both intervals due to the compressional efforts. The base of the interval is marked by the FO of *Micula concava*, probing the reliability of this event for the identification of the Coniacian/Santonian boundary in this basin.

Sixth Interval. Coniacian (8340–8350 to 8690–8700 ft).

As in the upper interval, this one is a repetition of the Coniacian succession above the fault. Assemblages are similar in species composition, but show worst preservation and richness. Even absolute abundance is low; in most of the samples, the value is rare. Species richness is less than five. Watznauerias are the most consistent species together with *Micula adumbrata* and *M. staurophora.* The first has its LO near to the top of the interval and the latter its FO at the base. Thicknesses of this interval and of the Coniacian interval above the fault are 410 and 233 ft, respectively. This difference of more than 170 ft for the interval below the fault would indicate that in the upper interval, the fault passes over the base of the Coniacian.

Seventh Interval. Early to late Turonian (8730–8740 to 8950–8960 ft).

This interval is characterized by moderate to poorly preserved assemblages with moderate to low richness and abundance. The dissolution-resistant Watznauerias are still the most important species in the assemblages, together with *Micula adumbrata*, which has its FO at the base of the interval. This event is also here highlighted for biostratigraphic correlation in the MMVB. The lack of good markers hinders the possibility of separate the interval between middle and late Turonian. For that reason, the entire interval is correlated with the UC7 to UC9 biozones. This interval correlates also with the *Helvetoglobotruncana helvetica* and *Marginotruncana sigali* (*Pars*) planktonic foraminifera biozones (Navarrete et al. this volume).

Eight Interval. Early Turonian (9230 to 9350–9360 ft).

In this interval, nannofossil assemblages are very poorly preserved and have low abundance. Only some dissolved Watznauerias could be recognized, and one sample was barren. Age interpretation is based on foraminiferal assemblage that corresponds to the *Whiteinella archaeocretacica* planktonic foraminifera biozone (Navarrete et al. this volume). The study finished at 9350–9360 ft, and the lowermost part of La Luna Formation was not reached.

Turonian–Santonian calcareous nannofossil Biozonation. A tentative biozonation based on calcareous nannofossils for the MMVB is suggested (Fig. 4). This biozonation is based on the nannoevents observed at Chuira-2 ST-1 well, but they are also compared to data collected in other localities (wells and outcrops) of this basin. Data from the other localities is not presented here as it is held in confidence by the oil and gas companies for which these studies had been done. Our data is also compared to nannoevents reported in the de Romero et al. (2003) biozonation for La Luna Formation in the Maracaibo Basin, Venezuela. Time correlation corresponds to Ogg and Hinnov (2012) calibration. Recognition of early, middle, and late Ages is difficult and doubtful since there are no clear nannoevents that could aid in these subdivisions. Future works might allow a better constraint inside each Age. Six biozones are proposed in chronostratigraphic order.

***Micula adumbrata* Interval Biozone**

Author: Pérez Panera, herein.

Definition: From the FO of *Micula adumbrata* until the FO of *Eprolithus moratus*.

Range: Within early Turonian.

Remarks: The species *Micula adumbrata* showed to be one of the most reliable Polycyclolithaceae for biostratigraphic purposes in the La Luna Formation both because of its abundance and continuous presence in the assemblages. According to Burnett (1998) and Corbett et al. (2014), its FO is at the base of UC9 biozone, upper Turonian (using Ogg and Hinnov 2012 calibration). However, we found this event to occur in older levels than the LO of *Eprolithus octopetalus*, which is middle Turonian (Burnett 1998; Bergen and Sikora 1999; de Romero et al. 2003) and within the *Helvetoglobotruncana helvetica* planktonic foraminifera biozone (Navarrete et al. this volume). Also, abundance of *M. adumbrata* is higher in the Turonian than in the Coniacian (Fig. 3). These give enough confidence to not consider this occurrence as contamination due to the drilling process. The FO of the marker *Quadrum gartneri* is also at the base of this biozone (early Turonian; Burnett 1998; Corbett and Watkins

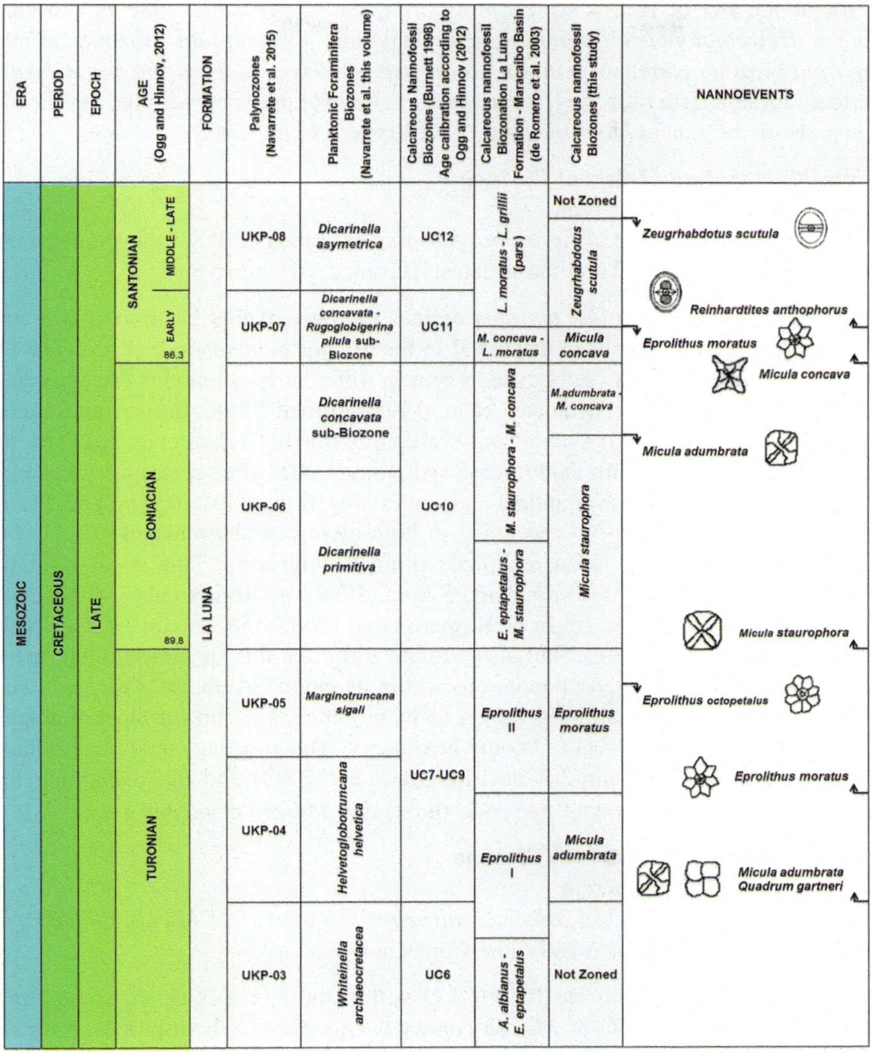

Fig. 4. Proposed calcareous nannofossil biozonation for La Luna Formation in the MMVB and major nannoevents. Correlation with the palinological biozones of Navarrete et al. (2015), the planktonic foraminifera biozones of Navarrete et al. (this volume), and the calcareous nannofossil biozones of de Romero et al. (2003)

2014a), but the continuity of this taxon along the succession is not so good, so it could be difficult to assess the real FO. The FO of *Eprolithus moratus* marks the top of this biozone. In this work, we use *Eprolithus moratus* according to the emendation of Burnett (1998). In her biozonation (Brunett 1998), its FO is used to define the UC6b sub-biozone of the early Turonian. In the Maracaibo Basin, de Romero et al. (2003)

also found the FO of *E. moratus* (*Eprolithus eptapetalus*) at the base of the early Turonian *Helvetoglobotruncana helvetica* planktonic foraminifera biozone. *Micula adumbrata* biozone correlates with the lower part of *Helvetoglobontruncana helvetica* planktonic foraminifera biozone (Navarrete et al., this volume) and with the middle and upper parts of the *Eprolithus* I biozone of de Romero et al. (2003).

***Eprolithus moratus* Interval Biozone**
Author: Pérez Panera, herein.
Definition: From the FO of *Eprolithus moratus* until the FO of *Micula staurophora.*
Range: Within middle Turonian to latest Turonian.

Remarks: FO of *Eprolithus moratus* defines the base of this biozone, and a consideration of this event has been discussed in the *Micula adumbrata* biozone. The FO of *Micula staurophora* is a widely used event in different biozonations (i.e., Sissingh 1977; Perch-Nielsen 1985; Bralower et al. 1995; Burnett 1998; Bergen and Sikora 1999; de Romero et al. 2003; Corbett and Watkins 2014a, b; Corbett et al. 2014) for the base of the Coniacian (recalibrated to Ogg and Hinnov 2012-time scale); so it is a high reliable event. This species is abundant and continuous through its range in the La Luna Formation and is quite easy to recognize in high-dissolved assemblages. The LO of *Eprolithus octopetalus* has been recorded within this biozone. This event is in the middle Turonian according to Bergen and Sikora (1999), de Romero et al. (2003), and Corbett and Watkins (2014a, b). In de Romero et al. (2003) biozonation, it defines the top of the *Eprolithus* I biozone, but also authors warn that this species is often rare in the assemblages and "relatively continuous within its range." At the MMVB, we found it to be rare and sporadic within its range, so its presence is useful for biostratigraphic analysis, but identifying its real LO could be difficult. This biozone could be correlated with the *Eprolithus* II biozone of de Romero et al. (2003) and the planktonic foraminifers *Helvetoglobotruncana helvetica* (*pars*) and *Marginotruncana sigali*.

***Micula staurophora* Interval Biozone**
Author: Pérez Panera, herein.
Definition: From the FO of *Micula staurophora* to the LO of *Micula adumbrata.*
Range: Early Coniacian to early late Coniacian.

Remarks: *Micula adumbrata* has its LO within the late Coniacian, some stratigraphic levels below the FO of *Micula concava.* This event is here proposed for the definition of this biozone due to the consistency of *M. adumbrata* along its range and the easy taxonomic identification. This biozone is correlated with the planktonic foraminifera *Dicarinella primitiva* biozone and lower part of *Dicarinella concavata* subbiozone of Navarrete et al. (this volume). It also correlates with the *Eprolithus eptapetalus–Micula staurophora* and *Micula staurophora–Micula concava* (*pars*) biozones of de Romero et al. (2003) for the Maracaibo Basin.

***Micula adumbrata*–**Micula concava Interval Biozone
Author: Pérez Panera, herein.
Definition: From the LO of *Micula adumbrata* until the FO of *Micula concava.*
Range: Late Coniacian.

Remarks: FO of *Micula concava* has been long used as a marker within the early Santonian (Perch-Nielsen 1985; Burnett 1998). Age calibration of this event according to Ogg and Hinnov (2012) is at the base of the Santonian, as observed by de Romero et al. (2003). *Micula concava* is an abundant and continuous species in the La Luna Formation which makes this event reliable for local correlation. This biozone could be correlated with the upper part of the *Dicarinella concavata* planktonic foraminifera sub-biozone (Navarrete et al., this volume) and the upper part of *Micula staurophora–Micula concava* of de Romero et al. (2003).

***Micula concava* Interval Biozone**
Author: Pérez Panera, herein.
Definition: From the FO of *Micula concava* to the LO of *Eprolithus moratus.*
Range: Early Santonian.

Remarks: According to the revision of Corbett and Watkins (2014b) *Eprolithus moratus* went extinct in the late Turonian. However, many authors use the LO of *E. moratus* for the early Santonian. Taxonomic concept of the species could explain these discrepancies, but Corbett and Watkins (2014b) definition of *E. moratus* coincides with the one we use in this work and the one of Burnett (1998). In the latter, the LO of *E. moratus* is within the early Santonian and used as a sub-biozonal marker. In the biozonation of de Romero et al. (2003) for the Maracaibo Basin, a coeval biozone is defined by the FO of *Micula concava* to the LO of *Lithastrinus moratus* (Fig. 4), but it is not clear which species concept the authors used for *Lithastrinus moratus.* It seems that they refer this species for the valid *Lithastrinus septenarius,* mostly because they state they followed, among others, the taxonomy of Varol (1992). But according to the photograph of *L. moratus* (de Romero et al. 2003: Fig. 6H), the specimen is so badly preserved that could actually be an *Eprolithus moratus*. However, it is most probable that they used the synonym *Eprolithus eptapetalus* Varol 1992 for *Eprolithus moratus.* If this is the case, de Romero et al. (2003) found the same chronostratigraphic distribution for *E. moratus* and for *L. septenarius* than Corbett and Watkins (2014b). This is a problem for local correlation between the MMVB and the Maracaibo Basin and needs a deep review on the *Eprolithus* and *Lithastrinus* morphotypes of both basins. The FO of *Reinhardtites anthophorus* could also be used to identify the top of this biozone. This event has been questioned by Burnett (1998) due to problems of identification in bad preserved assemblages and due to the earlier evolution of this species at high latitudes. However, for low paleolatitudes, it seems to be a reliable event. This biozone is correlated with the planktonic foraminifera sub-biozone *Dicarinella concavata–Rugoglobigerina pilula* of Navarrete et al. (this volume).

***Zeugrhabdotus scutula* Interval Biozone**
Author: Pérez Panera, herein.
Definition: From LO of *Eprolithus moratus* to LO of *Zeugrhabdotus scutula.*
Range: Early Santonian to late Santonian.

Remarks: *Reinhardtites anthophorus* first occurs at the base of this biozone. *Zeugrhabdotus scutula* has its LO within the late Santonian (Burnett 1998), and it proves to be a reliable event for the MMVB. This event lies within the *Dicarinella asymetrica* planktonic foraminifera biozone, which is early to late Santonian. Assemblages of this biozone are the most abundant and had the highest values of species richness in the La Luna Formation. Accompanying taxa of this biozone, which did not often occur in other biozones in La Luna Formation, are: *Chiastozygus synquadriperforatus*, *Cribrosphaerella ehrenbergi*, *Cyclagelosphaera reinhardtii*, *Gartnerago obliquum*, *Prediscosphaera ponticula*, *Retecapsa ficula*, *Zeugrhabdotus bicrescenticus*, *Z. diplogrammus*, *Z. embergeri*, and *Z. trivectis*. This biozone corresponds to the lower part of the *Lithastrinus moratus–Lithastrinus grillii* biozone of de Romero et al. (2003).

5 Conclusions

The calcareous nannofossils of La Luna Formation in the MMVB are strongly affected by dissolution, and biostratigraphic analysis is compromised due to lack of classic biozonation marker species. In this work, the study of the calcareous nannofossil assemblages of the Chuira-2 ST-1 well allows the identification of eleven nannofossil events, mostly based on dissolution-resistant species. Those events were analyzed within the same well, comparing the data from below and above the fault, with data from other sites, and with local biozonations for La Luna Formation in the MMVB (planktonic foraminifera) and the Maracaibo Basin (calcareous nannofossils). Six local biozones for the lower to the upper part of the La Luna Formation in the MMVB are proposed. *Micula adumbrata* Interval Biozone is defined from the FO of *Micula adumbrata* until the FO of *Eprolithus moratus*, which is early Turonian; *Eprolithus moratus* Interval Biozone, from the FO of *Ephrolithus moratus* until the FO of *Micula staurophora* comprising the middle Turonian to latest Turonian; *Micula staurophora* Interval Biozone, from the FO of *Micula staurophora* to the LO of *Micula adumbrata* ranging from the early Coniacian to the early late Coniacian; *Micula adumbrata–Micula concava* Interval Biozone, defined from the LO of *Micula adumbrata* to until the FO of *Micula concava* of Late Coniacian age; *Micula concava* Interval Biozone, from the FO of *Micula concava* to the LO of *Eprolithus moratus*, for the early Santonian; and *Zeugrhabdotus scutula* Interval Biozone, defined from LO of *Eprolithus moratus* to LO of *Zeugrhabdotus scutula* which corresponds to early Santonian to late Santonian.

Acknowledgements. We thank Gran Tierra Energy Colombia Ltd., who acquired Petronorte-UT Midas, for allowing us to publish these results and to Paleosedes EU consulting Colombian company, for funding this work. We also thank Dr. D. Watkins (University of Nebraska, USA) for his useful comments in the review of the MS.

Appendix—List of Species Recognized in This Study

Axopodorhabdus albianus (Black 1967) Wind and Wise 1983
Biscutum magnum Wind and Wise in Wise and Wind 1977
Braarudosphaera *africana* Stradner 1961
Broinsonia parca (Stradner 1963) Bukry 1969 ssp. *expansa* Wise and Watkins in Wise 1983

Broinsonia furtiva Bukry 1969 *emend* Hattner, Wind and Wise 1980
Calculites obscurus (Deflandre 1959) Prins and Sissingh in Sissingh 1977
Calculites sp.
Chiastozygus synquadriperforatus Bukry 1969
Cribrosphaerella ehrenbergii (Arkhangelsky 1912) Deflandre in Piveteau 1952
Cyclagelosphaera reinhardtii (Perch-Nielsen 1968) Romein 1977
Eiffellithus eximius (Stover 1966) Perch-Nielsen 1968
Eiffellithus parallelus Perch-Nielsen 1973
Eiffellithus turriseiffelii (Deflandre in Deflandre and Fert 1954) Reinhardt 1965
Eprolithus moratus (Stover 1966) Burnett 1998
Eprolithus octopetalus Varol 1992
Gartnerago obliquum (Stradner 1963) Reinhardt 1970
Gartnerago segmentatum (Stover 1966) Thierstein 1974
Helicolithus compactus (Bukry 1969) Varol and Girgis 1994
Helicolithus trabeculatus (Górka 1957) Verbeek 1977
Lithraphidites sp.
Micula adumbrata Burnett 1997
Micula concava (Stradner in Martini and Stradner 1960) Verbeek 1976
Micula staurophora (Gardet 1955) Stradner 1963
Prediscosphaera ponticula (Bukry 1969) Perch-Nielsen 1984
Quadrum gartneri Prins and Perch-Nielsen in Manivit et al. 1977
Quadrum sp.
Quadrum svabenickae Burnett 1997
Reinhardtites anthophorus (Deflandre 1959) Perch-Nielsen 1968
Retecapsa crenulata (Bramlette and Martini 1964) Grün in Grün and Allemann 1975
Retecapsa ficula (Stover 1966) Burnett 1998
Retecapsa surirella (Deflandre and Fert 1954) Grün in Grün and Allemann 1975
Rhagodiscus angustus (Stradner 1963) Reinhardt 1971
Rhagodiscus asper (Stradner 1963) Reinhardt 1967
Rhagodiscus reniformis Perch-Nielsen 1973
Rhagodiscus splendens (Deflandre 1953) Verbeek 1977
Thoracosphaera sp.
Tranolithus gabalus Stover 1966
Uniplanarius gothicus (Deflandre 1959) Hattner and Wise 1980
Watznaueria barnesae (Black in Black and Barnes 1959) Perch-Nielsen 1968
Watznaueria biporta Bukry 1969
Watznaueria fossacincta (Black 1971a) Bown in Bown and Cooper 1989
Watznaueria sp.
Zeugrhabdotus bicrescenticus (Stover 1966) Burnett in Gale et al. 1996
Zeugrhabdotus diplogrammus (Deflandre in Deflandre and Fert 1954) Burnett in Gale et al. 1996
Zeugrhabdotus embergeri (Noël 1958) Perch-Nielsen 1984
Zeugrhabdotus scutula (Bergen 1994) Rutledge and Bown 1996
Zeugrhabdotus trivectis Bergen 1994

References

Barrero D, Pardo A et al (2007) Colombian sedimentary basins: nomenclature, boundaries and petroleum geology, a new proposal. ANH publication, Bogotá

Bergen JA, Sikora PJ (1999) Microfossil diachronism in Southern Norwegian North Sea chalks: Valhall and hod fields. In: Jones RW, Simmons MD (eds) Biostratigraphy in production and development geology, vol 152. Geological society of London, Special Publications, pp 85–111

Boesi T, Galea FA et al (1988) Estudio estratigráfico del Flanco Norandino en el sector Lobatera–El Vigía: III Simposio Bolivariano, Exploración Petrolera de las Cuencas Subandinas, Caracas, Memoria, Tomo I, pp 1–41

Bralower TJ, Leckie RM et al (1995) An integrated cretaceous microfossil biostratigraphy. In: Berggren WA et al (eds) Geochronology, time scales and global stratigraphic correlation: society of economic paleontologists and mineralogists (SEPM), vol 54. Special Publication, pp 65–79

Bramlette MN, Sullivan FR (1961) Coccolithophorids and related nannoplankton of the early Tertiary in California. Micropaleontology 2:129–188

Burnett JA (1998) Upper cretaceous. In: Bown PR (ed) Calcareous nannofossil biostratigraphy. Klewer Academic Publishers, Dordrecht, pp 132–199

Cooper M, Addison F et al (1995) Basin development and tectonic history of the Llanos Basin, Eastern Cordillera, and Middle Magdalena Valley, Colombia. AAPG Bull. 79:1421–1442

Corbett MJ, Watkins DK (2014a) Cenomanian through basal Coniacian Calcareous Nannofossil biostratigraphy of the Mancos Shale reference section, Mesa Verde National Park, CO. Stratigraphy 11:97–108

Corbett MJ, Watkins DK (2014b) Transitional forms in the *Eprolithus-Lithastrinus* lineage: a taxonomic revision of Turonian through Santonian species. Micropaleontology 60:175–193

Corbett MJ, Watkins DK et al (2014) A quantitative analysis of calcareous nannofossil bioevents of the Late Cretaceous (Late Cenomanian—Coniacian) Western Interior Seaway and their reliability in stablished zonation schemes. Mar Micropaleontol 109:30–45

Crux J (1982) Upper Cretaceous (Cenomanian to Campanian) calcareous nannofossils. In Lord AR (ed) A stratigraphical index of calcareous nannofossils: british micropalaeontological society, Ellis Horwood, Chichester, p 81–135

Davis C, Pratt L et al (1999) Factors influencing organic carbon and trace metal accumulation in the Upper Cretaceous La Luna formation of the western Maracaibo Basin, Venezuela. In: Barrera E, Johnson CC (eds) Evolution of the Cretaceous Ocean climate system: geological society of America special paper, vol 332. Geological Society of America, Boulder, pp 203–230

de Romero LM, Galea-Álvarez FA (1995) Campanian Bolivinoides and microfacies from the La Luna Formation, westernVenezuela. Mar Micropaleontol 26:385–404

de Romero LM, Odreman OE (1996) Correlation of the Upper La Luna Formation, western Venezuela (abs.). In: Abstract volume, fifth international cretaceous symposium and second workshop of Inoceramids, Freiberg, p 20

de Romero LM, Truskowski I et al (2000) An integrated calcareous microfossil biostratigraphic framework for the La Luna Formation (abs.). In: Extended abstracts, society of economic paleontologists and mineralogists research conference: paleogeography and hydrocarbon potential of the la luna formation and related anoxic systems CD, Caracas, p 3

de Romero LM, Truskowski I et al (2003) An integrated calcareous microfossil biostratigraphic and carbon-isotope stratigraphic framework for the La Luna Formation, western Venezuela. Palaios 18:349–366

Farías A, Crux J et al (1995) Estudio bioestratigráfico de nanofósiles calcáreos de la Formación La Luna y sus equivalentes laterales en el occidente de Venezuela (abs.). In: XV Coloquio de Bioestratigrafiaa, Los Teques, Venezuela, unpaged

Farías A, Pilloud A et al (2000) Biostratigraphic study of calcareous nannofossils of the La Luna Formation and its lateral equivalents in western Venezuela (abs.). In: Extended abstracts, society of economic paleontologists and mineralogists research conference: paleogeography and hydrocarbon potential of the la luna formation and related anoxic systems CD, Caracas, p 5

Farías A, Pilluod A et al (1996) Biostratigraphic and lithostratigraphic study of the La Luna Formation and its lateral equivalents in western Venezuela (abs.). In: Abstracts, sixth North American paleontological convention, Washington, D.C., p 117

Galvis-Portilla HA, Higuera-Diaz I et al (2014) Regional sequence stratigraphy of the upper cretaceous la Luna formation in the Magdalena Valley Basin, Colombia. In: Proceedings unconventional resources technology conference, Denver, Colorado, 25–27 August 2014, Society of Exploration Geophysicists, American Association of Petroleum Geologists, Society of Petroleum Engineers, pp 2721–2730

Gardet M (1955) Contribution à l´étude des coccolithes des terrains néogènes de l´Algérie, vol 5. Publications du Service de la Carte Géologique de l´Algérie (Nouvelle Série), pp 477–550

Garner AH (1926) Suggested nomenclature and correlation of geological formations in Venezuela. In: Petroleum development and technology in 1925: papers presented at symposiums of the petroleum division, American Institute of Mining and Metallurgical Engineers, pp 677–684

Hedberg H, Sass L (1937) Sinopsis de las formaciones geológicas de la parte occidental de la Cuenca de Maracaibo, Venezuela. Bol Geol Min 1:77–120

Konyukhov AI (2012) Oil source rocks at Mesozoic and Cenozoic continental margins: communication 2. Oil source rocks at continental margins in the second half of the cretaceous and in the Cenozoic. Lithol Min Resour 47:419–436

Liddle RA (1928) The geology of Venezuela and Trinidad. J.P. MacGowan, Forth Worth, 552 p

Lorente MA, Durán I et al (1996) Late Cretaceous in western Venezuela: a new biostratigraphical approach (abs.): abstracts second American association of petroleum geologists/Sociedad Venezolano de Geólogos international congress and exhibition, Caracas. Am Assoc Pet Geol Bull 80:1309–1310

Melinte MC, Lamolda MA (2002) Calcareous nannofossils around the Coniacian/Santonian boundary interval in the Olazagutia section (N. Spain). In: Wagreich M (ed) Aspects of cretaceous stratigraphy and palaeobiogeography, vol 15, pp 351–364

Morales LG (1958) General geology and oil occurrences of middle Magdalena valley, Colombia: South America. In: Weeks LG (ed) Habitat of oil. American Association of Petroleum Geologists-AAPG, pp 641–695

Navarrete R E, Parra FJ et al (2015). Maastrichtiense a Mioceno temprano, correlación entre 19 escenarios en el sector norte del Valle Medio del Magdalena mediante palinoestratigrafía y foraminíferos. XV Congreso Colombiano de Geología. Bucaramanga, Colombia. pp 1

Navarrete RE, Parra-Navarrete FJ et al (2016) Foraminíferos, nanofósiles calcáreos y palinomorfos del Cretácico Superior—Daniense del Valle Medio del Magdalena (VMM), Colombia: Bioestratigrafía y Paleoambientes, 11° Congreso de la Asociación Paleontológica Argentina 2016. Roca, Argentina. Resúmenes, p 130

Notestein FB, Hubman FB et al (1944) Geology of the Barco concession, republic of Colombia, South America. Geol Soc Am Bull 55:1165–1216

Ogg JG, Hinnov LA (2012) Cretaceous. In: Gradstein F et al (ed) The geologic time scale 2012, vol 2. Elsevier, pp 813–853

Perch-Nielsen K (1979) Calcareous nannofossil zonation at the Cretaceous/Tertiary boundary in Denmark. In: Symposium on the cretaceous/tertiary boundary events; I. The Maastrichtian and Danian in Denmark. University of Copenhagen, Copenhagen, Denmark, pp 115–135

Perch-Nielsen K (1985) Mesozoic calcareous nannofossils. In: Bolli HM et al (eds) Plankton stratigraphy. Cambridge University Press, Cambridge, pp 329–426

Petters V (1955) Development of upper cretaceous foraminiferal faunas in Colombia. J Paleontol 29:212–225

Prince M, Cáceres C et al (2008) Acordionero oil field discovery, Middle Magdalena Basin, Colombia: an unusual large stratigraphic trap within the Paleocene Lisama Formation. Asociación Colombiana de Geólogos y Geofísicos del Petroleo, pp 1–19

Renz O (1959) Estratigracía [sic] del Cretáceo en Venezuela occidental. Bol Geol Ministerio Minas e Hidrocarburos Venezuela 5:3–48

Renz O (1977) The lithologic units of the Cretaceous in western Venezuela: Memorias, V Congreso Geológico Venezolano, Caracas, Tomo I, pp 45–58

Renz O (1982) The cretaceous ammonites of Venezuela. Birkhäuser Verlag, Basel, 132 p

Rodríguez M, Cáceres C et al (2106) An approximation to the understanding of natural fractured reservoirs in La Luna Formation. In: A study case of from the northern Middle Magdalena Basin, Colombia, XII Simposio Bolivariano -Exploración en Cuencas Subandinas: Bogotá, Colombia, ACGGP, pp 1–8

Rolón L (2004) Structural geometry of the jura-cretaceous rift of the Middle Magdalena Valley Basin—Colombia, thesis M.sc., Department of Geology and Geography, West Virginia University

Sissingh W (1977) Biostratigraphy of Cretaceous calcareous nannoplankton. Geol Mijnbouw 56:37–65

Sutton FA (1946) Geology of Maracaibo Basin, Venezuela. Bull Am Assoc Petrol Geol 30:1621–1741

Tchegliakova N, Pérez V (1995) Registro de la Zona bioestratigráfica *Dicarinella concavata* (Foraminiferida) en el Extremo noroccidental de Suramérica (Colombia). Geol Colomb 19:131–143

Truskowski I, Galea-Alvarez FA , Sliter W (1995) Cenomanian hiatus in Venezuela: geological society of America abstract with programs, 27, p. A-303

Truskowski I, Galea-Álvarez FA et al (1996) Late Cretaceous biostratigraphy of the La Luna Formation, Maracaibo Basin: abstracts, second American association of petroleum geologists/Sociedad Venezolana de Geólogos, international congress and exhibition, Caracas. Am Assoc Petrol Geol, Bull 80:1341

Varol O (1992) Taxonomic revision of the Polycyclolithaceae and its contribution to Cretaceous biostratigraphy. Newsl Stratigr 27:93–127

West J (1994) International petroleum encyclopedia. Penn Well Publishing Company

Navarrete RE, Parra-Navarrete FJ et al (this volumen) Turonian-Campanian Foraminifera Zonation for La Luna Formation, Middle Magdalena Valley Basin, Northern Colombia

Turonian-Campanian Foraminifera Zonation for the La Luna and Lower Umir Formations, Middle Magdalena Valley Basin, Northern Colombia

Rosa E. Navarrete-Parra[1(✉)], Francisco J. Parra[1,2,3], Juan P. Pérez Panera[4,5], Diana Daza[6], Christian Sánchez[6], Mario Prince[6], and Manuel Rodriguez[6]

[1] PALEOSEDES E.U., Departamento de Bioestratigrafía, Transversal 27 #57-49, 111311 Bogotá, Colombia
paleosedes@gmail.com
[2] Géosciences-Environnement Toulouse, Université de Toulouse, UPS (SVT-OMP), CNRS, IRD, 14 Avenue Édouard Belin, 31400 Toulouse, France
[3] Laboratorio de Bioquímico-Estratigrafía, Facultad de Geociencias, Universidad Nacional de Colombia, Edificio Manuel Ancizar (425), Ciudad Universitaria, Bogotá D.C., Colombia
[4] Laboratorio de Bioestratigrafía, Gerencia de Geociencias, YPF Tecnología S.A./ CONICET, Avenida del Petróleo Argentino s/n entre 129 y 143, 1923 Berisso, Buenos Aires, Argentina
[5] Cátedra de Micropaleontología, División Paleozoología Invertebrados, Museo de La Plata, Universidad Nacional de La Plata, Paseo del Bosque s/n, 1900 La Plata, Buenos Aires, Argentina
[6] Gran Tierra Energy Colombia Ltd., Calle 113 No. 7-80 Piso 17, Bogotá, Colombia

Abstract. Well-site biostratigraphy using planktonic foraminifera, palynomorphs, and calcareous nannofossils was performed in Chuira-2ST well, Middle Magdalena Valley Basin, Colombia. Biostratigraphic analysis was focused in the latest Early Campanian marly and sandy-shale deposits of the Umir Formation and the limestone and shale deposits of the La Luna Formation. In this contribution, foraminifera results are discussed and correlated with palynomorphs and calcareous nannofossil data. A foraminifera zonation for the Middle Magdalena Valley Basin is proposed, which is based on the Chuira-2 and Chuira-2ST wells and other exploration wells and outcrop sections in the area. In the La Luna Formation, a total of six globotruncanids zones were established: *Whiteinella archaeocretacea*, *Helvetoglobotruncana helvetica*, *Marginotruncana sigali*, *Dicarinella primitiva*, *Dicarinella concavata*, *Dicarinella concavata–Rugoglobigerina pilula*, and *Dicarinella asymetrica* (earliest part). In the Umir Formation, *Globotruncanita elevata* (latest part), *Globotruncana ventricosa*, and *Globotruncanita calcarata* zones were recognized. The *Dicarinella asymetrica* and *Globotruncanita elevata* zones are not developed in its total range since there is a regional unconformity in the Santonian/Campanian boundary, which corresponds to the La Luna and the

G. C. Cusminsky et al. (eds.), *Advances in South American Micropaleontology*,
Springer Earth System Sciences, https://doi.org/10.1007/978-3-030-02119-1_4

Umir Formations contact and involves a 4.4 Myr hiatus. In the La Luna Formation, paleoenvironments correspond to a marine inner to outer neritic, epicontinental sea with oxygen-depleted bottom waters and episodic upwelling events. Anoxia was more intense at the Middle Turonian and the Coniacian/Santonian boundary. In the Umir Formation, a shallowing trend is observed that ends in littoral to marginal marine paleoenvironments.

Keywords: Planktonic foraminifera · Cretaceous biostratigraphy
Colombian biostratigraphy · Cretaceous South America · La Luna Formation
Umir Formation

1 Introduction

This work presents the results of the foraminifera recovered on a well-site biostratigraphy (WSB) performed at Chuira-2ST well. The wells were drilled in the northeastern area of the Middle Magdalena Valley Basin, (MMVB) Colombia. WSB advanced technology is a tool, which results to assist with the placement of corings, casings, and terminal depth (TD), aids in the reduction of drilling risk and uncertainty assisting in the stratigraphic positioning while supports the caving analysis, to identify stratigraphic position of borehole instability. WSB was achieved by means of integrated biostratigraphy using foraminifera (planktonic and benthonic), calcareous nannofossils and palynomorphs. WSB verified inverse faulting between 8230 and 8263 ft. The fault causes Upper Cretaceous to repeat (Rodríguez et al. 2016a), and WSB reveled the Santonian and Coniacian strata above and under the fault. The biostratigraphy and zones here discussed eliminates 310 ft, which corresponds to the fault zone (8130–8440 ft). Thus we use two intervals: Campanian-Late Coniacian, interval (7880–8130) and Early Coniacian to Early Turonian, interval (8440–9365 ft TD). The 310 m thick of La Luna Formation includes pelagic calcareous deposits rich in planktonic foraminifers also bearing calcareous nannofossils and marine palynomorphs, mostly dinocysts, ranging in age from the Turonian to the Santonian. Overlying, Campanian strata of the base of the Umir Formation also yield calcareous microfossils and palynomorphs dominated by pollen and spores. As exposed by Schiebel (2002) the calcareous tests of planktonic foraminifera produce a substantial amount of carbonate in the world oceans. They just began their radiation in the Early Aptian and contribute 32–80% of the total deep marine carbonates production, therefore planktonic foraminifera figure as one of the most important carbonate producers within the La Luna Formation. Calcareous and mixed rocks are very rich in pelagic components, also including nannofossil and dinocysts.

One-hundred and thirty-three samples from the La Luna Formation were analyzed using foraminifera, calcareous nannofossils (see Pérez Panera et al., this volume) and palynomorphs. These were used to gain insight about ages and hiatus, and also were used to establish a first approximation to the recognition of Oceanographic Anoxic Events (OAE) characterized by the widespread deposition of pelagic sediments rich in organic matter such as black shales that are considered as key mechanisms for organic carbon burial (Arthur et al. 1990; Schlanger and Jenkyns 1976). A dominance of these black shales in shallow water settings in Atlantic and Caribbean regions was noted by some authors (see Arthur and Schlanger 1979; Jenkyns 1980; Hofmann and Wagner 2011). All samples were analyzed in well site. A total of 3785 specimens (17 genera with 47 species) of planktonic foraminifera were counted and classified. The foraminiferal vertical distribution was calibrated with dinocysts (Navarrete-Parra et al. 2015, 2016a, b) and calcareous nannofossils (Pérez Panera et al., this volume). The proposed zonation in this contribution is also tied to the geological time scale based on absolute ages and data from previous studies conducted on two cored boreholes and three outcrop sections distributed across the MMVB. They are Colón-2, Colón-3. Chuira-1 and Chuira-2 wells, and surface outcrops of the La Luna Formation in Aguablanca creek, El Salto creek, and Cuesta El Diablo section (Fig. 1). Additionally, Turonian and Coniacian faunas were also correlated with ammonite biostratigraphy from Aguablanca and El Salto. Previous works on the La Luna Formation showed that Turonian/Coniacian and Coniacian/Santonian boundaries are challenging to establish. This difficulty is due to low recovery of double-keeled planktonic foraminifera using conventional process and picking methods. Biostratigraphic markers as *Dicarinella*, *Marginotruncana* and *Globotruncana* genera are scattered, either missing or very rare. However, those difficulties were overcome by the achievement of total and rapid disintegration of rock obtaining an excellent recovery of specimens from these groups. Additionally, data was also improved using key dated age ranges of non-complex planktonic foraminifera and thin section analyses. The results from thin-slides analyzed are in full-agreement with routine methods. Standard foraminifer's zones of Caron (1985) were identified and *Dicarinella concavata* and *Dicarinella asymetrica* sensu Caron were lightly modified. Two subzones are introduced: *D. Concavata* sub-zone [*D. concavata* Zone Caron (1985) modified], and *Dicarinella concavata–Costellagerina* (*Rugoglobigerina*) *pilula* association sub-zone. This sub-zone has been previously recognized in other scenarios and although it has not been published before, it has been used in the private industry. By means of this sub-zone, the O and G have gained better resolution in stratigraphical interpretations. The Santonian/Campanian boundary is missing due to an hiatus and the late early Campanian shows condensed sedimentation.

The aim of this article is to introduce a zonal scheme for the La Luna and Umir formations that represents a first approximation to the vertical distribution of the deposits of these formations and even further, a review of the spatial distribution of organic-rich sediments in the MMVB (Colombia). The planktonic foraminifera studies allow us to propose this first biostratigraphic framework from the La Luna Formation in the MMVB, which can be applied in Rancheria and Catatumbo basins (Colombia) with high reliability, and to other equivalents as the Villeta Group in the Upper Magdalena Valley, and "Villeta informal group" in the Putumayo Basin. This paper is a first contributive effort using well-site Cretaceous tied with geochronological datings. This will improve our understanding of the biostratigraphy based on microfossils, and helps us to constrain the age of the units and the hiatus. Furthermore, briefly introduce us by means of foraminifera a first vision of the Cretaceous Oceanographic Anoxic Events during sedimentation in the MMVB (Colombia), which are implicit objectives in O and G Exploration.

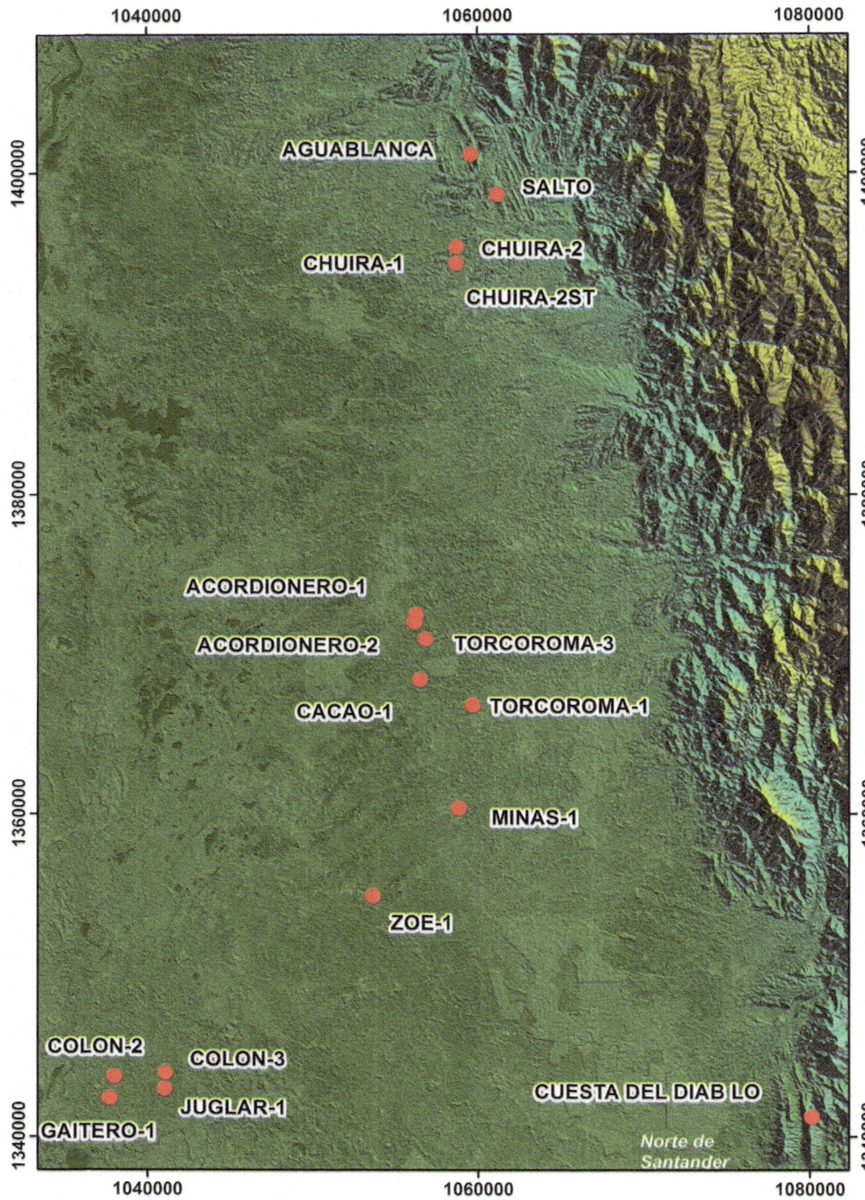

Fig. 1. Localization of Chuira-1, Chuira-2, and Chuira-2ST wells in the northern Middle Magdalena Valley Basin and surface outcrops more referred in the text

The well-site biostratigraphy in Chuira-2ST disclosed a Turonian to earliest Late Santonian age for the La Luna Formation in the MMVB and a hiatus spanning 4.4 Myr in the contact with the overlying latest Early Campanian condensed strata of the lower Umir Formation (Navarrete-Parra et al. 2016c). The very base of the La Luna Formation was not drilled in Chuira-2ST, but in Aguablanca and El Salto sections its base (Salada Member) is of Cenomanian to Early Turonian age.

1.1 Middle Magdalena Valley Basin Background

The Andes Mountains of South America consist of three distinct ranges in Colombia: The Western, Central, and Eastern Cordillera. Each one represents an accretionary and/or deformational event resulting from various instances of rifting, oblique collisions, and transpressional/transtensional tectonics, occasioned by the movements and interactions of the Caribbean, the Nazca and the South America tectonic plates (Barrero et al. 2007). The modern Middle Magdalena Valley Basin (MMVB) (Fig. 2) is an asymmetric, south-north trending intermountain depression in the middle of the Central and Eastern Cordilleras of the Colombian Andes extending over 30,000 km^2. The MMVB according to Barrero et al. (2007) is bound on the north and south by the Espiritu Santo fault system (ESFS) and the Girardot folded belt (GFB), respectively. The Bucaramanga-Santa Marta fault system (BSMF) marks the northeast boundary, and the Bituima and La Salina Fault Systems (BSFS) mark the southeast boundary. The western limit is delineated by the onlap of Neogene sediments over the Serranía de San

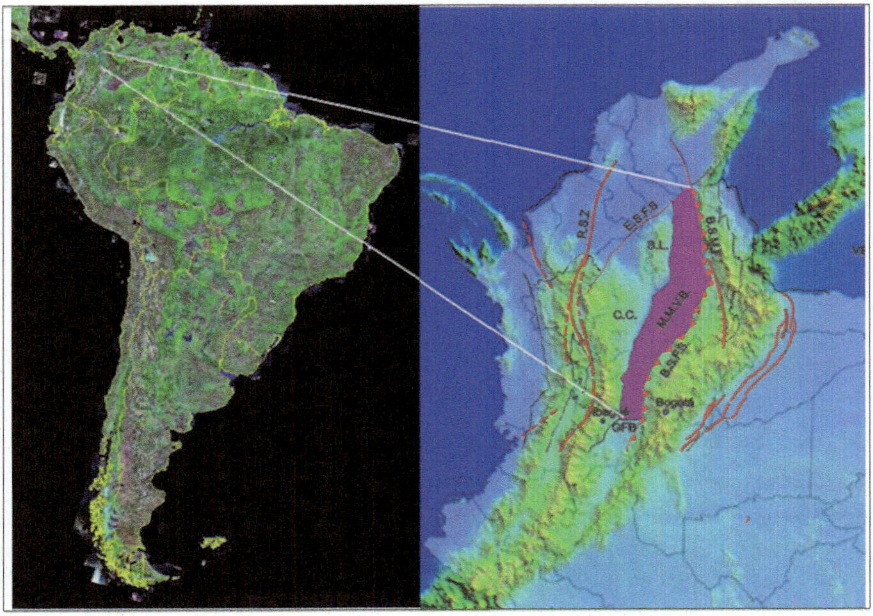

Fig. 2. Colombia in South America, structural boundaries from Middle Magdalena Valley Basin (abbreviations description in text)

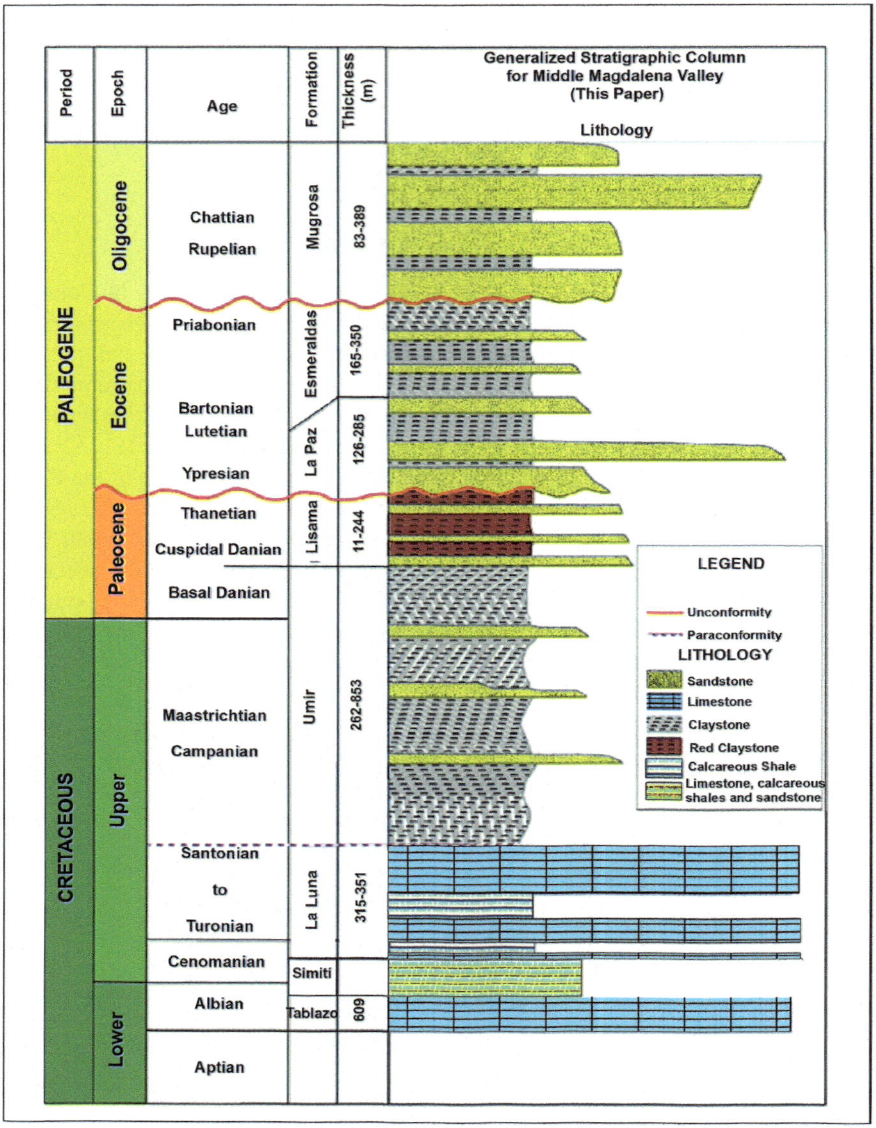

Fig. 3. Generalized stratigraphic column in Middle Magdalena Valley Basin MMVB

Lucas (SL) and Central Cordillera (CC) basement boundary. The Magdalena Valley Basin separates the Central and Eastern Cordillera; however, the underlying basin represents a major regional sedimentary basin that received sediment from the Triassic through the Cenozoic (Fig. 3).

According to Cooper et al. (1995) from the Triassic to the middle Miocene, the MMVB was part of a greater regional sedimentary basin including Eastern Cordillera and Llanos basins. The separation of North and South America during the Triassic and early Cretaceous triggered the development of an extended rifting phase in the Proto-Caribbean. This marked the beginning of a large transgression and deposition into what became the sedimentary basins of Colombia. The MMVB expansion began with an epicontinental transgression deposited in the rifts, related to the separation of North and South America and the opening of the Proto-Caribbean Sea (Cediel et al. 2011). The sedimentary record shows a succession of pre-Cretaceous continental conglomerates deposits. These deposits correspond to Jordan and Girón formations (Cooper et al. 1995). The pre-Cretaceous in MMVB is overlaid by calcareous and siliciclastic rocks deposited in a shallow marine shelf during Lower Cretaceous. These deposits correspond to Cumbre, Rosablanca, Paja and Tablazo formations (basal group) which resemble a marine platform wider than the modern MMVB (Cooper et al. 1995).

Basin development continued into the Cretaceous in a back-arc setting at the east of the Andean subduction zone. The back-arc sedimentation was mainly shallow marine and produced the excellent source rock of the La Luna Formation during the Turonian-Coniacian. Albian to Campanian period includes sedimentation in a shallow marine basin controlled by eustatic changes [(Simití and the La Luna formations: Rolon (2004)]. Marine deposition was abruptly interrupted during the early Maastrichtian due to the final accretion of the Western Cordillera (Cooper et al. 1995). During this time, mafic magmatism associated with the accretion of the Western Cordillera overheated the middle- and outer-shelf shales (Cediel et al. 2011). Deposition was re-established in the early pre-Andean foreland basin after accretion of the Western Cordillera in the late Maastrichtian (Cooper et al. 1995). Paleocene is represented by the siliciclastic Umir Formation. The Paleogene sequence is made up of siliciclastic rocks. These deposits consist of alluvial, coastal plain and estuarine deposits Eocene deformation, caused by the uplifting of the Central Cordillera, temporarily interrupted deposition into the basin once again subsidence and deformation eventually re-established sedimentation in the basin, into which late pre-Andean foreland basin sediments were deposited. These Cenozoic deposits represent environments similar to those in the early pre-Andean foreland basin; coal-rich alluvial plain, coastal plain, and estuarine deposits (Cooper et al. 1995). The Paleogene to early Neogene sedimentation includes continental deposits (i.e. fluvial, lacustrine) which correspond to La Paz, Esmeraldas, Mugrosa, and Colorado formations (Cooper et al. 1995; Caballero et al. 2010). By the middle Miocene, transpressional deformation of the Eastern Cordillera fully isolated the MMVB from adjacent basins. The resulting accommodation space was filled with the Andean foreland basin megasequence (Cooper et al. 1995). This megasequence corresponds to the Real Group and subsequent sedimentation (Caballero et al. 2010). Marine mudstones are present in the lower part of the megasequence; however, fluvial sands and conglomerates dominate the upper portions of the continental interior sequence (Cooper et al. 1995; Cediel et al. 2011).

1.2 Overview of the La Luna Formation

The La Luna Formation was originally named "La Luna limestone" by Garner (1926) in the La Luna creek in the Perijá range (Zulia state, Venezuela). It was formally described as a Formation by Hedberg and Sass (1937) as follows: "It consists of thin-bedded and laminated dense dark gray to black carbonaceous-bituminous limestone and calcareous shale. The limestone beds vary from a few inches to about a foot and one-half in thickness. The formation is particularly characterized by hard black ellipsoidal and discoidal limestone concretions ranging from a few inches to several feet in diameter. Black Chert is common as seams and nodules. The uppermost part of the formation is sometimes lighter in color than is the bulk of the formation. When freshly broken, the limestone frequently has a strong petroleum odor." Later, Notestein et al. (1944) introduced it in the "Barco Concession" (Catatumbo Basin, Colombia). Subsequently, the La Luna Formation was described by Morales (1958) in the MMVB and divided it into three members (from base to top): Salada, Pujamana and Galembo. These members have been described by various authors (e.g., Morales 1958; Reyes et al. 2000; Rangel et al. 2000a, b; Torres et al. 2012). According to Kuuskraa et al. (2013), best organic-rich layers (3–12% TOC) are in the 150 m thick Salada Member, which consists of hard, black, thinly bedded and finely laminated limy shales (40% $CaCO_3$), along with thin interbeds of black fine-grained limestone. Pyrite veins and concretions are common, as are planktonic (but not benthonic) foraminifera and radiolarians. The lower-TOC Pujamana Member consists of gray to black, thinly bedded and calcareous shale (43% $CaCO_3$). The Galembo Member has moderate TOC (1–4%) and consists of black, thinly bedded, calcareous shale, but with only thin argillaceous limestone interbeds. The Galembo also has abundant blue to black chert beds (Kuuskraa et al. 2013). In Venezuela and Colombia, organic-rich, cyclic strata such as laminated black shale-limestone couplets prevailed for most of the Coniacian–Santonian up to the Lower Campanian (de Romero et al. 2003; Rey et al. 2004). Near the Chuira-2ST Rodríguez et al. (2016a) found that in the Cuesta el Diablo, El Salto and Aguablanca creeks outcrops La Luna Formation develops a large stratigraphic sequence, with 211.5, 328 and 382 m of thickness, respectively. Recently, Galvis-Portilla et al. (2014) introduced a detailed description for the La Luna Formation. Unfortunately, great part of their description does not permit to differentiate objectively between the three members as previously described. This is due to disparity in lithological descriptions from the La Luna Formation and, mainly, because the three member sensu Morales (1958) are not always recognized. In this work, we use the scheme for the La Luna Formation as occurring in field sections cited by Rodríguez et al. (2016a) and in Chuira-2ST, without differentiating in the tree members.

The detailed and new biostratigraphical analyses carried out by Paleosedes E.U. biostratigraphic consultant (1987 to present), shows that, indeed, limy shales from the Salada Member correspond to the lower part of globigerinids-mud, with calcareous nannofossils, dinocysts, radiolarians and pithonellids of Cenomanian age (e.g., Cuesta El Diablo). The Chuira-2ST well did not drill the Salada Member lower part, but reached to a stretch of the upper part of this member. The upper hard part consists in black, limy shales with occasional black fine-grain limestones, very rich in calcareous and non-calcareous dinocysts, coccolithophorids and non-keeled foraminifers, frequent calcisphaerulids and rare radiolarians of earliest Turonian age.

In Chuira-2ST, the La Luna is an organic-rich Upper Cretaceous formation, consisting of dark gray to black, white gray and brown-gray laminated and massive calcareous rocks. From base-up, it discloses fossiliferous wackestones and packstones (micrite, biomicrite and spate-micrite) strata, interbedding with calcareous mudstone and subordinate siliceous mudstone and calcareous concretions. This conjunct is rich in foraminifera, calcareous nannofossils, dinocysts, pithonellids, and ammonites. The middle part consists, in general, of white gray and brown-gray colored limestone (micrite and biomicrite), dark brown to black colored calcareous and siliceous mudstone (margas), with benthonic foraminifera richness, in less proportion but abundant planktonic foraminifera, calcareous nannofossils and dinocysts. Bivalves are abundant and ammonites are present. Subordinately, wackestones, packstones and thin sandstones beds, also occur. The upper strata consist of brown gray, dark brown and dark green limestones in wackestones and packstones (micrite, biomicrite and spatemicrite), interbedded with white gray limestones and calcareous mudstones, with subordinated siliceous mudstone. Big and abundant calcareous concretions are characteristic of these levels. It also yields finely disseminated tuffaceous material. Toward the top, bluish and blackish chert beds are common. At the top, mature glauconite and hardgrounds are abundant. The La Luna Formation, marine unit rich in black calcareous shale and limestone with organic-rich facies, was formed in a widespread epicontinental sea. The environments and facies distribution were controlled by the paleo-topography, eustasy, accommodation space, oceanographic events, and climate change while post-depositional tectonics and diagenesis altered its thickness and obliterates its paleontological content in most places. The upper contact with the Umir Formation is well defined in electric logs. The gamma ray log shows a peak at the top of the La Luna Formation due to the glauconitic layers. The Resistivity log abruptly increases due to calcareous concretions and high organic matter contents. Glauconite and hardgrounds are recognized both in surface and subsurface.

1.3 The La Luna Formation in Oil and Gas Reservoir

The MMVB is a mature basin in terms of oil exploration and production and its activity has been largely focused on Cenozoic conventional hydrocarbon reservoirs (Prince et al. 2011). Since the first giant oil field discovery in 1918 (Cira-Infantas Oil Field), more than 100 oil fields had been discovered, adding more than 3990 MMboe of recoverable reserves and cumulative production (Rodríguez et al. 2016a). These proven conventional oil accumulations confirm an active and effective petroleum system. As 40% of the generated hydrocarbons are retained in the source rock (Jarvie et al. 2007), there is a great potential for hydrocarbon accumulations in unconventional organic shales in the MMVB. Thus, this basin has begun to attract efforts in looking for these unconventional reservoirs. La Luna Formation is notoriously the great source rock in Colombia and Venezuela (Talukdar et al. 1985), and one of the most important worldwide (West 1996). Most recent assessment of technically recoverable, continuous shale oil and shale gas resources estimates 233 million barrels of oil (MMBO) and 0.87 trillion cubic feet of gas (TCF) within Cretaceous units in the MMVB (Schenk et al. 2017)

1.4 La Luna Formation Biostratigraphy Background

The northern area of Middle Magdalena Valley produces oil from the La Luna Formation and there is high expectative about a better understanding of its time and space distribution. It has been extensively studied by the Oil and Gas (O and G) industry, but little, or certainly nothing, of these works had been published. The age of the unit has not been accurately constrained or is simply very poorly understood. No planktonic foraminifer's zonation has been published for the Cretaceous of the Middle Magdalena Valley, neither for the La Luna Formation in Colombia. The local stratigraphy was based on benthonic foraminifers (Cushman and Hedberg 1941; Petters 1954, 1955). Petters (1954) introduced the first benthonic zonation that included the La Luna Formation at MMVB. In this zonation, he mentioned high abundance of small Globigerinidae, *Güembelina* spp. and few *Globotruncana* spp. and Radiolarians in the lower part of the La Luna Formation. He also described the ammonite assemblage in those levels. The *Globigerinella–Güembelina* assemblage, mainly composed by pelagic forms, is present in black to dark shales, limestones, and cherts. The ammonite assemblage reported contains the Lower Turonian genera *Fagesia*, *Mammites*, *Metoicoceras*, *Neoptychites*, and *Vascoceras*. Upper Turonian ammonites are characterized by the genera *Coelopoceras*, *Prionotropis*, and the Coniacian assemblage by *Baculites*, *Barroisiceras*, *Peronoceras*, and *Prionocycloceras*. The Upper Cretaceous planktonic foraminifera from Colombia were first studied by Gandolfi (1955), who concentrated his studies on the *Globotruncana* genera. Later local publications like Martinez (1989) on the Maastrichtian of the Catatumbo basin, Tchegliakova (1993, 1995, 1996) and Rodríguez et al. (2016a) on the Upper Cretaceous of the MMVB and Tchegliakova and Pérez (1995) in the Central Eastern Cordillera; resume all the published works on Cretaceous planktonic foraminifera of Colombia. An important contribution is the one of Tchegliakova and Pérez (1995) where the authors recognize for the first time the Coniacian *Dicarinella concavata* Zone. Most micropaleontological studies in the La Luna Formation are available within the literature of Maracaibo Basin (Venezuela). For instance, Boesi et al. (1988) examined nannofossils and foraminifera in surface sections of the southwestern part of the Maracaibo Basin. Davis (1995) examined the planktonic foraminifera from the Alpuf-6 well core in northwestern Maracaibo Basin, and de Romero et al. (2003) developed calcareous nannofossils zonation correlated with foraminifera and stable isotopes analysis for the La Luna Formation in the Maracaibo Basin. For the La Luna Formation in Colombia, although several foraminiferal, dinocysts and calcareous nannofossils biostratigraphical studies have been carried out recently published results correspond to short abstracts in congresses (Navarrete-Parra et al. 2016a, b, c). There are not descriptions of MMVB or Colombian Cretaceous Planktonic Foraminifera, probably due to the lithological characteristics of these units. Late Aptian to Campanian sedimentites are hard calcareous and recrystallized carbonates and have been very poorly and sporadically explored. Most biostratigraphic works were taken for petroleum private industry and due to the hardness and recrystallization of limestone and carbonate biostratigraphic analysis in the La Luna Formation have been improved through thin section analysis.

2 Materials and Methods

Forty-seven micropaleontological samples of La Luna Formation and six of the Umir Formation from Chuira-2ST well (N: 1′394.42504 m and E: 1′058.71599 m Magna Sirgas, Bogota origin) in the middle Magdalena Valley, northern Colombia were studied. The sampled interval (310 m thick) from the Luna Formation consists of massive dark grayish limestones (biomicrites and biosparites) with very rare laminated fine black calcareous mudstone. Sub-mature and highly mature glauconitic grains were observed at the top of the formation. The Umir Formation yields gray siliceous mudstones, calcareous mudstones, and discrete sandstone lamina interlayer. Mature reworked glauconitic grains, from the upper La Luna Formation, were observed in the lower levels. Sediment samples were processed using standard micropaleontological methods. The picking, although attentive on foraminifers, reported also other fossil occurrences as ostracods, echinoids, mollusks, radiolarians, and conspicuous minerals such as pyrite, glauconite, and phosphates, among others. Approximately, 300 foraminiferal tests were picked from the >63 μm fraction and identified under the stereomicroscope, while representative specimens were examined in detail under scanning electron microscope and microphotographs are included. The foraminiferal taxonomy follows classic Cretaceous papers (e.g., Gandolfi 1955; Caron 1985; Bolli 1966; Sliter 1989). The age was sharpened from correlation with surface or reference sections with independent age control, mostly planktonic foraminifers and isotopes. Consequently, our foremost control for the La Luna and Umir Formations is from strata types, other surface sections and multiples core samples. For this paper, we utilized especially the results from well-site biostratigraphy (Chuira-2 and Chuira-2ST), Colon-2 and Colon-3 core samples, and Aguablanca, El Salto and Cuesta El Diablo outcrop sections (Navarrete et al. 2015). These reference sites are located at northern MMVB, Colombia. The proposed zones are calibrated to time scale using 25 absolute dates in those sections and cores, and correlated to palynomorphs (mainly marine dinocysts) and calcareous nannofossils. As a result, foraminiferal biostratigraphy has been integrated with Palynostratigraphy and Calcareous nannofossils biostratigraphy to suggest a geological column within the temporal framework. This column might be helpful in further studies from La Luna Formation in the MMVB and the adjacent Rancheria Basin (northeastern Colombia) and probably in equivalent units across Colombia (e.g., Villeta Group in the Upper Magdalena Valley, "Villeta informal group") in Putumayo Basin, Chipaque in Cordillera Oriental, and Gacheta in Llanos Basin, northern South America sub-Andean basins, Napo Group in Equator and Chonta Formation in Peru, as cited by Mello et al. (1995). For effects of calibration age, we adopted the Gradstein et al. (2004) and Cohen et al. (2013) timescales.

3 The Data Set in Chuira-2 ST and Biostratigraphy

As it has been evident among micropaleontologist, foraminifera species from one basin to another present stratigraphic range differences and mostly range differences occur regionally, a foraminifer's species is infrequently synchronous worldwide. The standard planktonic foraminifer's zonal scheme of Caron (1985) was followed with modifications in the ranges of foraminifer's genera and species from Premoli-Silva and Sliter (1999) and Patterson et al. (2004), together with unpublished absolute dating and stratigraphic distribution of selected genera and species in the MMVB. In this paper, we use the stratigraphic ranges of Late Cretaceous planktonic foraminifera (Table 1) (Paleosedes E.U 2008 to 2016). Stratigraphic ranges are given as first appearance (FA) and last appearance (LA) in mega-annum age (Ma) for each taxon.

In the 1485 ft studied, seven calcareous nannofossil zones (UC-6 a UC-12) (Pérez Panera et al., this volume) and seven dinocysts zones (UKP-03 to UKP-7 and UKP-8 to UKP-10) range from Early Turonian to Early Campanian. La Luna Formation and lower part of Umir Formation (Navarrete-Parra et al. 2015, 2016a, b, c) were also identified and here used for correlation with foraminifera data (Fig. 4).

In the scarce 310-m-thick section of the La Luna Formation, six Globotruncanids zones were identified: *Whiteinella archaeocretacea* (53–54 m), *Helvetoglobotruncana helvetica* (84–85 m), *Marginotruncana sigali* (39–40 m), *Dicarinella primitiva* (67–68 m), *Dicarinella concavata* (22–23 m), *Dicarinella concavata–Costellagerina (Rugoglobigerina) pilula* (24–25 m) and *Dicarinella asymetrica* (earliest part) (7–8 m). According to that, sedimentation of the Late Cretaceous pelagic limestones started in the Turonian and last until the earliest Late Santonian. In the 54-m-thick section corresponding to the lower part of the Umir Formation, *Globotruncanita elevata* (latest part) (4–5 m), *Globotruncana ventricosa* (4–5 m), and *Radotruncana (Globotruncanita) calcarata* (44 m) zones were recognized.

Pelagic sediments from the succession of carbonates of Turonian to late Santonian age and the overlying siliciclastic succession of early late Campanian age are separated by a hiatus which includes the latest Santonian and earliest Campanian. This hiatus involves the latest part of *Dicarinella asymetrica* and earliest part of *Globotruncanita elevata* planktonic foraminifer's zones, as well as the calcareous nannofossil zones UC-13 and UC-14 and dinocyst zones UKP-9 and UKP-10.

Planktonic foraminifera, calcareous nannoplankton and palynomorphs mostly dinocysts, suggest an Early Turonian to earliest Late Santonian age for the studied deposits of the La Luna Formation in the Chuira 2ST well (7880–8130 to 8440–9365 ft) and a latest Early Campanian age for the overlaying deposits of the lower Umir Formation (7700–7915 ft). The planktonic foraminifera record allows differentiation of five intervals in the studied intervals of the Chuira-2ST well. They are here characterized in chronostratigraphic order (Fig. 5a–j).

Interval 9365–8660 ft; ≈210 m thick (17 samples) is a very hard biomicrite and spate-micrite, bearing abundant calcispheres (recrystallized planktonic foraminifer fragments and pithonellids) and high planktonic foraminifera species richness. Minor components are a few small-sized benthonic foraminifera (mainly recrystallized buliminids) and bad preserved radiolarians. Moderate preserved planktonic foraminifera are very abundant. They are commonly silicified and/or infilled with silica and/or recrystallized and always infilled with kerogen and/or black amorphous organic matter. The planktonic foraminiferal fauna is characterized by very abundant small hedbergellids, whiteinellids and heterohelicids abundant marginotruncanids, common dicarinellids, and rare helvetoglobotruncanids, *Globigerinelloides* and *Schackoina*. Most relevant species are *Planoheterohelix (Heterohelix) reussi* (very abundant), *Muricohedbergella (Hedbergella) delrioensis* and *M. (H.) planispira* (common),

Table 1. Stratigraphic ranges in mega-annum age (Ma) for each selected Late Cretaceous planktonic foraminifera species in the Northern Middle Magdalena Valley Basin and the Luna and Umir Formations (after PALEOSEDES 2016)

Late Cretaceous planktonic foraminifera species	Range (FA-LA) (Ma)
Archaeoglobigerina blowi Pessagno 1967	(86.3–65)
Archaeoglobigerina bosquensis Pessagno 1967	(87–83.6)
Archaeoglobigerina cretacea (d'Orbigny 1840) Pessagno 1967	(89.8–69.9)
Contusotruncana fornicata (Plummer 1921) Korchagin 1982	(85.2–67.4)
Costellagerina (*Rugoglobigerina*) *pilula* (Belford 1960)	(86.3–78)
Dicarinella canaliculata (Reuss 1854) Porthault 1970	(93–84.7)
Dicarinella concavata (Brötzen 1934) Porthault 1970	(87–83.9)
Dicarinella imbricata (Mornod 1949) Porthault 1970	(92.7–86.3)
Dicarinella primitiva (Dalbiez 1955) Porthault 1970	(89.8–85.6)
Falsotruncana maslakovae Caron 1981	(92.8–89.2)
Globigerinelloides asper (*asperus*) (Ehrenberg 1854) Pessagno 1967	(92.8)
Globigerinelloides ultramicrus (*ultramicra*) (Subbotina 1949) Caron 1985	(101.6–70.4)
Globotruncana ventricosa White 1928	(79.4–68.4)
Globotruncanella havanensis (Voorwijk 1937) Reiss 1957	(73–66)
Globotruncanita elevata (Brötzen 1934) Reiss 1957	(83.7–73.6)
Helvetoglobotruncana helvetica (Bolli 1945) Reiss 1957	(93–92.5)
Helvetoglobotruncana praehelvetica (Trujillo 1960) Reiss 1957	(93.9–90.7)
Marginotruncana coronata (Bolli 1945) Hofker 1956	(92.52–82.1)
Marginotruncana marginata (Reuss 1845) Bolli, Loeblich and Tappan 1957	(90.3–83.5)
Marginotruncana pseudolinneiana Pessagno 1967	(90.7–83.6)
Marginotruncana renzi (Gandolfi 1942) Hofker 1956	(93–85)
Marginotruncana sigali(Reichel 1950) Hofker 1956	(92.7–85.6)
Muricohedbergella (*Hedbergella*) *simplex* (Morrow 1934) Huber and Leckie 2011	(110.3–89.8)

(*continued*)

Table 1. *(continued)*

Late Cretaceous planktonic foraminifera species	Range (FA-LA) (Ma)
Muricohedbergella (*Hedbergella*) *delrioensis* (Carsey 1926) Huber and Leckie 2011	(117.5–85.4)
Muricohedbergella (*Hedbergella*) *holmdelensis* (Olsson 1964) Huber and Leckie 2011	(FA 89.8)
Muricohedbergella (*Hedbergella*) *planispira* (Tappan 1940) Huber and Leckie 2011	(118.1–89.8)
Planoheterohelix (*Heterohelix*) *moremani* (Cushman 1938) Georgescu and Huber 2009	(101.6–92.7)
Planoheterohelix (*Heterohelix*) *reussi* (Cushman 1938) Georgescu and Huber 2009	(91.8–83.6)
Planoheterohelix (*Heterohelix*) *globulosa* (Ehrenberg 1840) Georgescu and Huber 2009	(83.5–69.3)
Planoheterohelix (*Heterohelix*) *striata* (Ehrenberg 1840) Georgescu and Huber 2009	(80.3–66)
Radotruncana (*Globotruncanita*) *calcarata* (Cushman 1927) El-Naggar 1971	(75.4–72.1)
Rugoglobigerina rugosa (Plummer 1926) Brönnimann 1952	(80.3–65)
Whiteinella archaeocretacea Pessagno 1967	(93.9–86.3)
Whiteinella baltica Douglas and Rankin 1969	(94.7–85.2)
Whiteinella inornata (Bolli 1957) Pessagno 1967	(92.7–84.7)
Whiteinella paradubia (Sigal 1952) Pessagno 1967	(94.5–89.8/88.5)

ERA	PERIOD	EPOCH	AGE		FORMATIONS (SSW-NNE SECTION)	PALYNOZONES VMM Navarrete *et al.* (2015)	PLANKTONIC FORAMINIFERA ZONES Navarrete *et al.* (2016)	PLANKTONIC FORAMINIFERA ZONES Navarrete-Parra R.E., (this work)	CALCAREOUS NANNOFOSSILS ZONES Burnett (1998) Age calibration according to Ogg and Hinnov (2012)	TIME Ma Cohen *et al.* (2013)	GENERAL CONDITIONS MARINE BASIN (this work)
MESOZOIC	*CRETACEOUS*	*LATE* *SENONIAN*	Maastrichtian	LATE	UMIR	UKP-12	*G.g-A.m*	*G.g-A.m*	*UC19 - UC20*	66.0	Marine, middle to outer shelf
				EARLY		UKP-11	*Gt.a -Gta. c*	*Gt. aegyptiaca to Gtlla. havanensis*	*UC17 - UC18*		Quickly deepest Anoxia post-Upwelling
			Campanian	LATE		UKP-10		*Rtna. calcarata*	*UC15 - UC16*	72.1 ± 0.2	Marine, Inner to middle shelf
							Gt. Ventricosa - Gta. Elevata	***Gt. ventricosa*** / *Gta. elevata*	*UC14?*	80.3	Marine, inner -middle shelf Upwelling condensed sedimentation
			83.6± 0.2	EARLY		UKP-09	Sterile and rework	Starvation sediments	Starvation sediments		Hardground formation Sudden relative peack sea level fall
			Santonian	LATE	LA LUNA	UKP-08	*D. asymetrica*	*D. asymetrica*	*UC12*	84.7	Middle to middle shelf, sudden deepest pick, anoxic bottom OAE3
				EARLY		UKP-07	*D. concavata*	*D. concavata- C. (R.) pilula*	*UC11*		
			Coniacian	LATE				*D. concavata*	*UC10*	86.3 ± 0.5	
				EARLY		UKP-06	*D. primitiva*	*D. primitiva*			Inner to middle shelf, upwelling
			Turonian	LATE		UKP-05	*M. sigali*	*M. sigali*	*UC7 - UC9*	89.8±0.3	Marine, Middle shelf Anoxic bottom OAE2
				MIDDLE		UKP-04	*H. helvetica*	*H. helvetica*			
				EARLY		UKP-03	*W. archaeocretacea*	*W. archaeocretacea*	*UC6*		

ABBREVIATIONS: MMV: Middle Magdalena Valley, ICS: International Chronostratigraphic Chart, *G.g.: Gansserina gansseri,* *A.m.: Abathomphalus mayaroensis, Gt. a.: Globotruncana aegyptiaca, Gta. c.: Globotruncanita calcarata, Gtlla.: Globotruncanella* *Rtna c.: Radotruncana calcarata, D.: Dicarinella, M.: Marginotruncana, H.: Helvetoglobotruncana, C.: Costellagerina R.: Rugoglobigerina W.: Whiteinella*

Discontinuity
Sterile & rework

Fig. 4. Summary planktonic foraminifera zones versus palynomorphs (mainly dinocysts) and calcareous nannofossils in Chuira-2ST, Middle Magdalena Valley (northern Colombia)

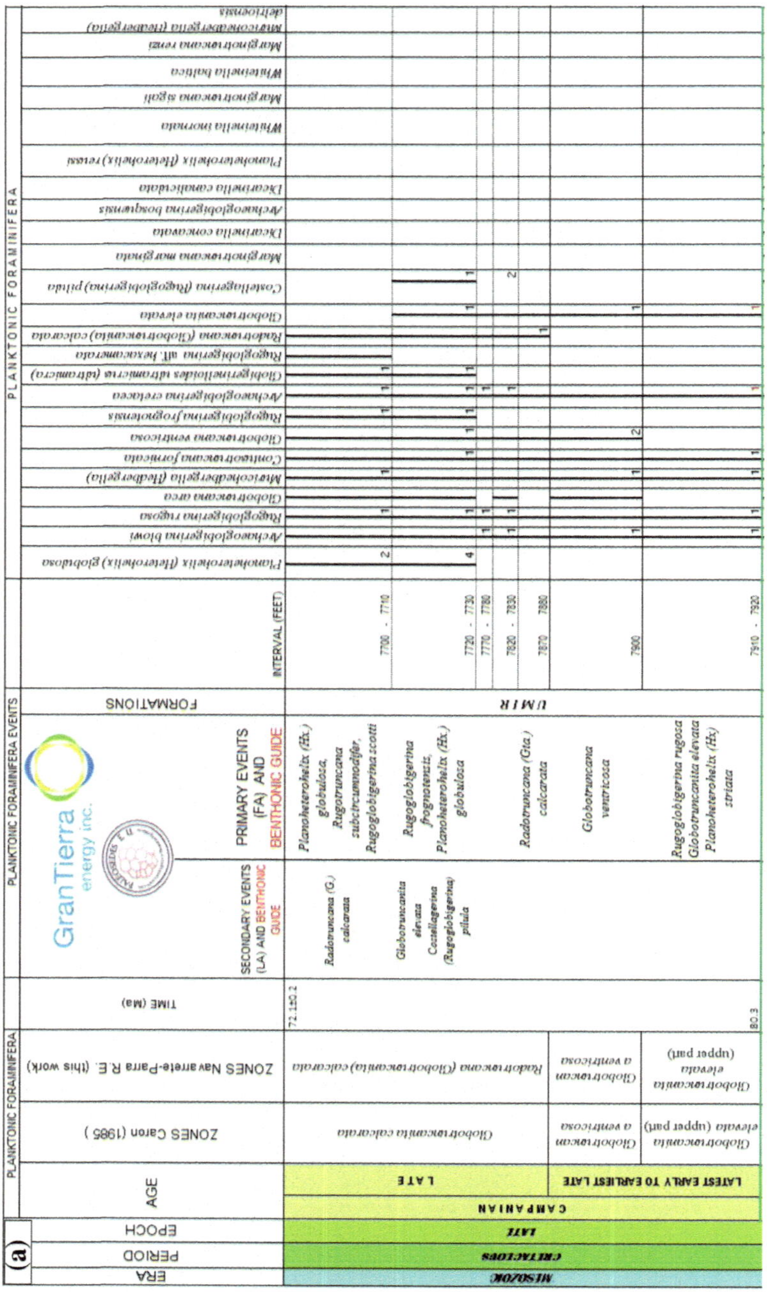

Fig. 5. **a–j** Range chart planktonic foraminifers from Chuira-2ST well in the northern Middle Magdalena Valley Basin (Colombia)

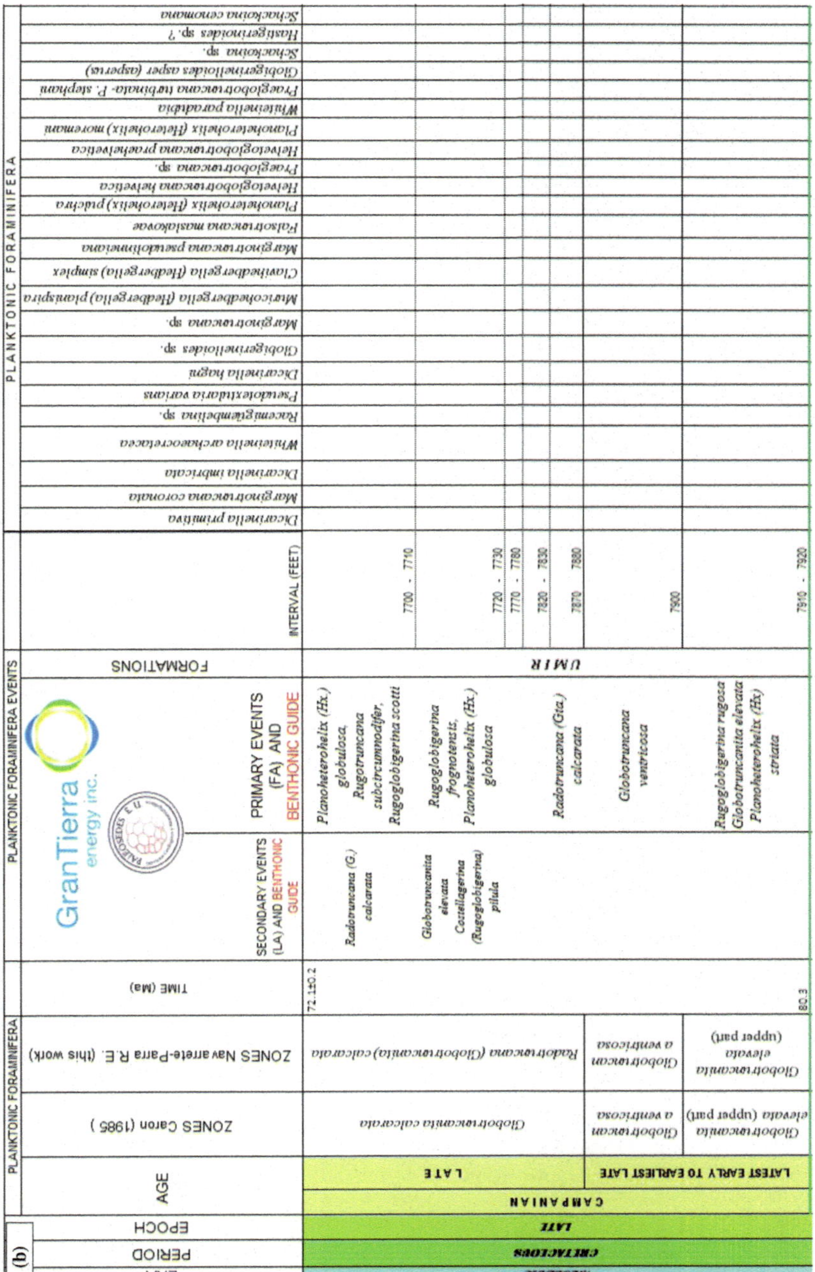

Fig. 5. (continued)

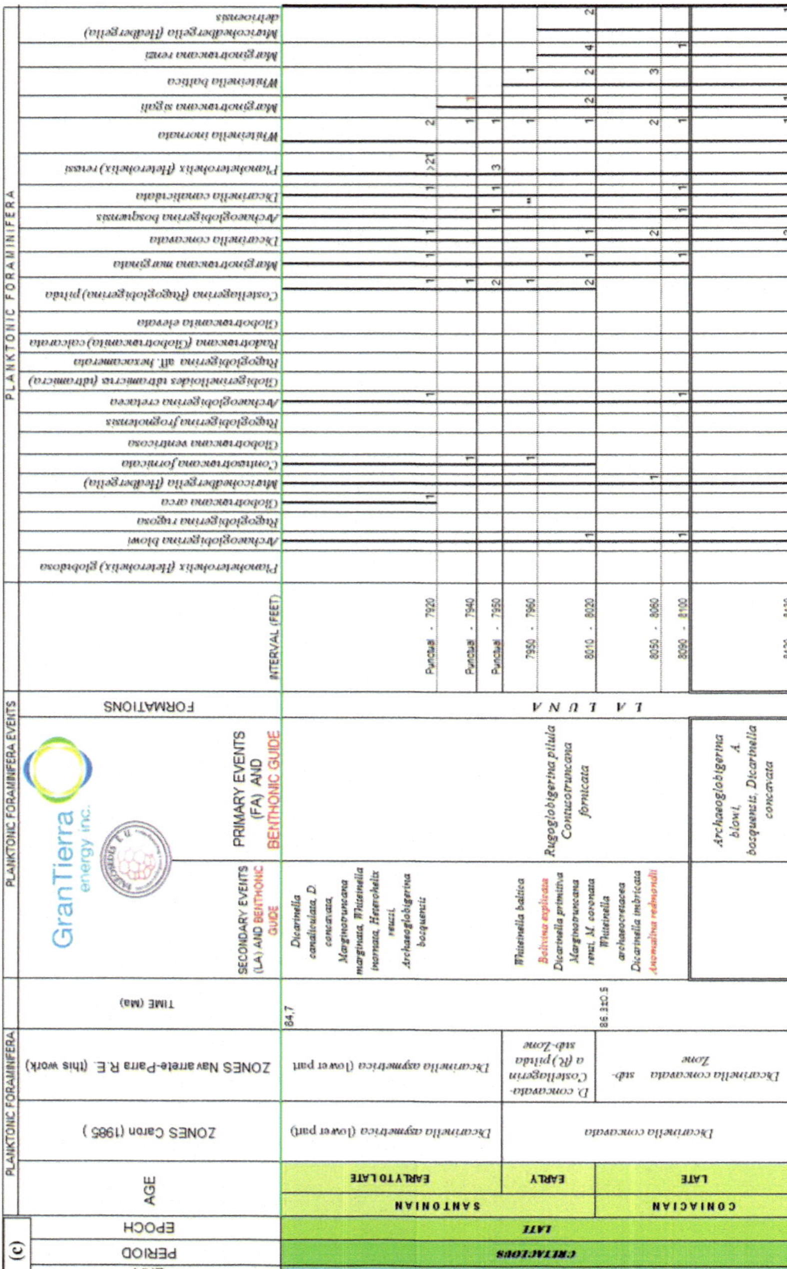

Fig. 5. (continued)

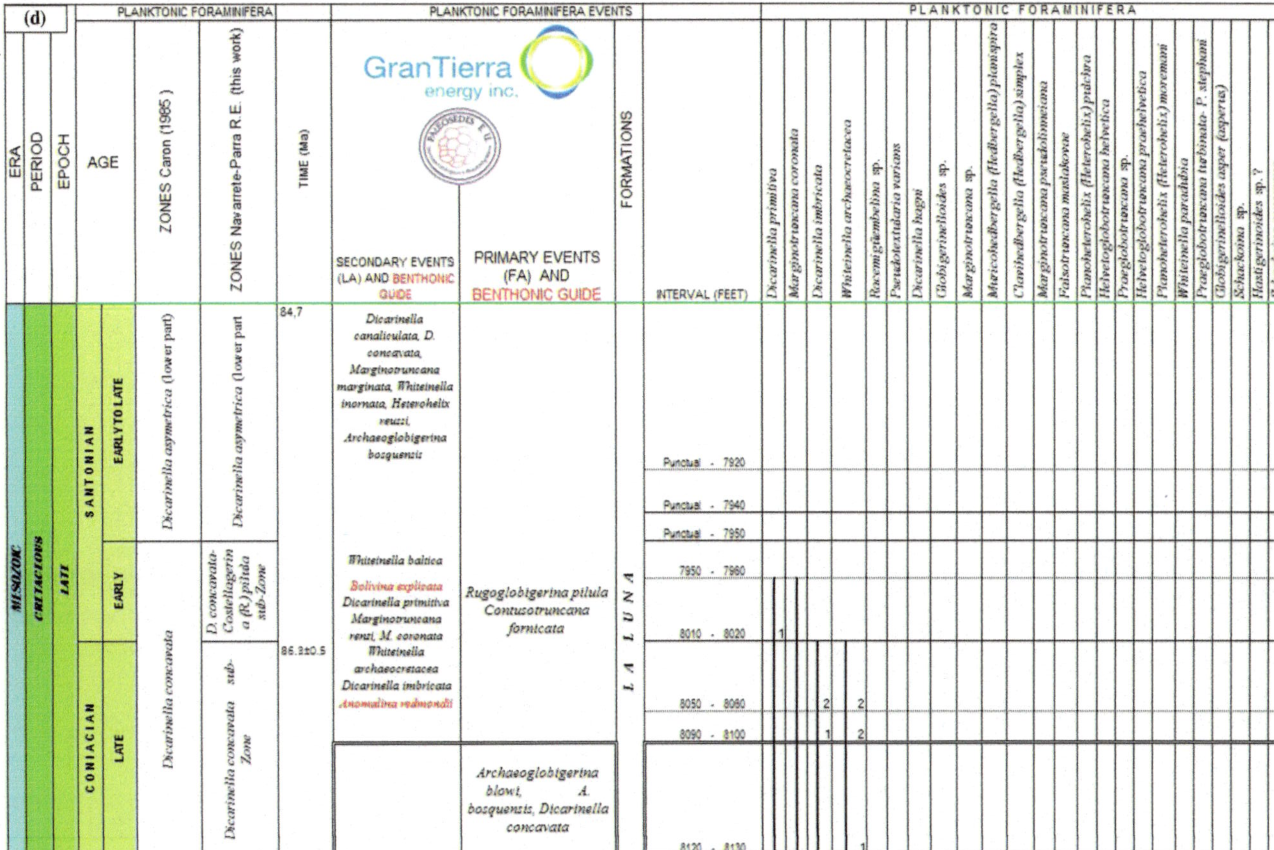

Fig. 5. (continued)

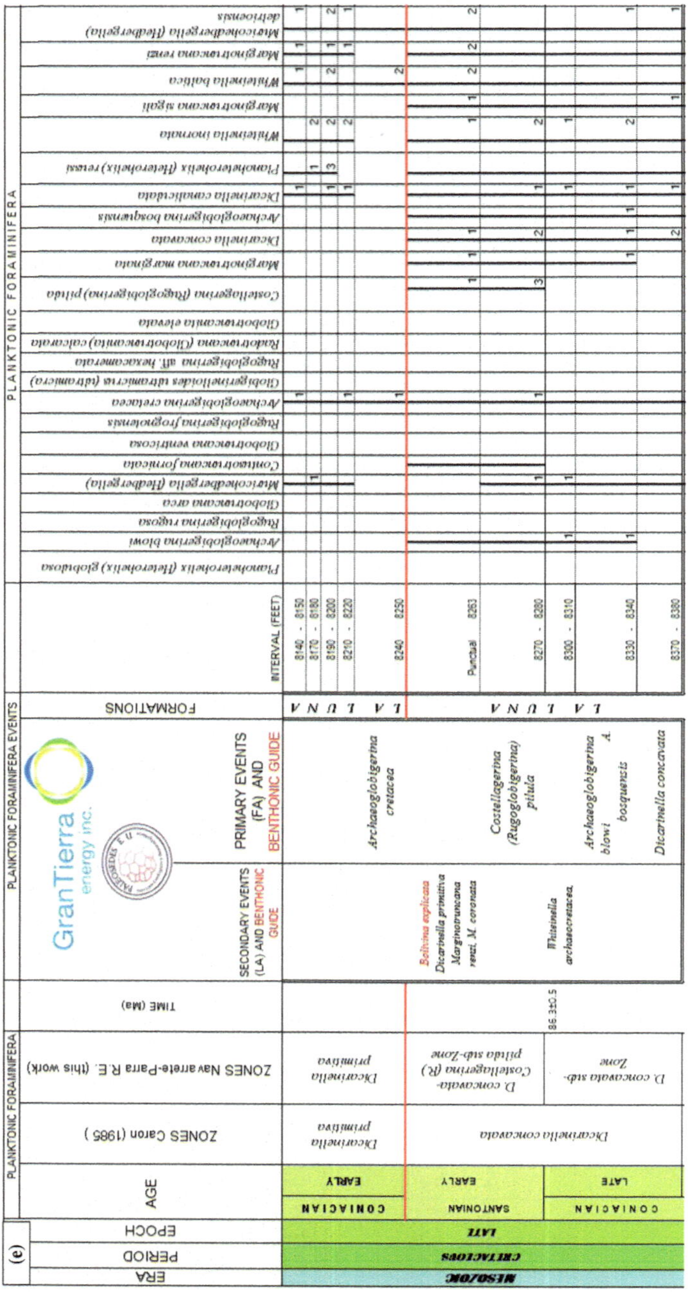

Fig. 5. (continued)

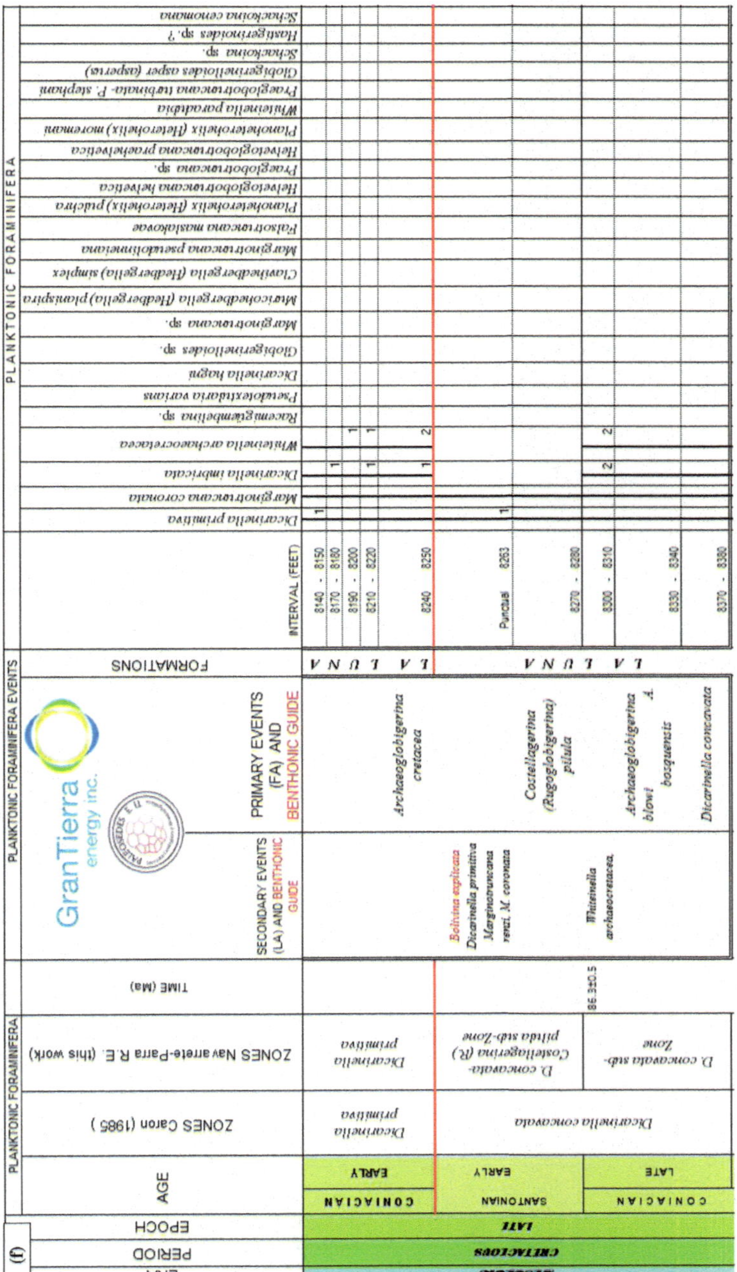

Fig. 5. (*continued*)

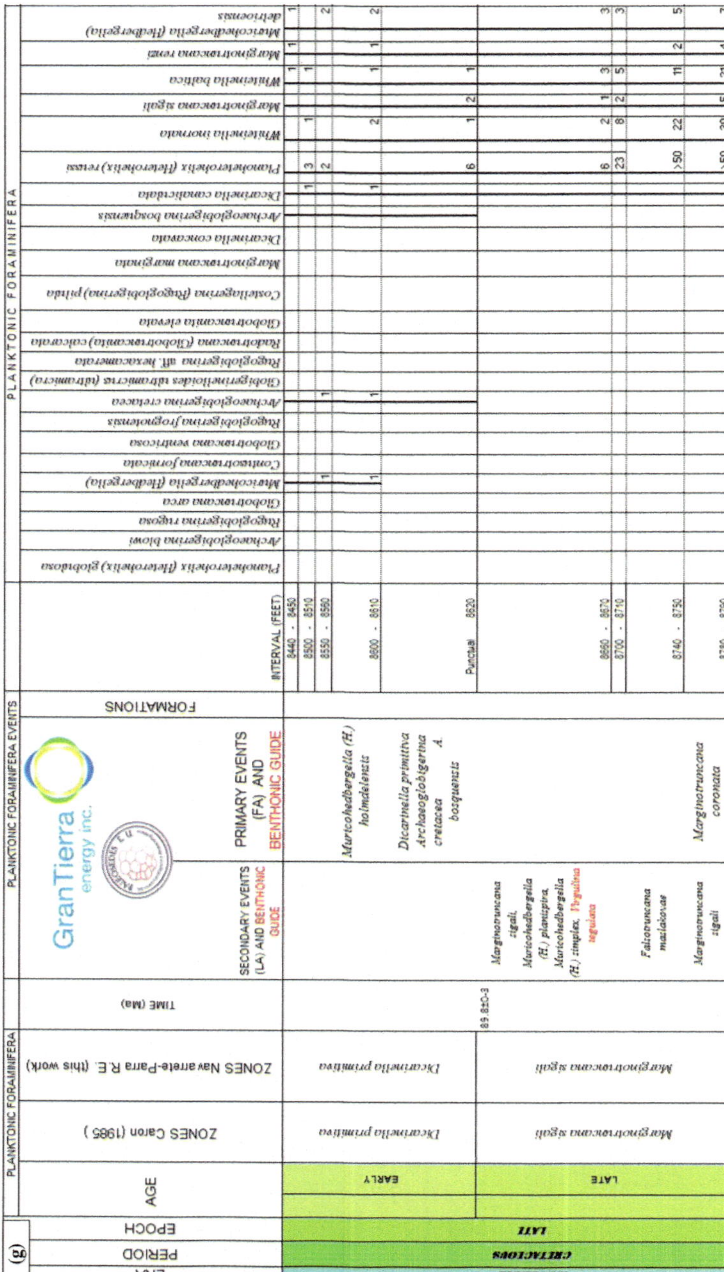

Fig. 5. *(continued)*

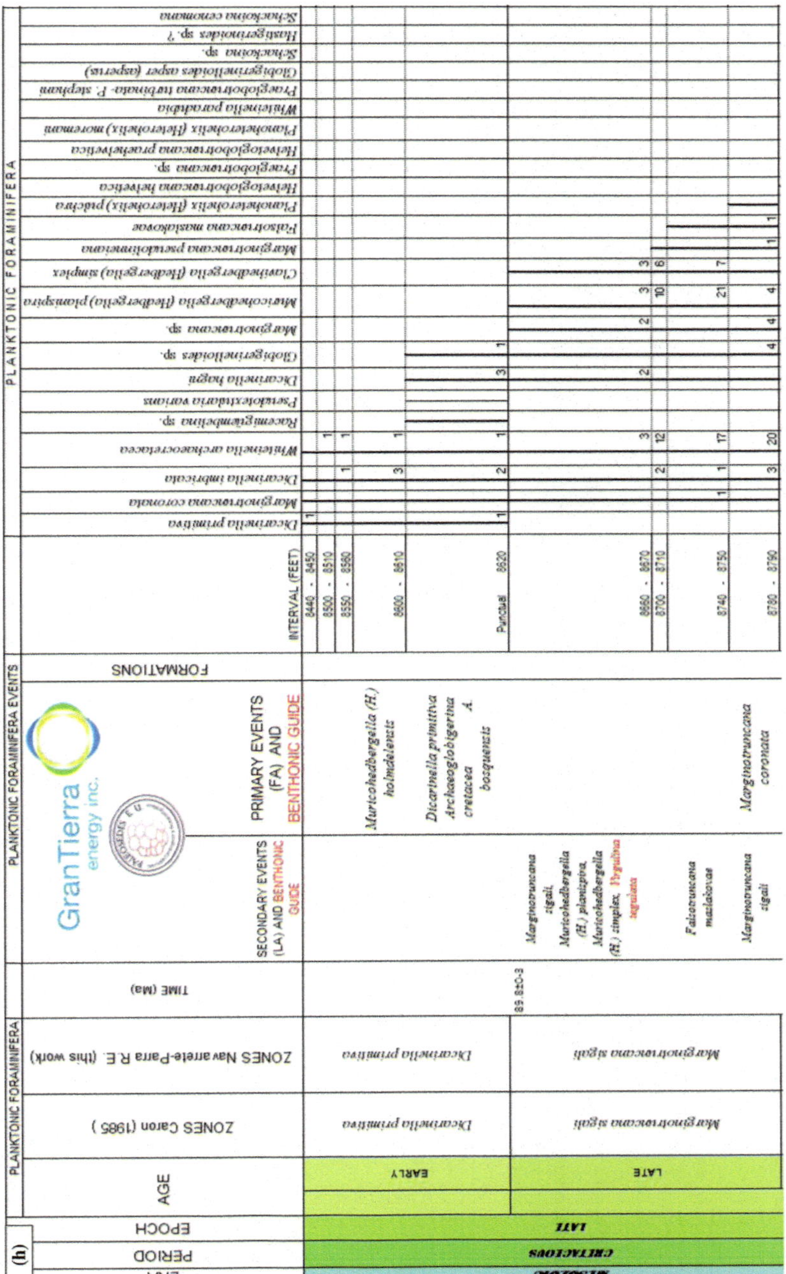

Fig. 5. (*continued*)

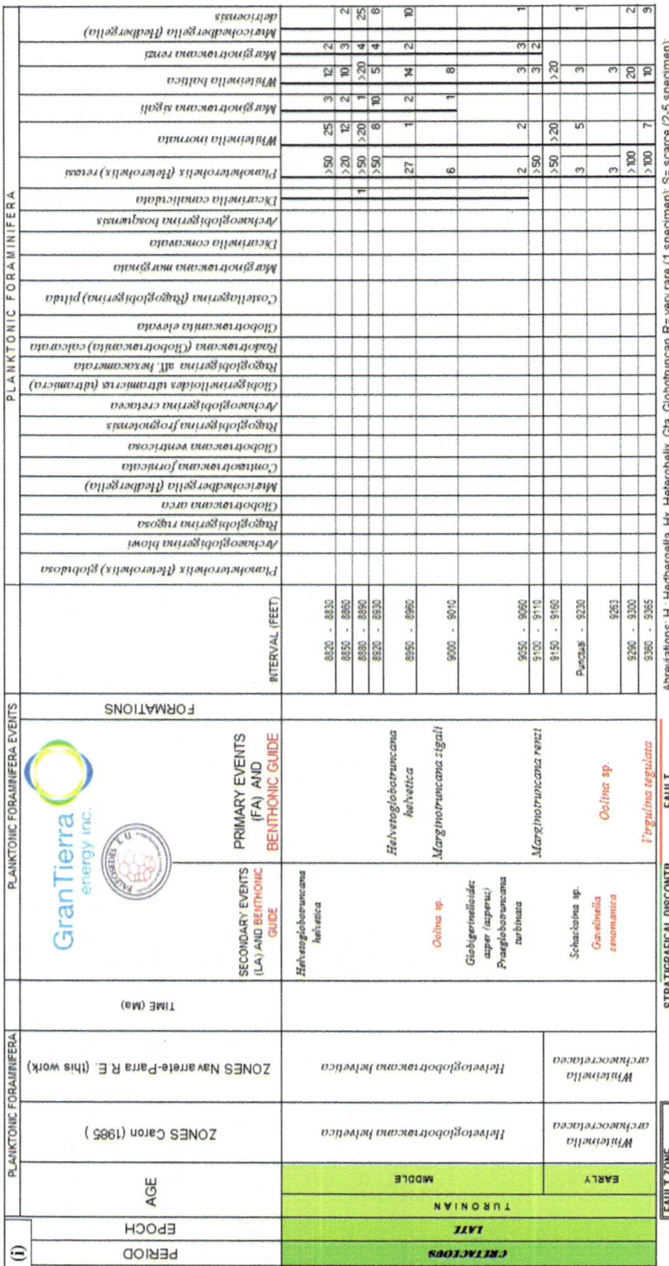

Fig. 5. (*continued*)

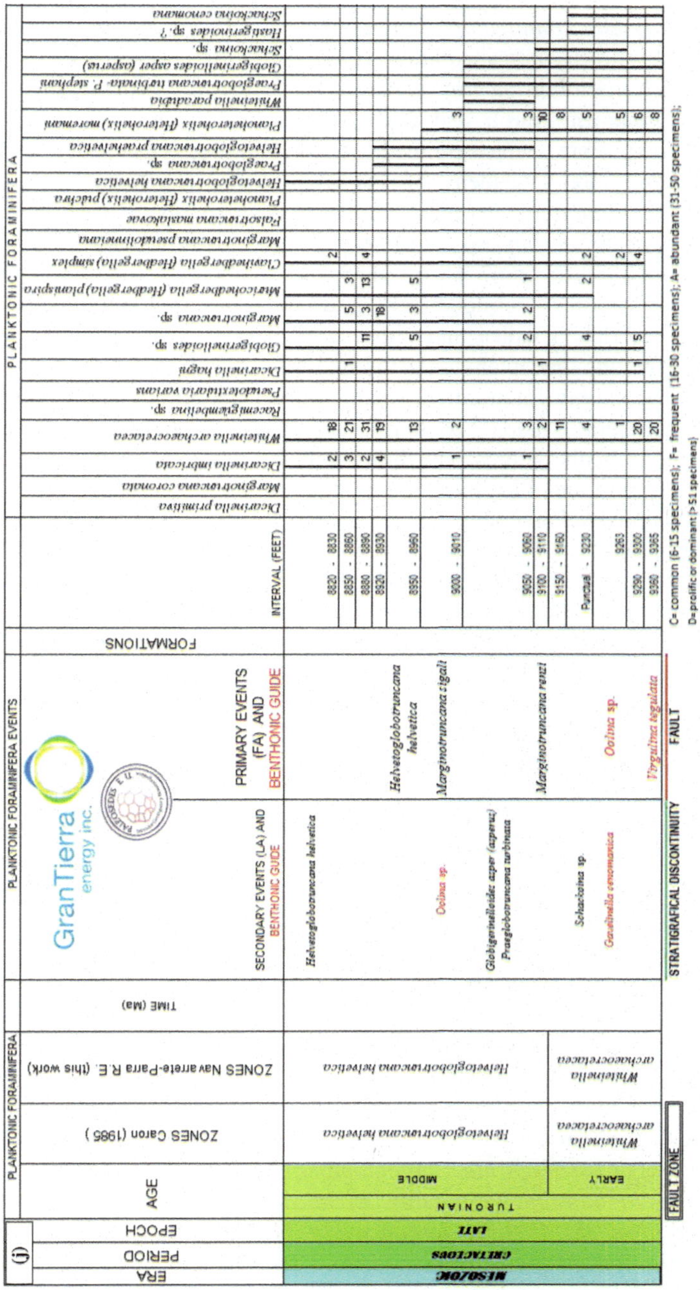

Fig. 5. *(continued)*

Clavihedbergella (*Hedbergella*) *simplex* (scarce), *Whiteinella archaeocretacea* (very abundant), *W. inornata* and *W. baltica* (abundant) and *Marginotruncana sigali*, *M. renzi*, and *Dicarinella canaliculata* (common). Other identified species are *Dicarinella imbricata* (few) *Marginotruncana* sp. (few), *M. coronata* (few), *M. pseudolinneiana* (rare), *Helvetoglobotruncana helvetica* (scarce), *H. praehelvetica* (rare), *Falsotruncana maslakovae* (rare), and *Planoheterohelix* (*Heterohelix*) *pulchra* and *Praeglobotruncana* sp. (rare), which occur dispersed in the upper part only. The lower interval also yielded rare specimens of *Planoheterohelix* (*Heterohelix*) *moremani* (scarce), *Whiteinella paradubia*, *Globigerinelloides asper* (*asperus*), *Praeglobotruncana turbinata–P. stephani*, *Schackoina* sp., *Hastigerinoides* sp.? and *Schackoina cenomana*.

Intervals 8650–8380 ft; ≈67–68 m thick (5 samples) and 8130–8050 ft; ≈22–23 m thick (3 samples) are moderately hard biomicrite, spate-micrite and scarce and dispersed layers rich in silica and phosphates. The biomicrites yield abundant calcispheres, pithonellids, and planktonic foraminifera and abundant homeomorphic, very small-sized, benthonic foraminifera, mainly small buliminids and rotalids. The planktonic foraminiferal fauna contains abundant median and small-sized archaeoglobigerinids, hedbergellids, and heterohelicids, frequent whiteinellids, common marginotruncanids and dicarinellids and rare *Globigerinelloides*, *Pseudotextularia*, and *Racemigüembelina*. Most characteristic species are *Archaeoglobigerina cretacea* (common), *A. bosquensis* (few), *Dicarinella canaliculata* (few), *Planoheterohelix* (*Heterohelix*) *Heterohelix reussi* (few), *Whiteinella inornata* (few), *Marginotruncana sigali* and *Whiteinella baltica* (rare), *M. renzi*, *Muricohedbergella* (*Hedbergella*) *delrioensis* (rare), *Dicarinella primitiva*, *D. imbricata*, and *W. archaeocretacea* (few). *Archaeoglobigerina blowi* (few), *Dicarinella concavata* (few), *Muricohedbergella* (*Hedbergella*) *holmdelensis* (few) and *Marginotruncana marginata* (rare) only occur in the upper levels (8050–8100 ft). Minor components are rare *Racemigüembelina* sp., *Pseudotextularia varians*, *Dicarinella hagni*, and *Globigerinelloides* sp., which occur only at bottom interval (8620–8650 ft).

The interval 8050–7920 ft; ≈32–33 m thick (5 samples) is a very hard and moderately hard biomicrite and spate-micrite. Discrete mixed (calcareous-terrigenous) rocks are interbedded, and they yield phosphatized fish remains and pyrite. The top of this interval has mature glauconite. The biomicrite is rich in robust planktonic and benthonic foraminifera, moderately to poorly preserved and abundant calcispheres and pithonellids. Minor components are few small-sized ooliths and pellets. Planktonic foraminifera are commonly broken and/or infilled with silica, glauconite, kerogen, or amorphous organic matter. Only minor proportion is recrystallized. The planktonic foraminiferal fauna contains abundant heterohelicids, low abundance of hedbergellids, whiteinellids, dicarinellids, and rugoglobigerinids, and rare globotruncanids. The identified species are *Planoheterohelix* (*Heterohelix*) *reussi* (abundant), *Archaeoglobigerina bosquensis*, *Costellagerina* (*Rugoglobigerina*) *pilula*, *Dicarinella canaliculata*, and *Whiteinella inornata* (common); *Archaeoglobigerina blowi*, *A. cretacea*, *Contusotruncana fornicata*, *Dicarinella concavata*, *Muricohedbergella* (*Hedbergella*) *holmdelensis*, *Marginotruncana marginata* (rare). *Globotruncana arca* (rare) occurs in the upper part of the interval (7920 ft). Rare specimens of *Dicarinella primitiva*, *Muricohedbergella* (*Hedbergella*) *delrioensis*, *Marginotruncana coronata*, *M. renzi*, and *Whiteinella baltica* occurs in the lower part of the interval (8010–8050 ft).

Interval 7920–7700 ft; ≈54–55 m thick (6 samples) is a monotonous soft gray mudstone and dark greenish gray shale with subordinated fine sandstone interbedded. The lower strata had few calcareous content, abundant blackish organic matter and carbonaceous particles. Disseminated pyrite is frequent, and toward the base, highly mature reworked glauconitic grains are abundant. The shale and mudstone are poor in planktonic foraminifera, but moderately to poorly preserved benthonic are abundant. There are not calcispheres, which are very abundant in the oldest intervals. Scarce components are few small-sized pellets and bioturbation. Moderately preserved planktonic foraminifera are commonly broken and/or lack the shell. Inner molds and crusts are always infilled with glauconite or kerogen and amorphous organic matter, and only a minor proportion of foraminifera is recrystallized. Specimens with regular preservation are scarce. The planktonic foraminiferal fauna contains rare heterohelicids, hedbergellids, rugoglobigerinids, and globotruncanids. The identified species are *Radotruncana* (*Globotruncanita*) *calcarata*, *Globotruncanita conica*, *G. elevata*, *Globotruncana arca*, *G. aegyptiaca*, *G. ventricosa*, *Contusotruncana fornicata*, *Globotruncanella citae*, *Rugoglobigerina rugosa*, *R. scotti*, *R. frognotensis*, *R.* aff. *hexacamerata*, *Costellagerina* (*Rugoglobigerina*) *pilula*, *Rugotruncana subcircumnodifer*, *Archaeoglobigerina blowi*, *A. cretacea*, *Globigerinelloides ultramicrus*, *Pseudogüembelina excolata*, *Planoheterohelix* (*Heterohelix*) *globulosa*, and *Muricohedbergella* (*Hedbergella*) *holmdelensis*.

4 Foraminiferal Zonation for the La Luna Formation (Turonian-Santonian) and the Overlying Lower Umir Formation (Campanian)

We studied the foraminifera and palynomorphs (mainly marine dinocysts) from six localities in the MMVB. Upper Cretaceous was recognized in stratigraphical profiles including the sections of Aguablanca and El Salto streams Cuesta El Diablo section, Chuira-1, Chuia-2, and Chuira-2ST wells. Chuira-2ST well spans great part of the stratigraphy of the La Luna Formation, and calcareous nannofossils were also studied in well-site (Pérez Panera et al., this volume). We also include absolute dates (Paleosedes unpublished Reports 2010; 2015 for Petrocaribe, Petronorte and Petronorte-UTMidas) of the most significant taxa. The zones were compared with the calcareous nannofossils (Pérez Panera et al., this volume) and dinocysts results for calibration purposes (from Chuira-2, Chuira-2ST) well-site integrated biostratigraphy and Navarrete-Parra et al. (2015). The outcrops and core samples were used to estimate error by caving occurred in normal drilling processes. Private Biostratigraphical Program (Paleosedes 2015) was used to improve the framework. Chuira-2ST completed a total depth 9365 ft and Cenomanian stage was not drilled. The constrained biostratigraphy spans from the Turonian to Early Campanian stage. The selected taxa for the proposed biozonation have wide-world distribution, are relatively common, and were recognized in each analyzed profile (Figs. 6a, b and 7a, b) and microphotographs are included. Keeled foraminifera were relatively scarce, whereas unkeeled were very abundant, as it commonly occurs in epicontinental seas. When transcontinental zonal

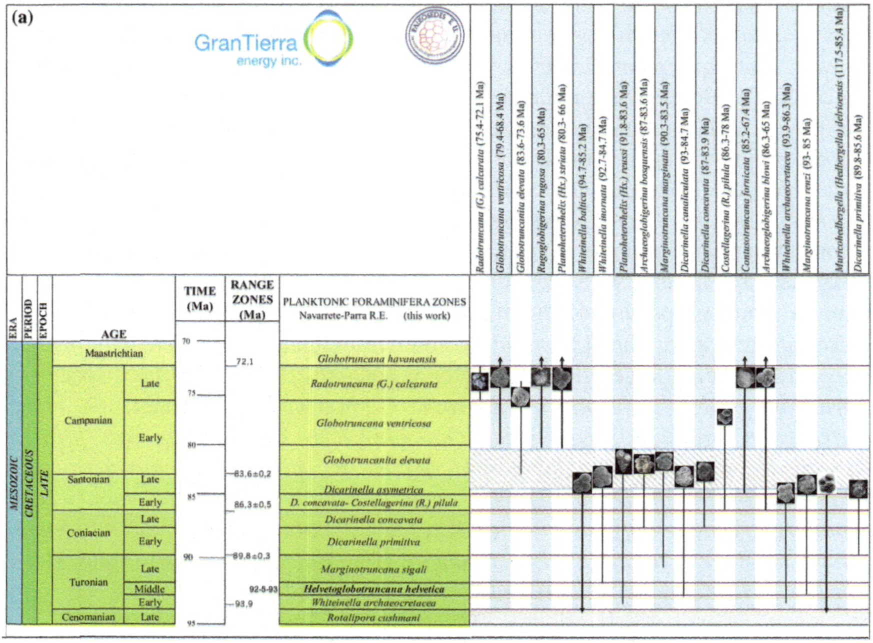

Fig. 6. **a**, **b** Stratigraphic ranges, in mega-annum age (Ma) for selected Late Cretaceous planktonic foraminifera species in the Northern Middle Magdalena Valley Basin and the Luna and Umir Formations

index keeled foraminifers were absent, they were replaced by other species with the same zonal value. Early Turonian and Early Coniacian zones were additionally constrained by means of ammonites from Aguablanca and El Salto sections according to Etayo-Serna (2010).

4.1 Turonian

High richness planktonic foraminiferal assemblages characterize the Turonian stage. Three foraminiferal zones Caron (1985) were identified: (1) *Whiteinella archaeocretacea*, (2) *Helvetoglobotruncana helvetica*, and (3) *Marginotruncana sigali* zones. The Turonian comprises the coeval zones UKP-03–UKP-05, *Florentina mantellii* interval dinocysts zones, of Turonian age (93.9–89.8 ± 0.3 Ma) (Navarrete-Parra et al. 2015), and the calcareous nannofossil UC-6 to UC-9 zones of Burnett (1998) and (Pérez Panera et al., this volume). These Turonian zones were also identified in the Aguablanca and El Salto creeks and Cuesta El Diablo section.

Whiteinella archaeocretacea **Partial Range Zone**. This zone was originally defined as the interval with *Whiteinella archaeocretacea*, from the last appearance (LA) of *Rotalipora cushmani* to first appearance (FA) of *Helvetoglobotruncana helvetica.* The age of this zone is Early Turonian (Bolli 1966; Caron 1985). According to Gradstein et al. (2004) and Cohen et al. (2013), it spans from 93.9 to 93 Ma.

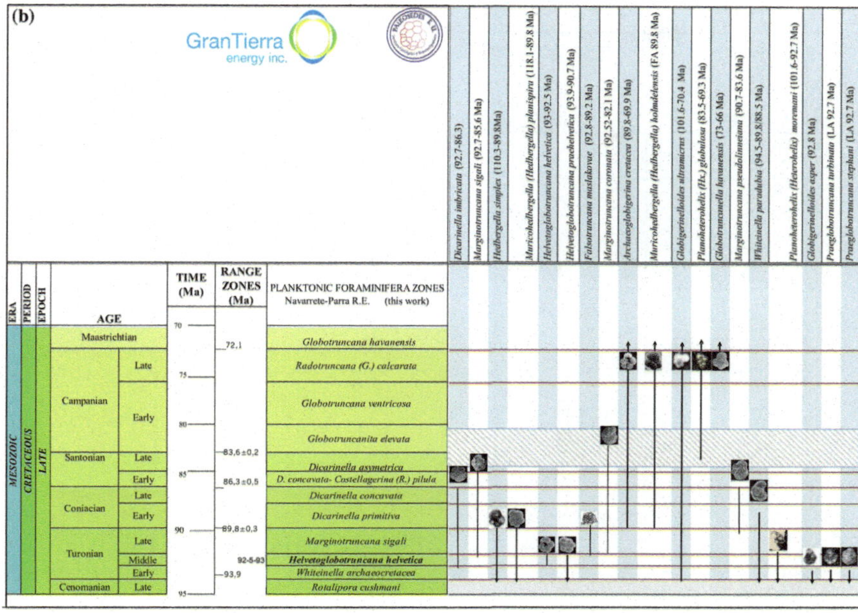

Fig. 6. (*continued*)

This partial range zone was identified in Chuira-2ST well (interval 9365–9150 ft), in Aguablanca and El Salto creeks and Cuesta El Diablo section. The bottom interval in Chuira-2ST well (9365–9150 ft) probably corresponds to the upper part of *Whiteinella archaeocretacea* Zone. This interval (9365 ft) yields abundant *Whiteinella inornata* (FA 93.1 Ma), *W. baltica*, and *Planoheterohelix* (*Heterohelix*) *reussi* (93.1–83.6 Ma), common *Whiteinella archaeocretacea* (93.9–86.3 Ma) and *Planoheterohelix* (*Heterohelix*) *moremani* (101.6–92.7 Ma). These suggest an age not older than 93.1 Ma, and it could be said that *W. archaeocretacea* Zone, whose duration is very brief (0.6 Ma), was reached until the 93.1 Ma, early Middle Turonian. The lower part was probably not reached by the perforation, since *R. cushmani* was not found. *Marginotruncana sigali* (92.7–85.6 Ma) and *M. renzi* (93–85 Ma) are present at 9110 ft, with their FAs constrained at 92.7 and 93 Ma, respectively, occur in the upper part of the zone. *Helvetoglobotruncana praehelvetica* occurs at 9060–9050 ft, and *Helvetoglobotruncana helvetica* (93–92.5 Ma) at 8960–8950 ft suggesting a Middle Turonian age. In addition, the dinocyst *Alterbidinium acutulum* (92.9–56 Ma) is present in the interval 9140–9250 ft, which indicates an age no older that early Turonian. Therefore, the early Turonian/Middle Turonian boundary is located at 9140 ft.

The precedent Early Turonian foraminifera were also found in the early part of *Whiteinella archaeocretacea* Zone in Aguablanca and El Salto creeks, accompanied by singular ammonite assemblages. According to Etayo-Serna (2010): ES-243 m sample contains *Benueites mosquerae* (Karsten 1858), *Kamerunoceras* cf. *seitzi* (Riedel 1933) and *Anomia* cf. *papyracea* (Deheé 1924?), and ES-245 m sample contains *Benueites*

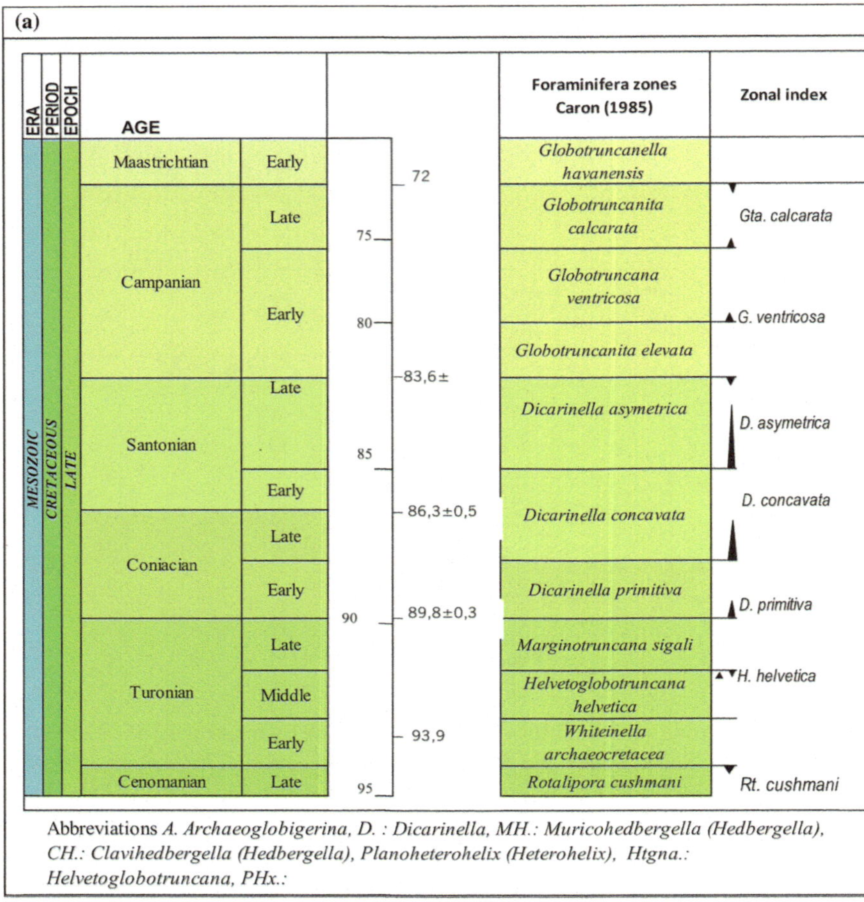

Fig. 7. **a**, **b** Schematic planktonic foraminifera zones (this paper) indicating respective First Appearance Datum (FAD) and Last Appearance Datum (LAD) compared to Caron (1985)

spinosus (Reyment 1954), all taxa from late Early Turonian age. The AB-305 m sample contains *Mammites nodosoides* (Schlueter 1871), from lower Early Turonian age. The ammonites found here confirm an Early Turonian age. The Cuesta El Diablo section ammonites were not found, but foraminifera span a coeval interval with *Whiteinella archaeocretacea* to *Helvetoglobotruncana helvetica* zones.

Helvetoglobotruncana helvetica **Total Range Zone**. This zone was originally defined as the total range interval of *Helvetoglobotruncana helvetica* and was assigned to the Middle Turonian (Sigal 1955; Caron 1985). According to Gradstein et al. (2004) and Cohen et al. (2013), it spans 93–92.5 Ma. *Helvetoglobotruncana helvetica* Zone is recognized in Chuira-2ST interval 9110–8820 ft. This was also recognized in Agua-blanca and El Salto creeks and Cuesta El Diablo section. *H. helvetica* was found in the 8960–8820 ft interval, but some taxa with coeval or close FA of occurs in lower levels.

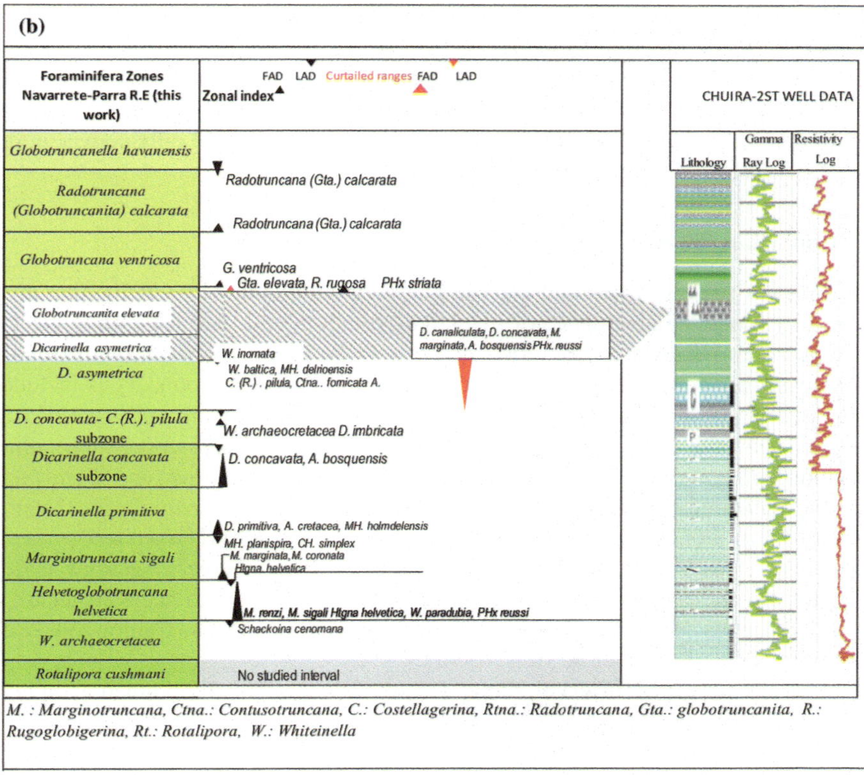

Fig. 7. (*continued*)

Those are *Helvetoglobotruncana praehelvetica* (93–92.5 Ma), *Marginotruncana renzi* (93–85 Ma), *M. sigali* (92.7–85.6 Ma), *Planoheterohelix* (*Heterohelix*) *moremani* (101.6–92.7 Ma), *Praeglobotruncana turbinata–P. stephani* (LA 92.7 Ma), *Whiteinella paradubia* (94.5–89.8/88.5 Ma), and *Globigerinelloides asperus* (LA 92.8 Ma). Therefore, we included *Helvetoglobotruncana helvetica* Zone in those lower levels.

Marginotruncana sigali **Partial Range Zone**. This Zone was originally defined as the interval with *Marginotruncana sigali*, from LA of *Helvetoglobotruncana helvetica* to FA of *Dicarinella primitive*. Late Turonian age was assigned (Barr 1972; Caron 1985). According to Gradstein et al. (2004) and Cohen et al. (2013) timescales, it spans from 92.5 to 89.8 Ma. The lower boundary is placed at the LA of *Helvetoglobotruncana helvetica* (92.5 Ma) at 8820 ft and the upper at the FA of *Dicarinella primitiva* (89.8 ± 0.3 Ma) at 8620 ft. This zone also comprises the LA of *Muricohedbergella* (*Hedbergella*) *planispira* and *Clavihedbergella* (*Hedbergella*) *simplex* and an unusual record of *Falsotruncana maslakovae*, which is very rare in the Middle Magdalena Valley. It also comprises the FA of *Marginotruncana sinuosa*, *M. marginata*, and *M. coronate*. *Planoheterohelix* (*Heterohelix*) *reussi*, *Whiteinella baltica*, and *W. archaeocretacea* are common. The *Marginotruncana sigali* sensu *strictu* was identified in Aguablanca and El Salto creeks; and a coeval interval was recognized in the Cuesta El Diablo section, although in the absence of *Marginotruncana sigali* zone marker species, Late Turonian age is revealed by the co-occurrence of *Clavihedbergella (Hedbergella) simplex* and *Planoheterohelix (Heterohelix) reussi*, and also based in the presence of *Trichodinium castanea* (dinocysts), which only coexist at that stage (see Turonian-Coniacian boundary).

4.2 Coniacian (89.8–86.3 Ma)

The Coniacian stage in Chuira-2ST intervals (8620–8440 and 8130–8020 ft) is characterized by moderate abundance of planktonic foraminiferal, and the planktonic foraminiferal zones *Dicarinella primitiva* and *Dicarinella concavata* were identified. The *D. concavata* has been here modified, and Coniacian/Santonian boundary is included. We divided it into two sub-zones; they are *Dicarinella concavata* sub-zone of Late Coniacian age and *Dicarinella concavata–Costellagerina (Rugoglobigerina) pilula* association sub-zone of Early Santonian age. *Dicarinella primitiva* and *Dicarinella concavata* sub-zones correlates to UKP-06 *Dinogymnium microgranulosum*, interval dinocyst zone from Coniacian age, (89.8 ± 0.3–86.3 ± 0.3 Ma) (Navarrete-Parra et al. 2015), which was established in Aguablanca and El Salto creeks sections and in Chuira-1 well, and also recognized in Chuira-2ST. The Coniacian stage correlates to UC-10 calcareous nannofossil zone of Burnett (1998) in the Chuira-2ST well, according to Pérez Panera et al. (this volume).

***Dicarinella primitiva* Interval Zone**. This zone was originally defined as interval from the FA of *Dicarinella primitiva* to the FA of *Dicarinella concavata*, and an Early Coniacian age was assigned (Caron 1978, 1985). According to Gradstein et al. (2004) and Cohen et al. (2013) time scale, it spans from 89.8 to 87 Ma). The lower boundary FA of *Dicarinella primitiva* (89.8 Ma) occurs at 8620 ft and the upper FA of *Dicarinella concavata* (87 Ma) at 8440 ft in Chuira-2ST. This zone comprises the FA of *Archaeoglobigerina cretacea* and *Muricohedbergella (Hedbergella) holmdelensis;* the LA of *Falsotruncana maslakovae* (98.8–89.2 Ma) occurs close to the lower boundary of the zone. The *Dicarinella primitiva* zone is also characterized by abundant heterohelicids with *Planoheterohelix (Heterohelix) reussi* dominant and high diversity of whiteinellids and common calcispheres and pithonellids. This interval zone was also recognized in the Aguablanca and El Salto creeks and Cuesta El Diablo section.

***Dicarinella concavata* Interval Zone (modified)**. *Dicarinella concavata* Zone was originally defined as the interval from the FA of *Dicarinella concavata* to the FA of *Dicarinella asymetrica* and a Late Coniacian to Early Santonian age was assigned (Sigal 1955; Caron 1985). According to Gradstein et al. (2004) and Cohen et al. (2013) this zone spans from 87 to 85 Ma. As cited, this zone includes the Coniacian/Santonian boundary (Chron 86.3 ± 0.5 Ma). The *Dicarinella concavata* Zone sensu Sigal (1955) occurs in the interval 8130–7950 ft. *Dicarinella asymetrica* is not present in the analyzed samples, but its FA is replaced by coeval events. Those are the LA of *Whiteinella baltica* and LA of *Muricohedbergella (Hedbergella) delrioensis* (85 Ma) which occurs at (7950–7960 ft in Chuira-2ST). Within this zone, as recorded in outcrops and wells of the MMVB, it also occurs the FAs of *Costellagerina (Rugoglobigerina) pilula* (86.3 Ma), *Contusotruncana fornicata* and *Archaeoglobigerina blowi* (86.3 Ma). Other LAs occur in this Zone: *Whiteinella archaeocretacea*, *Marginotruncana renzi*, *M. coronata*, *Dicarinella imbricata*, and *D. primitiva*. When recognized or established, the *D. concavata* Zone sensu Sigal (1955) does not reveal the Coniacian/Santonian boundary. Nevertheless, it is possible to constrain that boundary based on *Whiteinella archaeocretacea* (LAD 86.3 Ma) and *Dicarinella primitiva*, species marker of the Coniacian top, but also, based on the FA of *Costellagerina* (*Rugoglobigerina*) *pilula* and/or *Contusotruncana fornicata* and/or *Archaeoglobigerina blowi* (86.3 Ma); which co-occur within the *D. concavata* Zone sensu Sigal (1955) and are markers of the Santonian base. Thus, *D. concavata* Zone could be divided into two sub-zones (this paper): *D. concavata* interval sub-zone of Late Coniacian age, and *Dicarinella concavata–Costellagerina* (*Rugoglobigerina*) *pilula* association sub-zone of Early Santonian age. Additionally, the *Dicarinella concavata* sub-zone was calibrated at El Salto section and Aguablanca creek. In the latter outcrop, age was constrained by the presence of *Didymotis variabilis* (Gerhardt 1897) in AB182 sample, *Baculites maldonadi* (Karsten 1858) and *Paralenticeras spathi* (Reyment 1958) in HB14a sample; typical mollusks of Early Coniacian age (Etayo-Serna 2010).

***Dicarinella concavata* Interval sub-zone**. This sub-zone is defined as the interval with *D. concavata* from the FA of *Dicarinella concavata* (87 Ma) to the FA of *Costellagerina* (*Rugoglobigerina*) *pilula* (86.3 Ma). Age of this sub-zone is Late Coniacian (87–86.3 Ma) according to Gradstein et al. (2004) and Cohen et al. (2013) timescales. This sub-zone was defined in cores Colon-2 and Colon-3 by Navarrete et al. (internal report, Petronorte 2010). Subsequently, it has also been identified in Aguablanca and El Salto creeks, and Cuesta El Diablo section. In Chuira-2ST (interval 8130–8050 ft), this sub-zone comprises also the LAs of *Whiteinella archaeocretacea* and *Dicarinella imbricata*, which co-occur at (8050–8060 ft), it includes moreover the FA of *Archaeoglobigerina bosquensis* and *A. blowi* at (8130–8120 ft). *Whiteinella archaeocretacea* (LA) and *A. blowi* (FA) are marker species of the Coniacian/Santonian boundary, while *Archaeoglobigerina bosquensis* (FA) is a marker Late Coniacian base. Those taxa are excellent alternative regional markers when keeled *Dicarinella imbricata* and *Dicarinella concavata* are not present. The top of this sub-zone is coeval with the upper boundary of UKP-06 *Dinogymnium microgranulosum* Interval dinocysts zone [86.3 $\pm$ 0.3 Ma Navarrete-Parra et al. (2015)]; in addition, its base is correlated to the FA of the dinocyst *Dinogymnium acuminatum* in sample (8040–8050 ft). It also correlates with the LA of *Eprolithus moratus*, a calcareous nannofossil marker of the Santonian/Coniacian boundary (Pérez Panera et al., this volume) found at 8140–8130 ft in Chuira-2 ST (see Coniacian/Santonian boundary).

4.3 Santonian

The Santonian in Chuira-2ST (7920–8130 ft) is characterized by high richness of planktonic foraminiferal assemblages and was possible to identify one zone (*Dicarinella asymetrica* Zone) and to propose one sub-zone (*Dicarinella concavata–Costellagerina* (*Rugoglobigerina*) *pilula* sub-zone). These were also identified in Chuira-1 well, Colon-2, Colon-3, and Cuesta el Diablo section. They are coevals with the UC-11C and UC12 calcareous nannofossils zones (Burnett 1998), as recognized Chuira-2-ST (Pérez Panera et al., this volume), and with dinocysts interval zones

UKP-7/UKP-8 *Dinopterygium cladoides* (Navarrete-Parra et al. 2015) of Santonian age (86.3 ± 0.3–83.6 ± 0.2 Ma), in Chuira-2ST and in Aguablanca and El Salto creeks. However, in those localities, *D. asymetrica* Zone is not complete, and the last 1.1 Myr of the top is missing.

***Dicarinella concavata–Costellagerina (Rugoglobigerina) pilula* Association sub-zone**. This sub-zone was defined in cores Colon-2 and Colon-3 by Navarrete et al. (internal report, Petronorte 2010). Subsequently, it has also been identified in Aguablanca and El Salto creeks, and Cuesta El Diablo section. This sub-zone is defined in Chuira-2ST at 8050–8130 ft as the interval with *D. concavata* from the LA of *Whiteinella archaeocretacea* (86.3 Ma) and FA of *Costellagerina* (*Rugoglobigerina*) *pilula* (86.3 Ma) at 8130 ft; to the FA of *Dicarinella asymetrica* and/or LA of *Whiteinella baltica* (85 Ma) at 8050 ft. An Early Santonian age is assigned, and it spans from 86.3 to 85 Ma according to Gradstein et al. (2004) and Cohen et al. (2013) timescales. This sub-zone comprises multiples extinction events in its upper boundary, among them: *Whiteinella baltica*, *Muricohedbergella* (*Hedbergella*) *delrioensis*, *Marginotruncana renzi*, *M. sigali*, *Dicarinella imbricata* and *Dicarinella primitiva*. It also comprises FAs of marker species of the Coniacian/Santonian boundary. First appearances of *Contusotruncana fornicata*, *Archaeoglobigerina blowi*, and *Costellagerina* (*Rugoglobigerina*) *pilula* (calibrated at 86.3 Ma) are excellent regional events for recognizing the Santonian/Coniacian boundary. This sub-zone correlates with the UC11c calcareous nannofossils zone of Burnett (1998) as it has been recognized in Chuira-2ST (Pérez Panera et al., this volume). The *Dicarinella concavata–Costellagerina* (*Rugoglobigerina*) *pilula* Association sub-zone proposed here has been recognized along diverse sections in Colombian basins.

***Dicarinella asymetrica* Total Range Zone**. This zone was originally defined as the total range of *Dicarinella asymetrica*, and a latest Early Santonian to Late Santonian age was assigned (Postuma 1971; Caron 1985). Calibrated with Gradstein et al. (2004) and Cohen et al. (2013) timescales, its age is Late Santonian and it spans from 85 to 83.6 Ma. In this survey, *Dicarinella asymetrica* is not present; nevertheless, a coeval interval with its early part was identified due to the LA of *Whiteinella baltica* (85 Ma), at 7950–7960 ft. This event occurs at the same time that the FA *Dicarinella asymetrica* (85 Ma). Just indicating the upper boundary of the *D. concavata–Costellagerina* (*Rugoglobigerina*) *pilula* sub-zone. Thirty feet up, in sample 7920 ft, multiple LAs of marker species near the top of *Dicarinella asymetrica* Zone co-occur. They are the LA of *Whiteinella inornata* (84.7 Ma), *Archaeoglobigerina bosquensis* (87–83.6 Ma), *Marginotruncana marginata* (90.3–83.5 Ma), *Dicarinella canaliculata* (93–84.7 Ma), *D. concavata* (87–83.9 Ma) and *Planoheterohelix* (*Heterohelix*) *reussi* (91.8–83.6 Ma). According to those events, the age at 7920 ft is not younger than 84.7 Ma. In resume, the interval 7950–7960 to 7920 ft correspond to the early part of the *Dicarinella asymetrica* zone, defined from the LA of *Whiteinella baltica* (7950–7960 ft) to the LA of *Whiteinella inornata* spans 85–84.7 Ma. At the top of this interval (9720 ft), the LAs of *Archaeoglobigerina bosquensis*, *Marginotruncana marginata*, *Dicarinella canaliculata*, *D. concavata*, and *Planoheterohelix* (*Heterohelix*) *reussi* also occur. The time interval documented (85–84.7 Ma) reveals the absence of a record of the latest Santonian (1.1 Myr).

4.4 Campanian

The Campanian in Chuira-2ST (7920–7700 ft) corresponds to the Lower Umir Formation. This stage is characterized by moderate abundance of planktonic foraminifera and high richness of benthonic foraminiferal assemblages. Three known planktonic foraminifer zones are described for Campanian stage: (1) *Globotruncanita elevata*, (2) *Globotruncana ventricosa*, and (3) *Radotruncana* (*Globotruncanita*) *calcarata* zones. Dinocysts and calcareous nannofossils did not allow the recognition the early or middle Campanian, but dinocysts UKP-10 *Dinogymnium heterocostatum* interval zone (Navarrete-Parra et al. 2015) and UC-12 calcareous nannofossils zone Burnett (1998) were identified in Chuira-2-ST (Pérez Panera et al., this volume) and allowed the recognition of Late Campanian. Chuira-1 and Chuira-2ST revealed the absence of the UKP-9 Isabelidinium cretaceum Zone (Navarrete-Parra et al. 2015), and new evidence in Chuira-2ST shows the absence of UC-13 to UC-14 calcareous nannofossils zones of Burnett (1998) (Pérez Panera et al., this volume) zones.

Globotruncanita elevata **Partial Range Zone**. This zone was originally defined as the interval with *G. elevata* from the LA of *Dicarinella asymetrica* to the FA of *Globotruncana ventricosa* and an Early Campanian age was assigned (Postuma 1971; Caron 1985). According to Gradstein et al. (2004) and Cohen et al. (2013) timescales, it spans between 83.6 and 79.4 Ma. In Chuira-2ST well, FA of *Globotruncanita elevata* (83.6 Ma) at 7920–7010 ft co-occur with the first record of *Rugoglobigerina rugosa* (80.3 Ma) and *Planoheterohelix* (*Heterohelix*) *striata* (80.3 Ma). These species constrain a range not older than 80.3 Ma. They post-date the LAs of multiple taxa whose extinction levels occur in the Santonian stage, and in Chuira-1 they are not younger than 84.7 Ma. This evidences that *Globotruncanita elevata* Zone is not complete; there is no record of their first 3.3 Ma. The zone is here defined as the interval with *G. elevata* from the LA of *Archaeoglobigerina bosquensis*, *Marginotruncana marginata*, *Dicarinella canaliculata*, *D. concavata*, and *Planoheterohelix* (*Heterohelix*) *reussi* at 7920 ft) to the FA of *Globotruncana ventricosa* at 7900 ft. As described, this zone was also identified in Cuesta El Diablo section and Chuira-1 well.

Globotruncana ventricosa **Interval Zone**. Originally defined as the interval from the FA of *Globotruncana ventricosa* to the FA of *Radotruncana* (*Globotruncanita*) *calcarata* and an age late Early Campanian to Late Campanian was assigned (Dalbiez 1955; Caron 1985). According to Gradstein et al. (2004) and Cohen et al. (2013) timescales, it spans from 79.4 to 75.4 Ma. The *Globotruncana ventricosa* Zone was also recognized in Cuesta El Diablo section. In Chuira-2ST (7900–7880), this zone is relatively very poor in planktonic foraminifera. Nevertheless, it is recognized sensu Dalbiez (1955), from the FA of *Globotruncana ventricosa* (7900 ft) to the FA of *Radotruncana* (*Globotruncanita*) *calcarata* (7870–7880 ft). *Globotruncanita elevata*, *Archaeoglobigerina blowi*, and *Muricohedbergella* (*Hedbergella*) *holmdelensis* are common in the assemblage.

Radotruncana **(*Globotruncanita*) *calcarata*** **Total Range Zone**. This zone was originally defined by total range of the *Radotruncana* (*Globotruncanita*) *calcarata*, and a latest Late Campanian age was assigned (Herm and Barthel 1962; Caron 1985). It occurs in Chuira-2ST interval 7880–7700 ft, spanning from 75.4–72.1 Ma, according to Gradstein et al. (2004) and Cohen et al. (2013) timescales, Late Campanian. In the

MMVB, this zone yields poor planktonic foraminifer assemblages, but the FAs of *Globotruncanella havanensis*, *Planoheterohelix* (*Heterohelix*) *globulosa* and LA of *Globotruncanita elevata* were recorded. The *Radotruncana* (*Globotruncanita*) *calcarata* Zone sensu *strictu* was also recorded in Chuira-1 well and Cuesta El Diablo section. In Chuira-2ST, a Late Campanian age has also been constrained with the identification of the UC-15 to UC-16 calcareous nannofossil zones of Burnett (1998), based on the FA of *Eiffellithus parallelus* at 7890–7880 ft and LA of *Broinsonia parca expansa* at 7690–7680 ft. It also correlated to the UKP-10 *Dinogymnium heterocostatum* dinocyst zone of Late Campanian age (Navarrete-Parra et al. 2015).

5 Stratigraphic Boundaries and Hiatus–Discussion

5.1 Turonian/Coniacian Boundary

The determination of the Coniacian/Santonian boundary is difficult to achieve through foraminifera, since the markers *Helvetoglobotruncana* and the *Marginotruncana* groups are either missing or scattered. That boundary has been successfully established using foraminifera and calcareous nannofossils. Turonian and Coniacian faunas contain Tethyan species, and taxa are well differentiated. This boundary comprises various extinction events of no complexes planktonic foraminifera, as *Clavihedbergella* (*Hedbergella*) *simplex*, *Muricohedbergella* (*Hedbergella*) *delrioensis* and *M.* (*H.*) *planispira*, and the double-keeled *Helvetoglobotruncana* group. Also, the FAs of simple forms like *Whiteinella* group, *Planoheterohelix* (*Heterohelix*) *reussi*, and other heterohelicids. Coniacian faunas are characterized by double-keeled *Dicarinella* and *Marginotruncana* groups. *Dicarinella primitiva*, *Archaeoglobigerina cretacea*, *Muricohedbergella* (*Hedbergella*) *holmdelensis* first appear in the Coniacian, and they are common to abundant taxa. The boundary is suggested at 8620 ft with the FA of *Dicarinella primitiva* of Coniacian age. The calcareous nannofossils revealed latest Turonian with LA *Eprolithus octopetalus* in interval (8780–8790 ft) and the FA of *Micula staurophora* at 8690–8700 ft, which occurs at the base of the Coniacian (Pérez Panera et al., this volume). The Turonian/Coniacian boundary was not identified with palynomorphs. At the boundary, they show abundant *Odontochitina costata*, *Dinogymnium digitus*, *Manumiella* sp., *Alterbidinium acutulum*, and *Palaeotetradinium silicorum.* This assemblage characterizes the UKP-03-UKP-06 dinocysts zones of Turonian–Coniacian age in Chuira-1 well, Aguablanca, El Salto creeks and Cuesta El Diablo section, without record of marker species that could identify the Turonian/Coniacian boundary. The bias in the Turonian/Coniacian boundary according to foraminifera and calcareous nannofossils would not be related strictly with temporary differences; it could be explained in minor proportion to very-bad preserved specimens and taxonomical difficulties.

5.2 Coniacian/Santonian Boundary

The Coniacian/Santonian boundary is difficult to locate since such an important stratigraphic marker as the *Dicarinella* and *Marginotruncana* groups are either missing or very rare. *Whiteinella*, *Rugoglobigerina*, and *Archaeoglobigerina* groups help to substitute those double-keeled taxa, whose records are poor and obscure due to preservation problems. The base and top of involved sediments are gradational in the wells and subsurface and high-resolution biostratigraphical studies show they represent conformable facies boundaries. The Coniacian/Santonian boundary is recognized by the LAs of *Whiteinella archaeocretacea* and *Dicarinella imbricata* at latest Coniacian and FAs of *Costellagerina* (*Rugoglobigerina*) *pilula*, *Contusotruncana fornicata*, and *Archaeoglobigerina bosquensis* at earliest Santonian. These events indicate the top of *Dicarinella concavata* sub-zone of Late Coniacian and base of *Dicarinella concavata*–*Costellagerina* (*Rugoglobigerina*) *pilula* sub-zone of Early Santonian, respectively. The top of the UKP-06 *Dinogymnium microgranulosum* Zone (86.3 ± 0.3 Ma) also indicates the Coniacian/Santonian boundary, which in Chuira-2ST is located at interval 8040–8050 ft, according to foraminifera, nannofossils, and dinocysts.

5.3 Santonian/Campanian Boundary

Campanian spans from 83.6 ± 0.2 to 72.1 ± 0.2 Ma and Santonian from 86.3 ± 0.5 to 83.6 ± 0.2 Ma (see Gradstein et al. 2004; Cohen et al. 2013). The earliest Santonian is 85–84.7 Ma, underlying to late Early Campanian (80.3–75.4 Ma). The *Dicarinella asymetrica* Zone and/or its coeval interval Santonian age reveals a hiatus of 1.1 Myr from the latest Santonian. The *Globotruncanita elevata* of Early Campanian age shows a hiatus of 3.3 Myr (see *Dicarinella asymetrica* and *Globotruncanita elevata* zones). The *Globotruncana ventricosa* Zone contains the Early Campanian/Middle Campanian boundary; therefore, they are not always sharply separated and are currently referred as Middle Campanian. Only the upper part of *Globotruncanita elevata* Zone is present in the La Luna Formation in MMVB (cf. Chuira-2ST, interval 7910–7920 ft) as it was shown by the dense sampling, multidisciplinary biostratigraphy and geochronology. According to the integrated biostratigraphy, at 7920 ft sediments are not older than 80.3 Ma. (see *Globotruncanita elevata* Zone). *Globotruncanita elevata* and *Globotruncana ventricosa* zones, from the FA of *Globotruncanita elevata*, *Rugoglobigerina rugosa*, and *Planoheterohelix* (*Heterohelix*) *striata*, to the FA of *Radotruncana* (*Globotruncanita*) *calcarata* (7760–7880 ft), comprises an age from 80.3 to 75.4 Ma. We found that it would be the lower part of *Globotruncanita elevata* Zone which is missing, so *Globotruncanita elevata* Zone only reveals the last 0.9 Ma of Early Campanian. When compared with standard geochronology, the Santonian/Campanian boundary (Chron 83.6 ± 0.2: Cohen et al. 2013), at Churia-2ST shows that the earliest 3.3 Myr are not present.

5.4 Santonian/Campanian Hiatus and Condensed Sedimentation

Foraminiferal analyses, palynology, calcareous nannofossils (Pérez Panera et al., this volume) and sedimentological evidence obtained from subsurface and outcrops (e.g., Cuesta El Diablo and Chuira-2 (Rodríguez et al. 2016a) and Chuira-2ST drill-down), revealed a condensed Campanian (interval 7920–7880 ft in Chuira-2ST) which resume the *Globotruncanita elevata* and *Globotruncana ventricosa* zones, and no record from latest Santonian (1.1 Myr) and Early Campanian (3.3 Myr), which excludes an abnormal distribution of those species. The distribution and frequency of *Globotruncanita elevata* and *Globotruncana ventricosa* are not abnormal, and their virtual absences and/or not differentiated apparition or poor record have diverse explanations. This could be due to epicontinental sea environment, where keeled foraminifera are always absent, or due to condensed sedimentation. In fact, in the interval 7880–7920 ft $\approx$ 9–10 m thick, the occurrence of autigenic glauconite-rich horizons (grains and infill), phosphates, pyrite, and high abundance of ostracods, microbivalves, tooth and another fish remains, suggests periods of slow, condensed sedimentation. Formation of glauconitic sediments occurs on the marine shelves during prolonged intervals of sediment starvation (Odin and Fullagar 1988). The Santonian/Campanian boundary in MMVB revealed a hiatus of 4.4 Myr approximately. This hiatus corresponds to the La Luna and Umir Formations contact and was calibrated in other localities, among them, Chuira-1, Chuira-2, Chuira-2-ST, and other adjacent Colombian sedimentary basins. Although, the hiatus has been identified regionally, but its details remain unpublished (see Rodriguez et al. 2016b, Fig. 19). The recognition of this hiatus and its duration is key for the oil and gas exploration, as it corresponds with a paraconformity that allow the recognition of the La Luna and Umir Formations contact in surface and most important, in subsurface. The hiatus reveals a glauconitic-rich layer, which is a regional guide horizon at the top of the La Luna Formation. This paraconformity is the surface for the development of a 10-m-thick condensed deposit of 4.9 Myr (Middle to Late Campanian) that correspond to the upper part of the *Globotruncanita elevata* and *Globotruncanita elevata* zones (79.4–75.4 Ma).

6 Paleoenvironmental Interpretations

A general overview on the sedimentation and paleoenvironmental conditions prevailing during the deposition of the La Luna Formation is possible to introduce (see Ecozones Fig. 8), after detailed micropaleontological studies in different sections and wells in the MMVB and adjacent basins (e.g., Catatumbo, Rancheria, and Upper Middle Magdalena Valley). During the sedimentation of La Luna Formation and equivalent units, benthonic foraminifera are scarce and always very small-sized (Fig. 9a, b). Across different profiles in the basin, most conspicuous benthonic could be grouped in robust buliminids (Type C: 80–120 μ), small-sized buliminids (Type B: 120–140 μ), and very small-sized buliminids (Type A: 140–200 and <200 μ). They are very diverse and constitute the most persistent benthonic group along the La Luna and Umir (lower part) formations. Thus, for instance, buliminids type A, with the very small *Virgulina tegulata* and *Oolina* sp. occur in Turonian stage. They are scarce at Early Turonian but abundant ending the Middle Turonian. During Coniacian, buliminids Type B occurs accompanied by small *Fursenkoina* and diverse rotalidae as *Anomalina redmondii* (abundant). This assemblage increases up, but precludes before ending the Santonian. For example, *Bolivina explicata* is a guide benthonic foraminifer along the upper La Luna Formation from northern corner of the MMVB to Venezuela. *Bolivina explicata* and *Anomalina redmondii* are cited sensu zonule by (Petters 1954) in the La Luna Formation (upper part). There are no other records of benthonic foraminifera in the Colombian literature of the La Luna Formation. This level is an important seismic

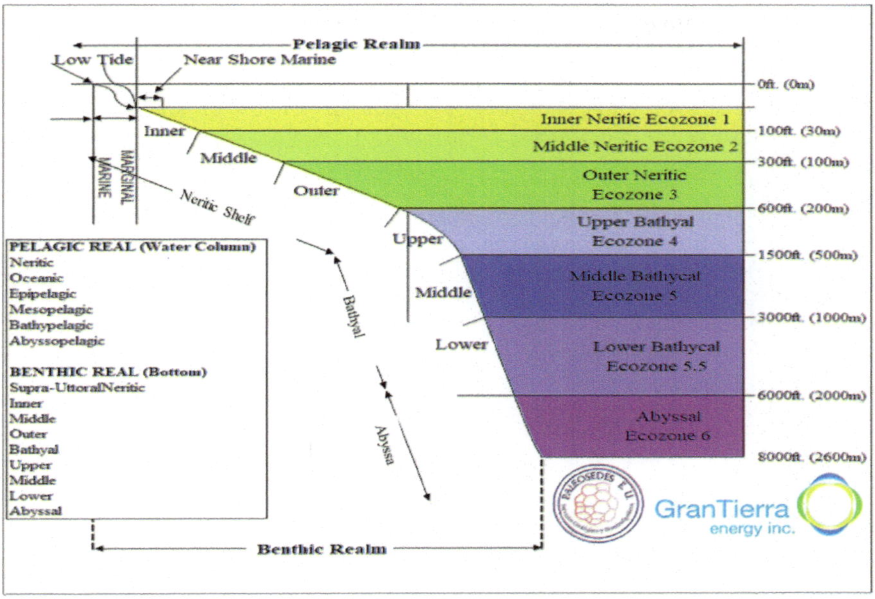

Fig. 8. Ecozones, shelf marine, and ocean profile used in this work

reflector at top of the La Luna Formation and is rich in mature glauconitic grains. This level coincides with the early part of *Dicarinella asymetrica* zone and an anoxic event in the Santonian. There are some turnovers in Santonian/Campanian boundary, for example, whiteinellids, dicarinellas, and marginotruncanas practically disappear, and *Planoheterohelix* (*Heterohelix*) *reussi* highest abundance records precede their extinction. The latest Early Campanian (in Umir Formation) shows a toward-up increase in buliminids Type C (robust). They are the main component clustered with a very diverse infauna and epifauna, which contain mainly robust specimens and high diversity and abundance. Calcareous rotalidae, *Siphogenerinoides*, *Bolivina*, diverse buliminids Type C and agglutinated foraminifera as *Ammobaculites*, *Haplophragmoides* spp. among very others, are common.

The La Luna Formation in the MMVB yields very abundant planktonic foraminifera, with noticeably affinities to the Tethyan realm according to biostratigraphical zonation from western areas of the Tethys [e.g., Gulf Coast Trinidad and Caribe Sea (Brönnimann and Brown 1958; Gandolfi 1955; Bolli et al. 1957; Bolli 1966)]. Planktonic foraminifera in the La Luna Formation range in age from early Turonian (*Whiteinella archaeocretacea* Zone), to early late Santonian (*Dicarinella asymetrica* Zone), while in the base of the Umir Formation, *Globotruncanita elevata–Radotruncana* (*Globotruncanita*) *calcarata* zones, indicates a Campanian age. In the La Luna Formation, simple globigerinidae dominates the assemblages, a characteristic shared with other basins in Colombia as the Rancheria and Catatumbo basins. Among them, heterohelicids are the most important in abundance, but with low diversity; followed by whiteinellids, abundant and high diversity; hedbergellids, abundant and moderate

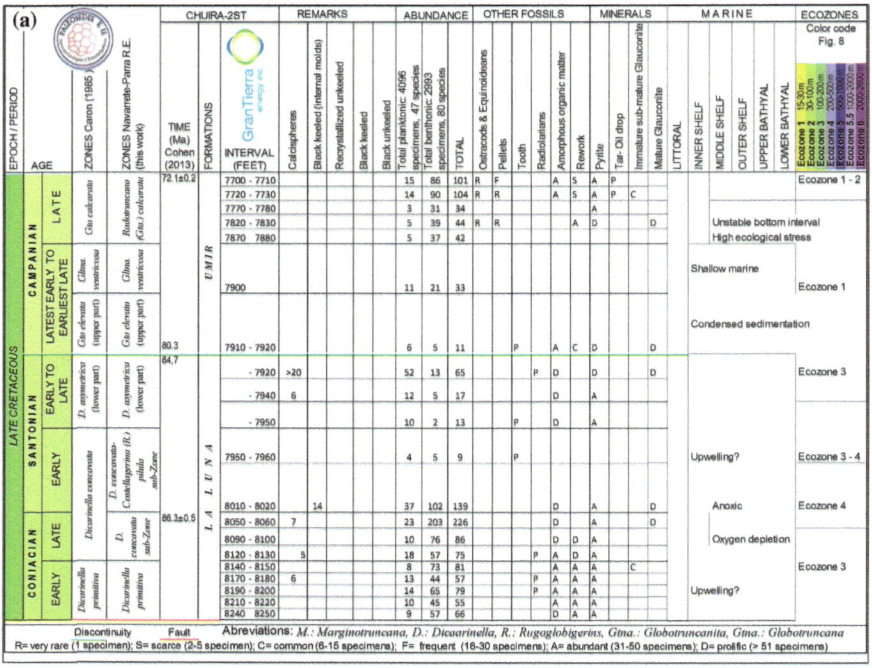

Fig. 9. **a**, **b** Chuira-2ST well in the northern Middle Magdalena Valley Basin, some conditions of deposit and paleobathymetry. Paleobathymetry and anoxic conditions strongly influenced the foraminiferal distribution

diversity. During the Turonian to the Coniacian/Santonian boundary, small keeled planktonic foraminifera had a moderate presence. The Planktonic/Benthonic ratio suggests that sedimentation occurred in a shallow sea, from middle to outer neritic marine environment, controlled by cyclic events of upwelling and depletion of oxygen content, from the Turonian to the Santonian, with important anoxic events in the Middle Turonian and Coniacian/Santonian boundary. These events could be associated with the OAE 2 and OAE 3 proposed by Jenkyns (1980). Paleobathymetry and anoxic conditions strongly influenced the foraminiferal distribution.

The lower part of Chuira-2ST (9365–8670 ft) corresponds to a 745 ft black to clear brown calcareous interval (biomicrite and spate-micrite). This interval is similar to the upper Salada and lower Pujamana members. This Turonian sequence shows a variation from inner to outer marine shelf (Ecozone 1 to Ecozone 3, Figs. 8 and 9a, b) as indicated by poorly developed associations of benthonic foraminifera of very small size and high abundance of small-sized opportunistic species as heterohelicids, hedbergellids, and whiteinellids (*Planoheterohelix* (*Heterohelix*) *reussi* disclose 80% planktonic fauna). This sequence begins with an upwelling event, very rich in black marine organic matter, quickly increasing in depth, ending in anoxic conditions. The Early Turonian is very poor in keeled foraminifera (*Praeglobotruncana turbinata*–*P. stephani* are present), and it is rich in small-sized whiteinellids, and hedbergellids,

(b)

EPOCH / PERIOD	AGE	ZONES Caron (1985)	ZONES Navarrete-Parra R.E. (this work)	CHUIRA-2ST: TIME (Ma) Cohen (2013)	CHUIRA-2ST: FORMATIONS	CHUIRA-2ST: INTERVAL (FEET)	REMARKS: Calcispheres	REMARKS: Black keeled (internal molds)	REMARKS: Recrystallized unkeeled	REMARKS: Black keeled	REMARKS: Black unkeeled	ABUNDANCE: Total planktonic: 4096 specimens, 47 species	ABUNDANCE: Total benthonic: 2993 specimens, 80 species	ABUNDANCE: TOTAL	OTHER FOSSILS: Ostracods & Equinoideans	OTHER FOSSILS: Pellets	OTHER FOSSILS: Tooth	OTHER FOSSILS: Radiolarians	OTHER FOSSILS: Amorphous organic matter	OTHER FOSSILS: Rework	MINERALS: Pyrite	MINERALS: Tar-Oil drop	MINERALS: Immature sub-mature Glauconite	MINERALS: Mature Glauconite	MARINE (LITTORAL, INNER SHELF, MIDDLE SHELF, OUTER SHELF, UPPER BATHYAL, LOWER BATHYAL)	ECOZONES Color code Fig. 8 (Ecozone 1 15-30m; Ecozone 2 30-100m; Ecozone 3 100-200m; Ecozone 4 200-500m; Ecozone 5 500-1000m; Ecozone 5.5 1000-2000m; Ecozone 6 2000-2600m)
	SANTONIAN EARLY	*Dicarinella concavata*	*D. concavato-Costellagerina (R.) pilula sub-Zone*		LA LUNA	8263		12				33	100	133					D		A	A		D		
						8270 - 8280	3					17	52	69					D		A	A		D	ANOXIC	
	CONIACIAN LATE		*D. concavata sub-Zone*	86.3±0.5		8300 - 8310						8	118	126								A		S	Oxygen depletion	Ecozone 4
						8330 - 8340	6					16	132	148								A		A		
						8370 - 8380						5	162	166					A		A	A				
	CONIACIAN EARLY	*Dicarinella primitiva*	*Dicarinella primitiva*			8440 - 8450	3					13	54	67					D	A	A		C			
						8500 - 8510	9					16	51	67					D	A	A		C			
						8550 - 8560	5		8			21	59	80					D		A		C			Ecozone 3
						8600 - 8610	4	3				22	28	50					D		A		S			
						8620	8		>20			56	10	66				C	D		A		S			Ecozone 2- 3
	TURONIAN LATE	*M. sigali*	*M. sigali*	89.8±0-3		8660 - 8670						28	15	43					D		A	A				Ecozone 3
						8700 - 8710				>20		97	20	117					D		A					
						8740 - 8750			>50	>50		246	12	258				S	D		A	A				
						8780 - 8790			>50	>50		258	2	260					D		A	D				Ecozone 4
	TURONIAN MIDDLE	*Helvetoglobotruncana helvetica*	*Helvetoglobotruncana helvetica*			8820 - 8830			>50	>50	>20	239	0	239					D		A	D				to Ecozone 3
						8850 - 8860	>20		>50	20	15	190	23	213		C		S	D		A	D			Sudden Anoxic pick	Ecozone 3- 4
						8880 - 8890	>20		>50	20	25	303	100	403					D		A	D				Ecozone 5
						8920 - 8930	>50		>20	>50	30	288	7	295					D		A	A				Ecozone 3
						8950 - 8960	20		20	10	15	153	9	162		P			D		A	A				
						9000 - 9010			18	4		44	18	62					D		A	A			Bottom cool-waters influx	
						9050 - 9060	10		20			63	4	67					D			D			Upwelling	Ecozone 2- 3
						9100 - 9110	>100		>100			269	6	275					D		A	D			ANOXIC	
	TURONIAN EARLY	*Whiteinella archaeocretacea*	*Whiteinella archaeocretacea*			9150 - 9160	>100		>100			309	1	310					D		A	A				Ecozone 3
						9230	>50		>50	20	15	172	3	175					D		A					
						9263	12					30	7	37					D		A	A			Bottom cool-waters influx	Ecozone 4
						9290 - 9300	>100		>50		>50	359	0	359					D		A	D				Ecozone 3 - 4
						9360 - 9365	>100		>100		>50	408	4	412					D		A	D				

Discontinuity Fault Abreviations: *M.: Marginotruncana, D.: Dicaarinella, R.: Rugoglobigerins, Gtna.: Globotruncanita, Gtna.: Globotruncana*
R= very rare (1 specimen); S= scarce (2-5 specimen); C= common (6-15 specimens); F= frequent (16-30 specimens); A= abundant (31-50 specimens); D= prolific (> 51 specimens)

Fig. 9. (*continued*)

and highly abundant on small *Planoheterohelix* (*Heterohelix*) *reussi*. Insulated cameras of planktonic recrystallized (calcispheres, pithonellids) are common and so are the scattered homeomorphic small benthonic foraminifera accompanied by microbivalves, ostracods, and phosphates.

The great abundance of whiteinellids and heterohelicids indicate high environmental stress and /or biotic stress maximum. According to Keller (2003), heterohelicids are extremely tolerant to low oxygen concentration. The dominance of heterohelicids (Pardo and Keller 2008) more precisely *Planoheterohelix* (*Heterohelix*) is an important global marker of OAE2. Moreover, the dominance of *Planoheterohelix* (*Heterohelix*) in any moment would indicate expansion of the oxygen minimum zone, normally associated to an increase of organic productivity.

Especially in the Atlantic and in the Western Tethys, numerous black shale horizons occur in regional and supraregional distribution. Given that some of these events can be correlated across a large area and various marine environments (e.g., deep water, marginal basins, and platform sediments), they were called global Oceanic Anoxic Events (OAEs) (Schlanger and Jenkyns 1976). These OAEs are defined by a massive deposit of organic matter in marine environments (Schlanger and Jenkyns 1976; Arthur et al. 1990). The OAE2 would be present in the lower part *Whiteinella archaeocretacea* Zone, interval 9365–9290 ft, when heterohelicids had its acme (80–90% on total planktonic fauna) and hedbergellids and whiteinellids are dominant, accompanied by rich black amorphous marine organic matter and rare benthonic with *Virgulina tegulata* and *Oolina*, typical of cold marine waters. The OAE3 would be present in the interval 8060–8010 ft at Chuira-2ST well, corresponding with Coniacian *Dicarinella concavata* sub-zone and Santonian *D. concavata–Costellagerina* (*Rugoglobigerina*) *pilula*, coincident with the sudden appearance of benthonic with high abundance and diversity of buliminids Type B, accompanied by double-keeled planktonic and very abundant amorphous marine organic matter. Only 30 ft up, before the end of Early Santonian, the inverse event occurs. Sudden fall of the precedent species and increase in phosphate, probably related to upwelling episodes, took place.

Diverse keeled foraminifera as *Helvetoglobotruncana helvetica*, *H. praehelvetica*, *Marginotruncana sp. M. coronata*, *M. pseudolinneiana*, *Dicarinella hagni*, *D. imbricata*, *D. canaliculata*, *Falsotruncana maslakovae*, *Marginotruncana sigali*, and *M. renzi*, first appear by mid-Turonian, during *H. helvetica* zone. They are taxa related to oceanic waters and reach high abundance. In a shallow marine environment, only few of these keeled species were found, probably brought by surface currents or by the action of strong tides. Their abundance and variety near the coast is more linked to oceanographic events and later burial histories, than with their position into the water column. The keeled forms also exhibit an excessively small size. Dissolution and recrystallization is the most common taphonomic process, which suggests offshore deposition. The eventual co-occurrence of *Oolina* spp., a benthonic form with cold water affinity, suggests an epicontinental environment, with cold-bottom waters that produced upwelling events. The great productivity of these groups associated to phosphatized fish remains and pellets also suggests a poorly stratified water column. Upwelling predates levels with very abundant buliminids Type A < 200 (e.g., *Virgulina tegulata*) and high abundance of not keeled planktonic and high diversity of keeled planktonic foraminifera. The high concentration of amorphous marine organic matter indicates anoxic bottoms and deepest environments during the increase of keeled forms at the Middle Turonian. The co-occurrence of buliminids Type A and *Virgulina* and *Oolina* sp. interbedded in at some levels, discloses disoxic conditions into the anoxic interval. The Middle Turonian interval (9110–8820 ft) shows variations from middle to outer shelf (Ecozone 2 to Ecozone 3, Figs. 8 and 9**a**, **b**), and a notable sudden deepest peak interval (8890–8850 ft), reaching outer shelf to upper bathyal (Ecozone 3 to Ecozone 5 $\approx$ 500–1000 m depth (Figs. 8 and 9**a**, **b**).

The Late Turonian reveals a decreasing-up in both, planktonic and benthonic foraminifera abundance. A lower diversity of keeled planktonic and bigger size of whiteinellids and *Planoheterohelix* (*Heterohelix*) *reussi* is recorded, together with a diminution in the radiolarians and buliminids Type A. Buliminids Type B first appear in these levels. These suggest a shallowing-up until Earliest Coniacian.

The second and third dated sequences (89.8–84.7 Ma) are represented in Chuira-2ST well by the intervals 8670–8440 and 8130–7920 ft. The first corresponds to 230 ft of calcareous mudstone (biomicrite and spate-micrite), black and blackish brown colored and a phosphate-rich interval with submature glauconite and pyrite at the base. This interval is similar to Pujamana Member of the La Luna Formation in MMVB. Planktonic foraminifera from this interval in Chuira-2ST range in age from Latest Turonian, upper part *Marginotruncana sigali* Zone, to Early Coniacian, *Dicarinella primitiva* Zone. The upper interval (8130–7920 ft) corresponds to 210 ft of bio-micritic and spate-micritic limestone with concretions, and gray and blackish clear brown interbedded chert levels toward the top. It reveals abundant mature glauconitic grains and euhedral pyrite at the top. This interval is similar to Galembo Member of the La Luna Formation in MMVB. Planktonic foraminifera in this interval in Chuira-2ST range in age from Late Coniacian, *Dicarinella concavata* sub-zone and *D. concavata* Zone, to Late Santonian, *Dicarinella asymetrica* Zone (lower part). Lithological changes between members are not necessarily a time line. The Early Coniacian is characterized by low abundance of planktonic and poor abundance in benthonic foraminifera, with buliminids type B and disperse anomalinids. Micro-plankton contains mainly calcispheres, pithonellids, and scarce foraminifera with scattered representation of the whiteinellids, hedbergellids, and heterohelicids. Keeled foraminifera are rare, and *Archaeoglobigerina* is present. Relatively quiet waters and variable depths, from inner-middle to outer marine shelf (Ecozone 2 to Ecozone 3, Figs. 8 and 9a, b) are interpreted for this sequence. The Late Coniacian is characterized by a dominance of benthonic over planktonic foraminifera. Type B buliminids dominates over small rotalids, accompanied by microbivalves and radiolarians remains. Keeled foraminifera are scattered and less diverse than in lower strata of the La Luna Formation, and the first small-sized *Archaeoglobigerina* appears. The Coniacian interval records some maximum flooded levels, developed in strata of transgressive facies and severe anoxic intervals; which are concurrent with minor abundance of planktonic foraminifera and proliferation of Type B buliminids. These characteristics are interpreted as sedimentation in an epicontinental sea, on the middle shelf (Ecozone 3 to Ecozone 2, Figs. 8 and 9a, b), relatively far from the coast and with oxygen depletion conditions. The Late Coniacian/Early Santonian boundary is characterized by a sudden planktonic foraminifera abundance (8010–8060 and 8330–8380 ft intervals), as well as by non-biogenic silica, pyrite, and black amorphous marine organic matter which increases from the middle part up. This record discloses an anoxic event in the transit of the Coniacian/Santonian boundary, probably related with OAE3 (sensu Jenkyns 1980). The latest Early Santonian to Late Santonian show a fall in the abundance of planktonic and benthonic foraminifera. A shallowing-up pattern is notable, sedimentation occurred in shallow marine, from middle to inner shelf (Ecozone 2 to Ecozone 1, Figs. 8 and 9a, b). Sedimentation ends at the beginning of the Late Santonian.

The Campanian stage is similar to the base of the Lower part of the Umir formation in MMVB (Navarrete-Parra et al. 2016b, c). Planktonic foraminifera in the Umir Formation range in age from latest Early Campanian, *Globotruncanita elevata* Zone, to Late Santonian, *Radotruncana* (*Globotruncanita*) *calcarata* Zone. Its transgressive base in Chuira-2 and Chuira-2ST consists of a calcareous mudstone (intra-biomicrite), with extraclasts of foraminiferic spate-micrite, black and dark brown colored, rich in mature glauconite, and rich in reworked blackish dark green glauconite, just in Chuira-2ST (7920–7910 ft). The base of the Umir Formation yields an abundant assemblage of benthonic foraminifera with allochthones from shallowest shelf; presence of phosphatic pellets, fish remains and echinoderm and bivalve fragments. Micro-bioturbation is common. It contains autogenic acicular and euhedral pyrite. Thus, the base of the Umir Formation is interpreted as a flysch-type sedimentation. It always reveals keeled planktonic foraminifera, tied with median and big size benthonic foraminifera (infaunal and epifaunal), which suggests stratified, oligotrophic surface waters and some dys-aerobic conditions with intervals characterized by high-stress ecologic conditions.

A shallow marine environment with unstable (depth, salinity, oxygen, and nutrients) bottom waters is evident by the alternation of levels with greater and lesser abundance of shallow bottom benthonic foraminifera and proliferation of opportunistic planktonic species. Although with few variations, a general depth between 100–200 m in the middle shelf (Ecozone 2, Figs. 8 and 9a, b), with high organic productivity is interpreted.

7 Conclusions

Planktonic foraminifera from Chuira-2ST, Upper Cretaceous La Luna, and Lower Umir Formations in the Middle Magdalena Valley Basin were analyzed. A detailed planktonic foraminiferal study allowed the identification of a Turonian to Santonian age for the La Luna Formation and a Campanian age to the lower part of the Umir Formation. The La Luna Formation in Chuira-2ST was reached at the Early Turonian at Terminal depth of 9365 ft, but the drilling did not reach its lower contact. In the Aguablanca and El Salto streams and Cuesta El Diablo section, the base of the La Luna Formation is Late Cenomanian to earliest Turonian. The base of the Umir Formation is Campanian.

The La Luna Formation in Chuira-2ST revealed planktonic foraminifera *Whiteinella archaeocretacea*, *Helvetoglobotruncana helvetica*, and *Marginotruncana sigali* zones from the Turonian stage. The Coniacian stage corresponds with planktonic foraminifera *Dicarinella primitiva* Zone, and *Dicarinella concavata* sub-zone into the *D. concavata* Zone. The upper part of the La Luna Formation discloses the *Dicarinella concavata–Costellagerina* (*Rugoglobigerina*) *pilula* association sub-zone into the *D. concavata* Zone of Early Santonian age. At the top of the La Luna Formation, *Dicarinella asymetrica* Zone was identified.

The Campanian was recognized with planktonic foraminifera *Globotruncanita elevata* zone (upper part), *Globotruncana elevata* and *Radotruncana* (*Globotruncanita*) *calcarata* zones, which occurs at the base of the Umir Formation in MMVB.

La Luna and Umir Formations are separated by a hiatus that comprises at least 4.4 Ma. It is associated with a turnover of diverse planktonic foraminifera as *Planoheterohelix* (*Heterohelix*) *reussi*, *Whiteinella archaeocretacea* and *W. baltica*, whose vertical distribution are truncated before reach their extinction levels. It discloses also the first record of *Globotruncanita elevata*, *Rugoglobigerina rugosa*, and *Planoheterohelix* (*Heterohelix*) *striata*. There is neither record of the upper part of *Dicarinella asymetrica* Zone, nor of the lower part of *Globotruncanita elevata* Zone. Sedimentary features reveal important presence of glauconitic grains exhibiting high maturity and hardground surfaces with phosphatic intraclasts and bioturbation.

Chuira-2ST reveals an unconformable relationship between the La Luna and Umir contact. This contact discloses two special events: the Santonian-Campanian hiatus and a Campanian condensed sedimentation. The base of the Umir Formation postdates the hiatus; its base corresponds to condensed sedimentation, 10 m spans (80.3–75.4 Ma), and the deposits are flysch type.

Environments with oxygen-depleted waters and episodic blackish marine organic matter deposits because of anoxic events and upwellings with nutrient input from the offshore characterizes the La Luna Formation. Foraminifera, dinocysts, and calcareous nannofossils assemblages indicate shallowing-up paleoenvironment from an outer shelf to marine littoral (transitional marine) setting. The Early Turonian disclose anoxic conditions linked with the OAE2. The Middle Turonian shows a deepest and sudden anoxic peak, probably an anoxic local event. The Coniacian/Santonian boundary also shows a peak of depth, corresponding to an outer-shelf environment with sedimentation associated with the OAE3. The sediments of the studied section were deposited during the late stage of the upper Cretaceous Turonian–Santonian relative sea-level rise in the Colombian Cretaceous Basin and a drop in the sea level is revealed at the end of the Santonian, at the La Luna and Umir Formations contact.

Acknowledgements. We thank Gran Tierra Inc, Colombia ltd, who acquired Petronorte-UTMidas, for allowing us to publish these results and Paleosedes Consulting Colombian Company, for funding this work. We also thank Jaime Andrés Navarrete for his help during edition. Special thanks to Isabella Premoli-Silva and Mercedes B. Prámparo for their reviews, which improved the manuscript.

References

Arthur MA, Schlanger SO (1979) Cretaceous "Oceanic Anoxic Events" as causal factors in development of reef reservoired giant oil fields. Am Assoc Petr Geol 63(B):870–885

Arthur M, Jenkyns HC et al (1990) Stratigraphy, geochemistry, and paleoceanography of organic carbon-rich Cretaceous sequences. In: Ginsburg RN, Baeudin B (eds) Cretaceous resources, events and rhythms: background and plans for research, vol 304. Kluwer Academic Publishers, Dordrecht, The Netherlands, pp 75–109

Barr F (1972) Cretaceous biostratigraphy and planktonic foraminifera of Libya. Micropaleontology 18(1):1–46

Barrero D, Pardo A et al (2007) Colombian sedimentary basins: nomenclature, boundaries and petroleum geology, a new proposal. Agencia Nacional de Hidrocarburos, vol 1. ANH and B and M Exploration Ltd Bogotá, pp 92

Boesi T, Galea F et al (1988) Estudio estratigráfico del flanco norandino en el sector Lobatera. 3er Simposio Bolivariano-Exploración Petrolera en las Cuencas Subandinas Soc. Ven. Geólogos, Caracas, Mem I, pp 1–41

Bolli HM (1966) Zonation of Cretaceous to Pliocene marine sediments based on planktonic foraminifera. Bol Inf Assoc Venez Geol Min y Petroleo 9:3–31

Bolli HM, Loeblich AR et al (1957) Planktonic foraminiferal families Hantkeninidae, Orbulinidae, Globorotaliidae and Globotruncanidae. US Nat Mus Bull 215:97–123

Brönnimann P, Brown NK (1958) *Hedbergella*, a new name for a Cretaceous planktonic foraminiferal genus. J Wash Acad Sci 48(1):15–17

Burnett J (1998) Upper cretaceous. Calcareous nannofossil biostratigraphy, pp 132–199

Caballero V, Parra M et al (2010) Levantamiento de la cordillera oriental de Colombia durante el Eoceno Tardío-Oligoceno temprano: Proveniencia sedimentaria en el sinclinal de nuevo mundo, cuenca valle Medio del Magdalena. Bol de Geol 32(1):45–77

Caron M (1978) Cretaceous planktonic foraminifers from dsdp leg 40, Southeastern Atlantic Ocean. In: Deep Sea Drilling Project Reports and Publications (ed) DSDP, vol 40, pp 651–678

Caron M (1985) Cretaceous planktic foraminifera. In: Bolli HM, Saunders JB, Perch-Nielsen K (eds) Plankton stratigraphy, vol 1. Cambridge University Press, Cambridge, pp 17–86

Cediel F, Leal-Mejia H et al (2011) Petroleum geology of Colombia: regional geology of Colombia. Fondo Editorial U. Eafit (ed) vol 1, pp 220

Cohen K, Finney S et al (2013) The ICS international chronostratigraphic chart. Episodes 36 (3):199–204

Cooper M, Addison F et al (1995) Basin development and tectonic history of the Llanos basin, Eastern Cordillera, and middle Magdalena Valley, Colombia. AAPG Bull 79(10):1421–1442

Cushman J, Hedberg H (1941) Upper cretaceous foraminifera from santander del norte, Colombia, SA. Contr Cushman Lab Foram Res 17(Pt 4):79–100

Dalbiez F (1955) The genus *Globotruncana* in tunisia. Micropaleontology 1(2):161–171

Davis CL (1995) Geochemical and carbon-isotopic stratigraphy of cretaceous marine sediments in a core from Maracaibo Basin, Western Venezuela. M.Sc. thesis. Indiana University, pp 1–114

de Romero LM, Truskowski IM et al (2003) An integrated calcareous microfossil biostratigraphic and carbon-isotope stratigraphic framework for the La Luna formation, western Venezuela. Palaios 18(4):349–366

Etayo-Serna F (2010) Identificación de Ammonites en columnas estratigráficas proyecto Tectonoestratigrafia Fm La Luna, VMM. Private Report Geostratos, pp 1–5

Galvis-Portilla HA, Higuera-Diaz I et al (2014) Regional sequence stratigraphy of the upper cretaceous La Luna formation in the Magdalena Valley Basin, Colombia. Paper presented at the unconventional resources technology conference, Denver, Colorado, pp 25–27

Gandolfi R (1955) The genus *Globotruncana* in northeastern Colombia. Bull Am Paleontol 36:1–118

Garner A (1926) Suggested nomenclature and correlation of geological formations in Venezuela. Amer Inst Min Metall Eng Trans 677–684

Gerhardt K (1897) Beitrag zur Kenntiniss der Kreidefbrmation in Columbien. Neues Jahrb Min Geol Palaeont Beit Bd 11:118–208

Gradstein FM, Ogg JG et al (2004) A geologic time scale. Cambridge University Press, Cambridge

Hedberg H, Sass L (1937) Sinopsis de las formaciones geológicas de la parte occidental de la Cuenca de Maracaibo, Venezuela. Bol Geol y Min 2(4):77–120

Herm D, Barthel KW (1962) Stratigraphische und mikropaläontologische Untersuchungen der Oberkreide im Lattengebirge und im Nierental. Bayer- Akad. Wiss Math-naturw Kl Abh., NF, Heft, vol 104, pp 1–119

Hofmann P, Wagner T (2011) ITCZ controls on Late Cretaceous black shale sedimentation in the tropical Atlantic Ocean. Paleoceanography 26(4):1–11

Jarvie DM, Hill RJ et al (2007) Unconventional shale-gas systems: The Mississippian Barnett Shale of north-central Texas as one model for thermogenic shale-gas assessment. AAPG Bull 91(4):475–499

Jenkyns HC (1980) Cretaceous anoxic events: from continents to oceans. J Geol Soc 137 (2):171–188

Karsten H (1858) Über die geognostischen Verhältnisse des westlichen Columbien, der heutigen Republiken Neu-Granada und Equador: Aus der Kaiserlich-Königlichen Hof-und Staatsdruckerei, pp 80–117

Keller G (2003) Biotic effects of impacts and volcanism. Earth Planet Sci Lett 215(1):249–264

Kuuskraa VA, Stevens S et al (2013) Northern South America. In: EIA/ARI World Shale Gas and Shale Oil Resource Assessment (ed) World shale gas resources: an initial assessment of 14 regions outside the United States Arlington, VA and Washington, DC, vol 17, pp 1–19

Martínez JI (1989) Foraminiferal biostratigraphy and paleoenvironments of the Maastrichtian Colon mudstones of northern South America. Micropaleontology 35(2):97–113

Mello M, Koutsoukos E et al (1995) The Napo formation, Oriente Basin, Ecuador: hydrocarbon source potential and paleoenvironmental assessment petroleum source rocks. Springer, Berlin, pp 167–181

Morales LG (1958) General geology and oil occurrences of Middle Magdalena Valley, Colombia: South America, pp 641–695

Navarrete-Parra RE, Parra-Navarete FJ et al (2015) Maastrichtiense a Mioceno temprano, correlación entre 19 Escenarios en el sector norte del Valle Medio del Magdalena Mediante palinoestratigrafía y foraminíferos. XV Congreso Colombiano De Geología, Bucaramanga Colombia, Memorias, pp 1033–1039

Navarrete-Parra RE, Parra-Navarrete FJ et al (2016a) Foraminíferos, nanofósiles calcáreos y palinomorfos del Cretácico Superior - Daniense del Valle Medio del Magdalena (VMM), Colombia: Bioestratigrafía y Paleoambientes. 11° Congreso de la Asociación Paleontológica Argentina 2016, Roca, Argentina. Memorias, p 130

Navarrete-Parra RE, Parra-Navarrete FJ et al (2016b) Palynostratigraphy and Meso-Cenozoic spores, pollen and dinocysts zonation for Middle Magdalena Valley, Colombia (NW South America). XIV International Palynological Congress and X international organization of palaeobotany conference, Salvador Bahía, Brazil Boletín de la Asociación Latinoamericana de Paleobotánica y Palinología (16):180

Navarrete-Parra RE, Parra-Navarrete FJ et al (2016c) Bioestratigrafía y análisis paleoambiental de la Formación Umir: pozos Chuira – 2 y Chuira-2ST-1, Cuenca del Valle Medio del Magdalena (VMM), Colombia. 11° Congreso de la Asociación Paleontológica Argentina 2016, Roca, Argentina. Memorias, pp 130–131

Notestein FB, Hubman CW et al (1944) Geology of the Barco concession, Republic of Colombia, South America. Geol Soc Am Bull 55(10):1165–1216

Odin G, Fullagar P (1988) Geological significance of the glaucony facies. In: Odin GS (ed) Green marine clays, oolitic ironstone facies, verdine facies, glaucony facies and celadonite-bearing rock facies—a comparative study, vol 45. Elsevier Science, Amsterdam, pp 295–332

Pardo A, Keller G (2008) Biotic effects of environmental catastrophes at the end of the cretaceous and early tertiary: *Guembelitria* and *Heterohelix* blooms. Cretac Res 29(5):1058–1073

Patterson RT, Fowler AD et al (2004) Evidence of hierarchical organization in the planktic foraminiferal evolutionary record. J Foramin Res 34(2):85–95

Pérez Panera JP, Parra Navarrete FJ et al (this volume) Turonian santonian zonation: calcareous nanoplancton and foraminifera, Dinoflagellates correlation in La Luna Formation—Middle Magdalena Valley Basin, northern Colombia

Petters V (1954) Tertiary and upper Cretaceous foraminifera from Colombia, South America, vol 5. Contributions from the Cushman Foundation of Foraminiferal Research, pp 37–41

Petters V (1955) Development of upper Cretaceous foraminiferal faunas in Colombia. J Paleontol 29(2):212–225

Postuma JA (1971) Manual of planktonic foraminifera. Elsevier Publishing Company, Amsterdam, p 420

Premoli-Silva I, Sliter WV (1999) Cretaceous paleoceanography: evidence from planktonic foraminiferal evolution. Special Papers-Geological Society of America, pp 301–328

Prince M, Acevedo R et al (2011) Upper Cretaceous in the Middle Magdalena Valley, Colombia, South America: a new exploratory target in an old mature basin. Search and discovery, Online Journal for E&P Geoscientists, pp 1–13. Retrieved from http://www.searchanddiscovery.com. Website: http://www.searchanddiscovery.com/documents/2011/10296prince/

Rangel A, Giraldo B et al (2000a) Estratigrafía química y facies orgánicas del Terciario Inferior y Cretácico Superior del Piedemonte Llanero y Valle Oriental del Magdalena. Internal report. ECOPETROL-ICP, pp 1–21

Rangel A, Parra P et al (2000b) The La Luna formation: chemostratigraphy and organic facies in the Middle Magdalena Basin. Org Geochem 31(12):1267–1284
Rey O, Simo J et al (2004) A record of long-and short-term environmental and climatic change during OAE3: La Luna Formation, Late Cretaceous (Santonian–early Campanian), Venezuela. Sediment Geol 170(1):85–105
Reyes J, Fajardo A et al (2000) Secuencia Calcárea del Cretáceo del Valle Medio Del Magdalena, Colombia. Una Nueva Frontera Exploratoria ACGGP, pp 1–6
Rodríguez M, Cáceres C et al (2016a) An approximation to the understanding of natural fractured reservoirs in La Luna formation: a study case of from the northern Middle Magdalena Basin, Colombia. XII Simposio Bolivariano-Exploración en Cuencas Subandinas, Bogotá, Colombia, p 179
Rodriguez M, Sanchez C et al (2016b) Acordionero oil field discovery, Middle Magdalena Basin, Colombia: an unusual large stratigraphic trap within the Paleocene Lisama formation. XII simposio bolivariano-exploración en cuencas subandinas, Bogotá, Colombia, p 180
Rolon LF (2004) Structural geometry of the Jura-Cretaceous rift of the Middle Magdalena Valley Basin–Colombia, Doctoral dissertation. West Virginia University Libraries, 117 pages
Schenk CJ, Brownfield ME et al (2017) Assessment of continuous oil and gas resources in the Middle and Upper Magdalena Basins, Colombia, 2017 (Fact sheet: 2017–3060). Retrieved from Reston, VA: http://pubs.er.usgs.gov/publication/fs20173060
Schiebel R (2002) Planktic foraminiferal sedimentation and the marine calcite budget. Glob Biogeochem Cycles 16(4):1065
Schlanger SO, Jenkyns HC (1976) Cretaceous oceanic anoxic events: causes and consequences. Geol Mijnbouw 55(3–4):179–184
Sigal J (1955) Notes Micropaléontologiques nord-africaines. I. Du Cénomanien au Santonien: zones et limites en facies pélagiques. Cr Somm Soc geol Fr 8:157–160
Sliter WV (1989) Biostratigraphic zonation for Cretaceous planktonic foraminifers examined in thin section. J Foramin Res 19(1):1–19
Talukdar S, Gallango O et al (1985) Generation and migration of hydrocarbons in the Maracaibo Basin, Venezuela. An integrated basin study. In: Leythaeuser D, Rullkötter J (eds) Advances in organic geochemistry 1985: organic geochemistry, vol 10, pp 261–279
Tchegliakova N (1993) Los foraminíferos y minerales antigénicos de la Formación Umir (sección quebrada la julia, Valle Medio del Magdalena): registro de una laguna costera a finales del cretácico superior (maastrichtiano). Geología Colomb 18:107–117
Tchegliakova N (1995) Los foraminíferos de la formación Umir (sección quebrada la julia): registro del cretácico superior cuspidal (maastrichtiano) en el Valle Medio del Magdalena, Colombia. Geología Colomb 19:109–130
Tchegliakova N (1996) Registro de las biozonas de foraminíferos planctónicos *Gansserina gansseri* y *Abathomphalus mayaroensis* (maastrchtiano medio y superior) en el extremo meridional del Valle Medio del Magdalena (Colombia, Sur América). Geología Colomb 20:67–80
Tchegliakova N, Pérez V (1995) Registro de la Zona bioestratigráfica *Dicarinella concavata* (Foraminiferida) en el Extremo noroccidental de Suramérica (Colombia). Geología Colomb 19:131–143
Torres E, Slatt R et al (2012) Characterization of the Cretaceous La Luna Formation as a Shale Gas System, Middle Magdalena Basin, Colombia. Paper presented at the Houston Geological Society conference on unconventional resource shales, Poster
West J (ed) (1996) International petroleum encyclopedia. Pennwell Publishing Co., Tulsa, pp 110, 186–187

Paleoenvironmental Interpretation of Late Miocene Deposits, from El Cajón Valley (Catamarca Province, Argentina) Based on Calcareous Microfossils

Belén Zamudio[1(✉)] and Susana Morton[1,2]

[1] Centro de Ecología Aplicada Del Litoral (CECOAL)-CONICET-Universidad Nacional Del Nordeste, Ruta 5 Km 2,5, 3400 Corrientes, Argentina
belenzamudio@live.com, lourdes_su@yahoo.com.ar
[2] Facultad de Ciencias Exactas Y Naturales Y Agrimensura (FACENA), Universidad Nacional Del Nordeste, Av. Libertad 5470, 3400 Corrientes, Argentina

Abstract. The Neogene sedimentary sequence from El Cajón Valley, Catamarca Province, Argentina, includes alloformations or sequences of three orders: Peñas Azules, Totoral, and Playa del Zorro, being this last one the one studied in this work. The samples come from the Río Totoral site, located in the southern portion of the valley. The calcareous microfossils associations, together with gastropods and bivalves, were found in levels of green and yellow-green fine sandstones and mudstones, with gypsum intercalations. The microfossils are represented by foraminifers of the genus *Ammonia*, together with Ostracoda, Cytherocopina, from which were recognized *Cyprideis* sp. *Perissocytheridea* sp., *Limnocythere* sp.1, and *Limnocythere* sp. 2. Foraminifers were found only at one level, where they constitute the 15% of the total specimens. The ostracods were dominant in all the associations, and they were represented by valves and carapaces of juveniles and adults. In many cases, they showed signs of compression and fragmentation. *Cyprideis* and *Perissocytheridea* were common in the lower levels of the outcrop; they are known as typical from brackish water. At the middle section level, the association was characterized by the dominant taxa of *Cyprideis* and *Limnocythere*. This association could indicate brackish waters with precipitations higher than evaporation and high alkalinity. At the upper section level, a low foraminifera diversity is represented only by *Ammonia* in association with *Cyprideis*. This fact would denote a shallow lacustrine and oligo–mesohaline paleoenvironment.

Keywords: Ostracods · Foraminifers · Brackish water · Neogene Paleoecology

1 Introduction

The Ostracoda (Arthropoda, Crustacea) is one of the main groups used in paleontology as paleoenvironmental and biostratigraphic indicators, since they provide data on different environmental parameters (e.g., salinity, depth) and on the nature of the substrate

G. C. Cusminsky et al. (eds.), *Advances in South American Micropaleontology*, Springer Earth System Sciences, https://doi.org/10.1007/978-3-030-02119-1_5

(Whatley and Mayburry 1990; Bertels Psotka 2000; Boomer et al. 2003; Laprida and Ballent 2007; Ferrero 2009). Another important characteristic of ostracods is that they are used for reconstruction of paleoecological conditions because of their high dispersal and their fast evolution in some cases (Martínez 2005; Laprida 2006). Along with other calcareous microfossils, such as foraminifers, they have a high fossilization potential. Furthermore, from a few grams of sediment it is possible to obtain hundreds of specimens, which can provide much of taxonomic and paleoecology information. As regards, the foraminifers, these organisms are sensitive to fast environmental changes (salinity, oxygen, and organic matter) (Alperin et al. 2008).

The Calchaquí Valleys are a system of mountains and valleys in the northwest part of Argentina, extended from north to south and embracing 520 km from the NE of the Catamarca Province. They also include the Neogene sedimentary sequence, which is integrated by the Santa María-Hualfín Basin (Catamarca and Tucumán Provinces), El Cajón Valley (Catamarca), and the Calchaquí Valley (Salta) (Herbst et al. 2000).

This study involves the area of El Cajón Valley, which is a tectonic depression lengthened in a north–south direction, which is narrowing to the north. This valley has a **V** inverted shape seen from above, and it separates the Puna and Sierras Pampeanas areas. El Cajón Valley is placed between Quilmes or El Cajón range to the east and El Chango Real range to the west (Fig. 1) (Malizzia et al. 1990). The Neogene of El Cajón Valley was constituted by three alloformations or sequences (Peñas Azules Alloformation (sequence I), the Playa del Zorro Alloformation (sequence II), and the Totoral Alloformation (Bossi et al. 1992) (Fig. 2). The section under study, Playa del Zorro, is characterized by deposits of fine lacustrine sediments, produced under reducing conditions and shallow water, in a semi-arid and seasonal climatic regime, although with torrential rains periods (Parra and Morales 2003). On the other hand, Bossi et al. (2000) through the paleocurrent analysis inferred how the drainage system was developed.

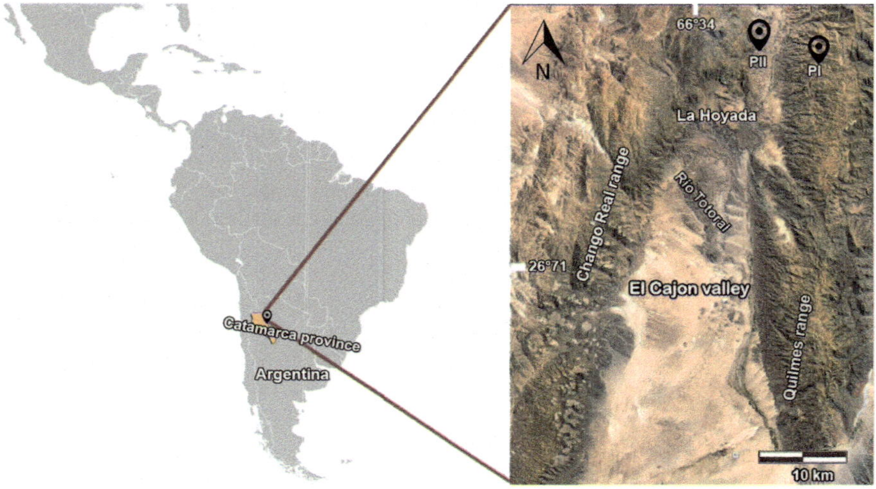

Fig. 1. Location of study area. Catamarca Province (left). Satellite image of El Cajón Valley from Google maps

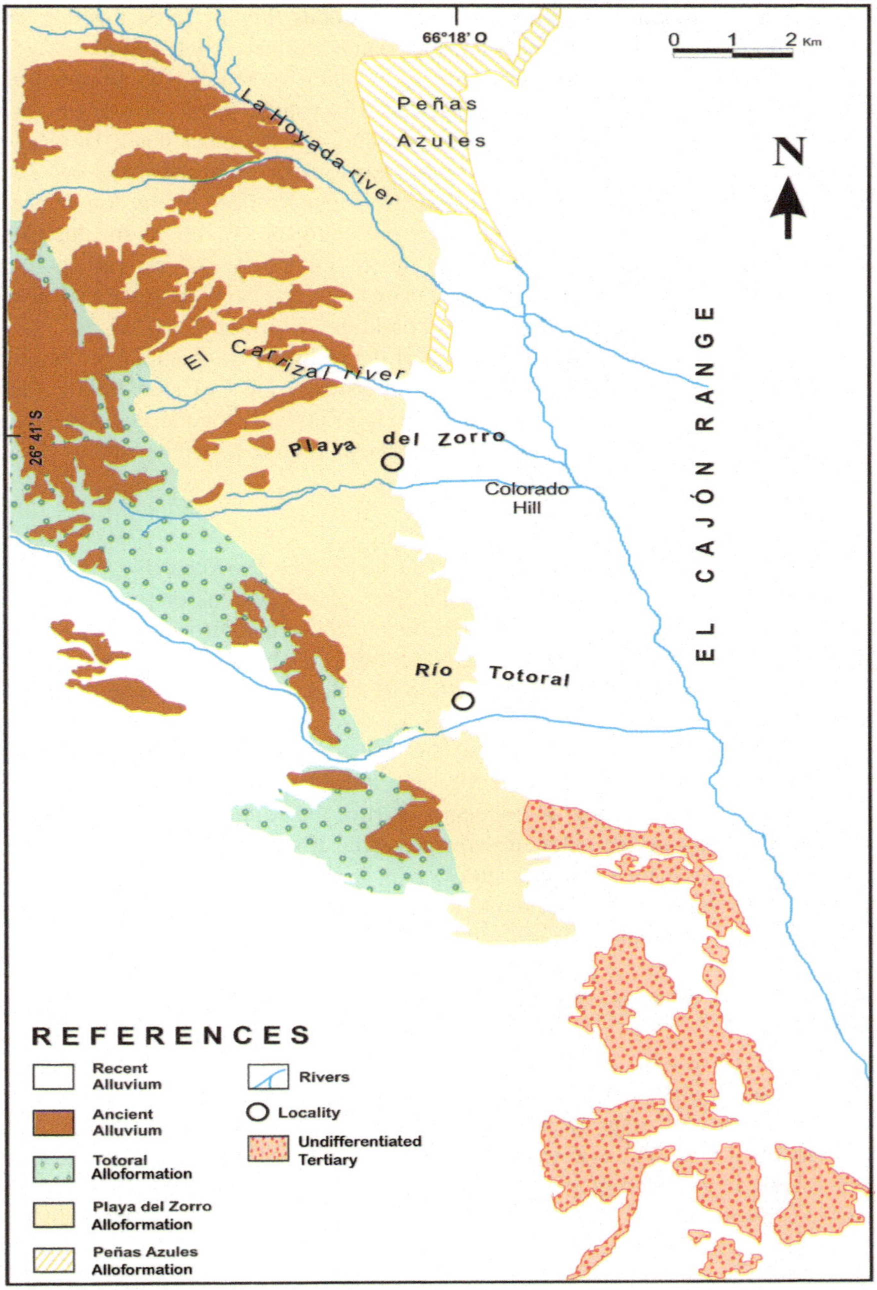

Fig. 2. Geological map from El Cajón Valley (taken from Morton 2004, modified from Bossi et al. 1992)

Here, the Playa del Zorro Alloformation shows sinuous fluvial facies, together with permanent flood stages, lakes formations, and scarce sediment contribution.

The micropaleontology studies from El Cajón Valley are scarce. García and Herbst (1997) analyzed the charophytes and suggested a lacustrine paleoenvironment, with salinity changes from scarcely saline to mesohaline. On the other hand, Horn et al. (2016) carried out an interesting study on the palynofacies from the Playa del Zorro Alloformation. The palynomorphs associations were characterized by the dominance of terrestrial elements, suggesting arid and warm paleoclimatic conditions for this place, being influenced by the aquatic environment, due to the presence of resistant spores of Zygnemataceae what reflects stressed water. However, the high frequency of *Botryococcus braunii* suggests oxygenated water conditions (Horn et al. 2016).

The current study focuses on the analysis of the Neogene associations of ostracods and foraminifera from the Playa del Zorro Alloformation, in El Cajón Valley, Catamarca Province. This information will help us to infer not only the paleoecological and paleoenvironmental conditions, but also where those microfossils inhabited.

1.1 Geological Settings

El Cajón Valley is located in the northwest of Catamarca Province (Fig. 1), and the Neogene of this valley is constituted by the three alloformations or sequences mentioned before. The age of the Playa del Zorro Alloformation was established as "Calchaquense," i.e., Miocene sensu lato, according to the nomenclature of Turner (1973). On the other hand, Bossi et al. (2000) dated a tuff, using the 40Ar/39Ar method, obtaining the following values: 5.38 ± 0.07 and 6.04 ± 0.07 Ma (average at 5.71 ± 0.04 Ma)". This value shows that the lacustrine section is slightly older than the tuff; however, it belongs to the Late Miocene. The Playa del Zorro Alloformation is the best known, and it is placed in the central part of the valley where it also displays good outcrops, being divided into subsequences (IIa, IIb, IIc, IId) (Bossi et al. 1992). The subsequence IIc, which provides study samples, is lacustrine in origin, and it has a rich fauna of vertebrates and plant remains (Bossi et al. 1992). In the lower section of this subsequence, siltstones and fine yellow and yellowish green sandstones predominate in thin beds, having parallel and ondulitic lamination. This section, with a thickness of about 80 m has also abundant levels of bivalves, gastropods, fish scales, and ostracods.

The upper section presents a grain-increasing succession, where fine sandstones are interbedded with crossed stratification toward the top of the sequence, becoming thicker and of conglomerate nature (Bossi et al. 1992). For this alloformation, Vides (2004) identified five sub-environments associated. Two of them were from lacustrine origin: 1. lake: represented by pelitic layers and thin sandstones, finely laminated with abundant gypsum; 2. lake fringe: sandstones, laminated and massive mudstones, with abundant pelecypods and gastropods.

2 Materials and Methods

The microfossils come from two stratigraphic profiles developed in the Río Totoral site (Playa del Zorro Alloformation), from siltstones and fine sandstones interbedded with gypsum at the base, and from grain increasing, having thick sandstones at the upper section. A total of 23 samples from two stratigraphic sections were studied:

1. Profile I, in Río Totoral (RT), at 66°20′33,9″W/26°42′40,1″S; with 67 m of thickness and two fertile levels, RT1 and RT6 (Fig. 3).
2. Profile II, Morterito (M), at 66°26′19,3″W/26°40′30,1″S; with 57 m of thickness and four fertile levels: M0, M2, M5, and M13 (Fig. 4).

The samples were processed according to the traditional methodology based on the disintegration/disaggregation, washing it through a sieve of 0.074 mm. The picking was carried out under a trinolocular Stereomicroscope, Motic SMZ 168, and specimens were picked. The microfossils recovered were observed under Motic SMZ 168 and Leitz Wetzlar Trinolocular Stereomicroscope. SEM images were obtained with a JEOL JSM-5800-LV Scanning Microscope of the Secretaría General de Ciencia y Tecnología de la Universidad Nacional del Nordeste from Corrientes, Argentina.

For the Ostracoda, we followed the classification of Moore and Pitrat (1961) and Van Morkhoven (1962), Zabert (1981), Zabert and Herbst (1986), Cusminsky et al. (1995), Meisch (2000), Laprida (2006), and Ferrero (2009) among others. For the Foraminifera the classification follows Loeblich and Tappan (1988) Calvo-Marcilese and Pratolongo (2009), Calvo-Marcilese et al. (2011), among others.

Several characteristics were accounted for the ostracods analysis: proportion of valves and carapaces, taphonomic features in the preservation of valves (either they were complete/or in fragments). In the case of the carapaces, the presence of compaction and/or recrystallization; and the population structure within each association (adults/juveniles) (Whatley 1988).

The abbreviations used in the text are: C = carapace, RV = right valve, LV = left valve, EV = external view, DV = dorsal view; SpV = spiral view, U = umbilical view.

The collected material was stored in the paleontological collection of the UNNE “Dr. Rafael Herbst,” under the acronym CTES-PMP (Preparados Micropaleontológicos). Repository: Lot No. 3.487-3.491.

3 Results

A total of 4.062 ostracods and 123 foraminifera specimens were obtained, from six fertile levels. In the profile I were obtained 2446 ostracods from four fertile levels. The collected associations showed low specific richness of the ostracod *Cyprideis* sp., together with *Perissocytheridea* sp., *Limnocythere* sp. 1, *Limnocythere* sp. 2 (Fig. 5). In the profile II were obtained 1.616 ostracods and 123 foraminifers. From two fertile levels were found *Cyprideis* sp., *Perissocytheridea* sp., and specimens of the foraminifer *Ammonia tepida* Cushman 1926 (Fig. 5). Both outcrops (profiles I and II) are represented by valves and carapaces of ostracods, denoting different states of

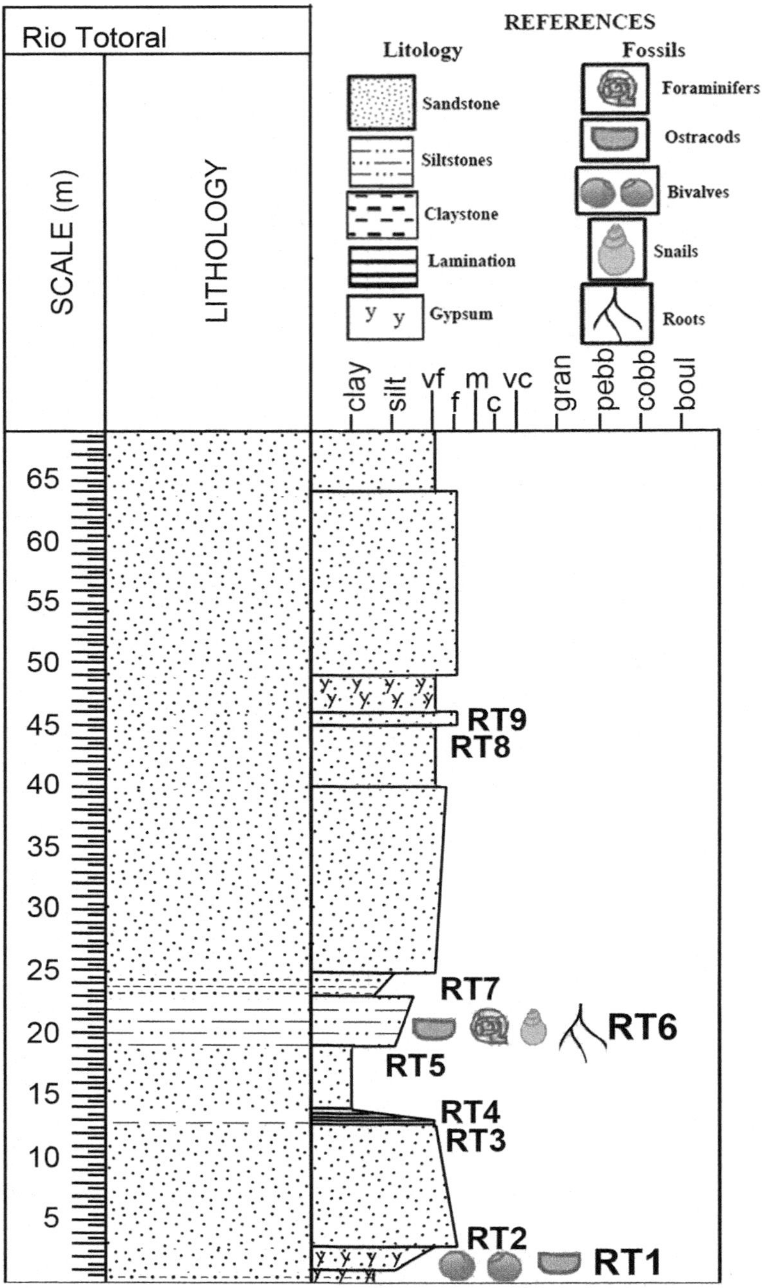

Fig. 3. Stratigraphic column of the Playa del Zorro Alloformation, at the Río Totoral site with indications of the study samples (Profile I)

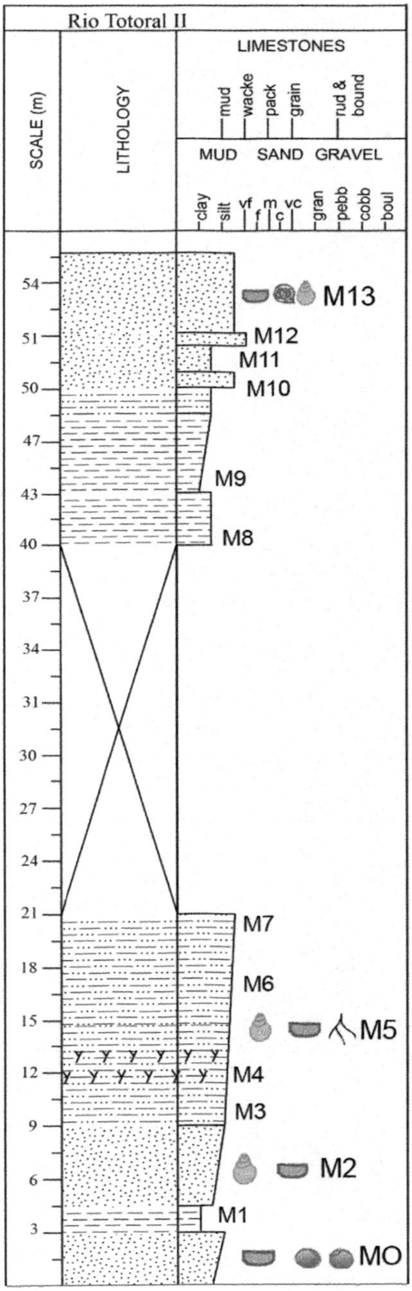

Fig. 4. Stratigraphic column of the Playa del Zorro Alloformation, at the Río Totoral site with indications of the study samples (Profile II)

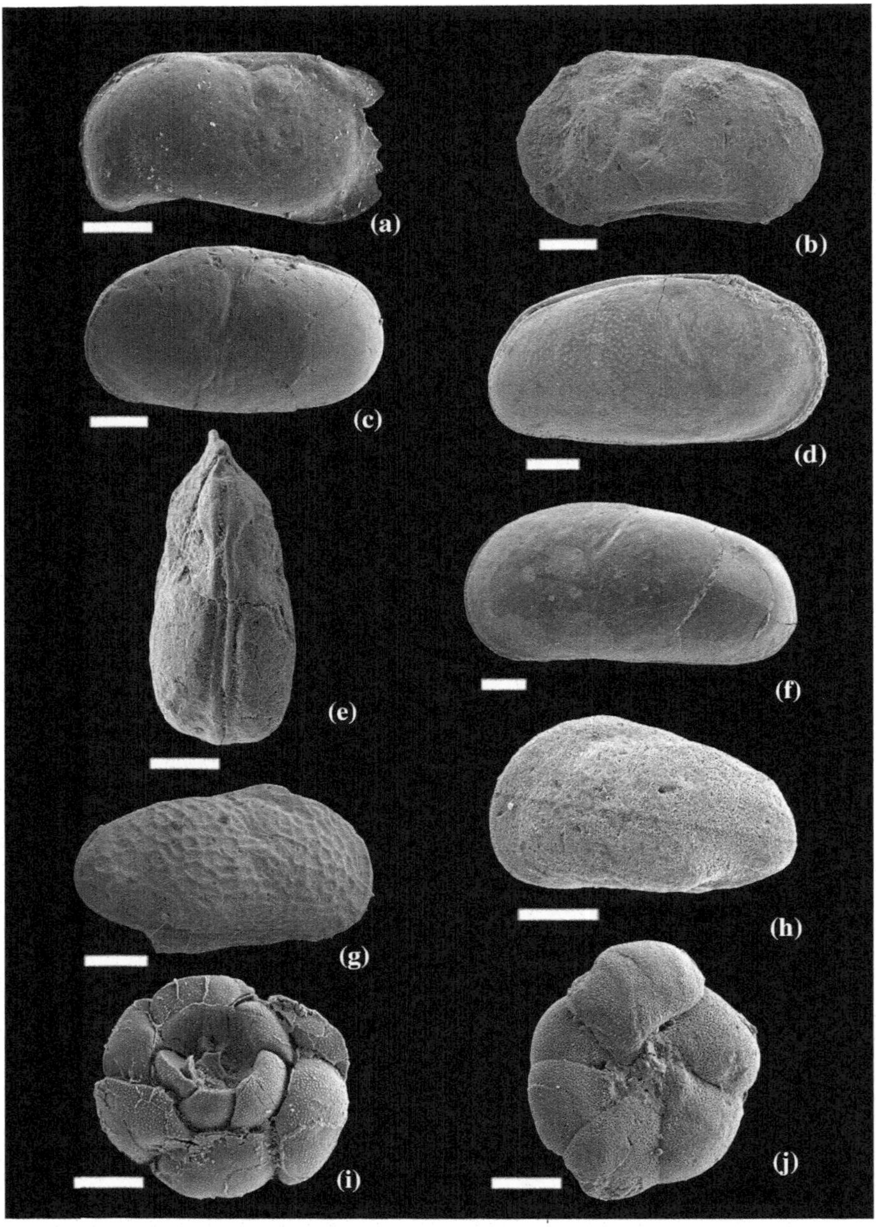

Fig. 5. Calcareous microfossils: **a** *Limnocythere* sp. 1. RV, EV, **b** *Limnocythere* sp. 2. LV, EV, **c d f** *Cyprideis* sp., **c f** LV, EV female, **d** RV, E male, **e g h** *Perissocytheridea* sp., **e** C, DVi, **g** C, RV female, **h** LV, EV, juvenile, **i**, **j** *Ammonia tepida*, **i** SpV, **j** UV, scale bar = 100 μm

preservation (Fig. 8). *Cyprideis* sp. was dominant in all the analyzed samples, from both profiles (Figs. 6 and 7).

In the profile I, *Perissocytheridea* sp. was present in less than 10% of the sample RT1 (Fig. 6). In this sample, only carapaces were found (Fig. 8) and juveniles specimens were scarce. Foraminifers were found only in the sample RT6, and many of them are preserved as molds of *Ammonia tepida*. They represent the 15% of the total specimens in the whole sample (Fig. 6).

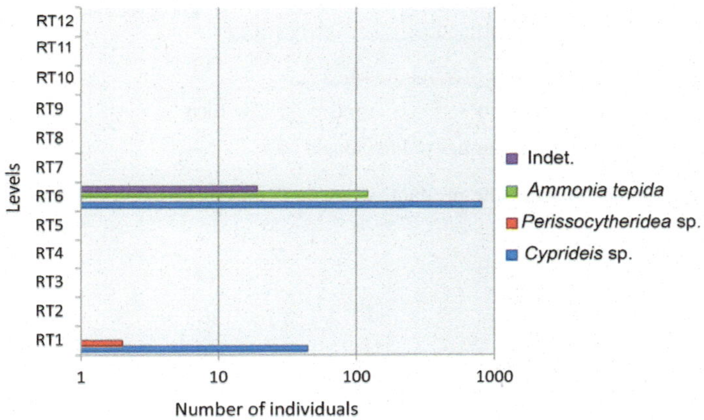

Fig. 6. Specific richness of the productive levels from the Río Totoral site. Profile I

As regards profile II, *Perissocytheridea* sp. was the second taxa more representative in the sample M0, following *Cyprideis* specimens with almost a 50% (Fig. 7). In this sample, the best-preserved carapaces and valves was observed, with a juvenile/adult ratio of 1:2. In this sample, both, *Limnocythere* sp. 1 and *L.* sp. 2, were scarce (Fig. 7). The sample M5 displayed an ostracod fauna represented by a monospecific association (Fig. 7) and a high number of compressed carapaces (Fig. 8). This fact would reveal some mechanical effort toward them, probably post-burial. The largest number of specimens is present in sample M13. They are represented by *Cyprideis* dominance and few *Perissocytheridea* specimens (Fig. 7). This is the only level where compression in carapaces was not found (Fig. 8). However, the 95% of the specimens are recrystallized, possibly consisting with fossil–diagenetic processes.

All the studied associations were represented mainly by adult carapaces. In both profiles, fragmented valves were very rare, between 15% or less (Fig. 8). This fact would evidence a high energy thanatocoenosis, with some postmortem disturbance produced by currents, waves, or biological activity (Boomer et al. 2003).

The associated macrofauna included few specimens of the gastropod *Heleobia* sp. and the pelecypod *Neocorbicula* sp., in particular *Heleobia*. This one was present in the sample M5 in a 15%, being scarcely present in the rest of the samples (less than 1%) on the other hand; *Neocorbicula* sp. was present in samples of base levels in both profiles.

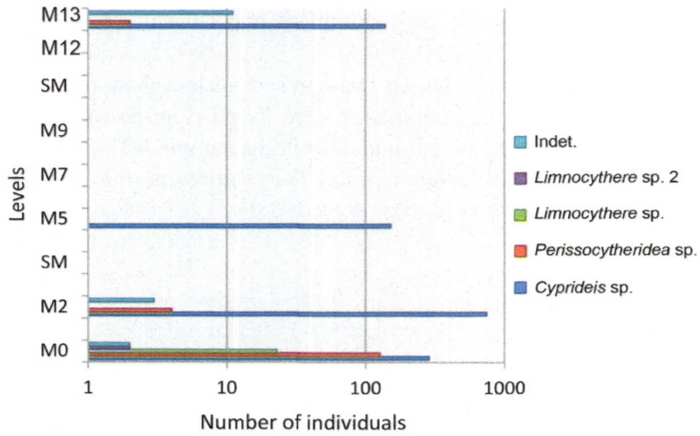

Fig. 7. Specific richness of the productive levels from the Río Totoral site. Profile II

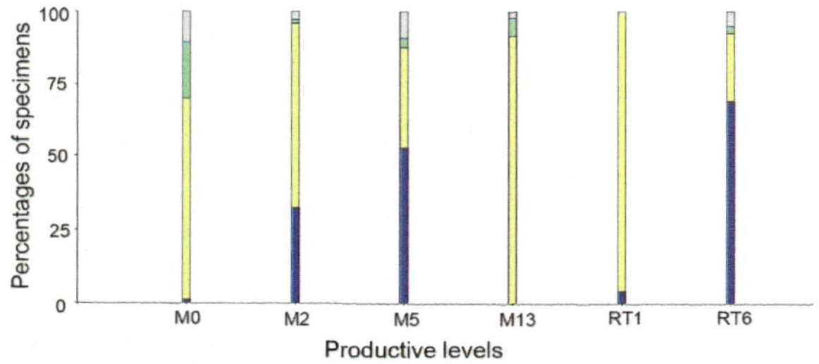

Fig. 8. Proportion of compressed carapaces (blue), carapaces without compression (yellow), valves (green), and fragmented material (gray) by productive levels, from *Cyprideis* ostracods

4 Discussion and Final Remarks

The *Cyprideis* genus has wide salinity tolerance, as it is able to adapt properly to the fluctuations of marginal marine environments (estuarine area, lagoons, etc.), brackish environment, and mixohaline lagoons (Maddocks 1983; Sandberg and Plusquellec 1974; Würdig 1983). It prefers a mud sandy or a mud substrate (Meisch 2000). The *Cyprideis*-dominant assemblage would indicate unstable environment with salinity fluctuations and dry conditions dominance (Laprida et al. 2009). This ostracod inhabits in mixed and purely continental saline water bodies, with low salinity to hypersaline coastal environments. It is also found in marine–marginal, estuaries, and marshes around the world (Walton and Sloan 1990; García and Herbst 1997).

In profile I, in the lower section (level RT1), *Cyprideis–Perissocytheridea* assemblage is considered as living in brackish water. These taxa predominance are typical of brackish environments (Evangelista Nogueira and Feijó Ramos 2016). On the other hand, *Perissocytheridea* Stephenson is a mixohaline genus that inhabits lagoons (Nicolaidis and Coimbra 2008; Evangelista Nogueira and Feijó Ramos 2016) and from estuarine environments (Laprida 2006). This genus can be found in relation with both, freshwater and marine species (Coimbra et al. 2007). In the upper section, level RT6 (Fig. 6), the *Ammonia* presence would indicate a salinity increase and a decrease in the energy level. The salinity increase could be due to the evaporation which results in precipitation of the evaporites present in some levels. This fact shows alternation of dry periods and floods with predominance of droughts. The *A. tepida* species is characterized by inhabiting marginal marine environments. This one has been registered in both hypersaline and hyposaline waters (Murray 2006; Márquez et al. 2016; Pint et al. 2017). This environmental interpretation is also supported by Madeira-Falcetta (1974), Boltovskoy (1990), and Geslin et al. (1998). The foraminifer *Ammonia* is often found associated with *Cyprideis*; they are considered as a typical microfauna from a brackish environment (Nicolaidis and Coimbra 2008).

As regards profile II, a gradual change in the level of the paleolake could be inferred. In the lower section (level M0), it was possible to identify the association with the highest specific diversity, with *Limnocythere* spp. presence. This fact is associated with strong alkalinity, abundant rainfall, and better conditions. This genus has been found in permanent shallow lakes of mesohaline waters, strongly alkaline, and little oxygenated (Ferrero 2009; Laprida and Valero-Garcés 2009). Furthermore, *Limnocythere* was used for reconstruction of several lake Quaternary settings from Pampean shallow lakes, being this one a good indicator of hydrological changes (Ramón Mercau et al. 2014). The associations would reveal unstable environments, suggesting strongly salinity fluctuations, probably determined by fast changes from wet to dry phases, but with predominant dry phases (Laprida et al. 2009). The *Cyprideis* Jones–*Limnocythere* Brady associations were found, generally of permanent shallow lakes with mesohaline and shallow waters, where there are more precipitations than evaporation and high levels of alkalinity (pH = 9.05) (Ferrero 2009). Those genera are characteristic of lacustrine environments with a wide salinity range (Calvo-Marcilese et al. 2016). Both of them are typical of shallow, mixed environments, connected to the sea and in continental saline water bodies. These species can also tolerate conditions of low salinity. Additionally, monospecific associations composed by *Cyprideis* Jones with a large number of specimens were identified in the middle and in the upper levels (M5 and M13). The presence of these associations makes possible to infer drier conditions and shallower water in these levels.

The associations studied here share similarities with other Miocene associations such as the Yecua Formation (Bolivia) (Nicolaidis and Coimbra 2008) and the Desencuentro Formation (La Rioja) (Cusminsky et al. 2006), due to the presence of *Cyprideis*, *Limnocythere*, *Perissocytheridea,* and *Ammonia*. The Yecua Formation shows evidence of calcareous microfauna from fresh to brackish water (Nicolaidis and Coimbra 2008). Cusminsky et al. (2006) described an ostracod microfauna associated with charophytes and inferred environment of variable salinity. This fact corresponds to lacunar bodies liable to desiccation or at the supply of rivers and streams, as well as, to bodies of water developed in a floodplain (Cusminsky et al. 2006).

The observed alternations between fine and sandy sediments in both profiles often suggest variations in the shallow water level, caused by seasonal climatic changes. The Playa del Zorro Alloformation would match to a sandy lake system, from hypohaline to oligohaline with high alkalinity during torrential rain periods (Parra and Morales 2003). It is also possible to infer that storms energy could have influenced the current energy, triggering associations with scarce juveniles. This situation is very common in continental and shallow environments (Whatley 1988). Adult carapaces were dominant in all the associations. The abundance of carapaces does not necessarily mean that the postmortem burial has been immediately, since *Cyprideis* has a complex hinge. For this reason, the valves would remain articulated longer than other species.

In relation to the mollusks association, *Heleobia* exhibits an important range of ecological tolerances, as it could live in both, freshwater and brackish water with or without vegetation (De Francesco et al. 2013). This gastropod prefers reduced coastal streams, as well as fast-flowing and well-oxygenated currents with silty clay bottoms. This genus is found in brackish water or freshwater, and it was never found in stagnant waters or in strong current areas (Biese 1944; Weyrauch 1963; Alonso and Wayne 1992).

Neocorbicula bivalves inhabit rivers and streams, either in firm sand or clay bottoms (Morton 2004).

The assemblages of microfossils would indicate that in the Playa del Zorro Alloformation, during the Late Miocene the specific diversity of the preserved microfauna was represented by a few genera *Cyprideis* sp., *Perissocytheridea* sp., *Limnocythere* spp., and *Ammonia tepida*.

Cyprideis–Perissocytheridea assemblage was found in profile I, in the lower section; they are a typical brackish water association. In the upper section, the presence of *Ammonia tepida* seems to indicate an increase in the salinity conditions, probably resulting from dry periods. On the other hand, in the lower section of profile II, the presence of *Limnocythere* specimens is associated with conditions of abundant precipitations and strong alkalinity. In the middle and upper levels, monospecific associations composed mainly by *Cyprideis* were identified with a large number of specimens. This circumstance suggests drier conditions and shallower water at these levels.

Acknowledgements. The authors are grateful to Ana P. Carignano for her comments, criticism, and advice in the improvement of this paper and to Verónica Espíndola for her critical review. We also thank the two anonymous reviewers and Gabriela Cusminsky for improving the manuscript with constructive reviews. We also express gratitude to Universidad Nacional Del Nordeste for the financial support (PICTO-UNNE-2011-0216 and SGCyT- UNNE. PI: 17/13F003).

References

Alonso RN, Wayne WJ (1992) Hallazgo de gastrópodos en la Formación Sijes (Mioceno superior), Salta, República Argentina. Estud Geol 48:179–186

Alperin M, Bernasconi E et al (2008) Asociaciones de foraminíferos bentónicos recientes de la Plataforma Continental Argentina (39°- 43° S y 58°- 65° O) analizadas con métodos estadísticos para datos composicionales. Ameghiniana 45(2):443–461

Bertels Psotka A (2000) Ostrácodos (Arthropoda, Crustacea) de hábitat límnico de la Formación Collón Cura, provincia de Chubut, Argentina. Ameghiniana 37:39–45

Biese WA (1944) Revisión de los moluscos terrestres y de agua dulce provistos de concha, de Chile. Boletin Museo Nacional Historia Natural de Chile 1(22):169–190

Boltovskoy E (1990) Morphological variations of benthic foraminiferal tests in response to change in ecological parameters: a review. J Paleontol 65:175–185

Boomer I, Horne DJ et al (2003) The use of Ostracodes in Paleoenvironmental Studies, or What can you do with an ostracod shell? Paleontol Soc Papers 9:153–180

Bossi GE, Vides M et al (2000) Análisis de las paleocorrientes y de la varianza de los componentes a tres niveles, Neógeno del valle del Cajón, Catamarca, Argentina. Asociación Argentina de Sedimentología 7:23–47

Bossi GE, Muruaga C et al (1992) Aloestratigrafía del Neógeno del Valle del Cajón (Provincia de Catamarca). In: Abstracts of the 4th Reunión Argentina de Sedimentología, La Plata, Buenos Aires, 5–9 October 1992

Calvo-Marcilese L, Ambrosio D et al (2016). In: Abstracts of the 11th Congreso de la Asociación Paleontológica Argentina, General Roca, provincia de Río Negro, Argentina, 17–21 October 2016

Calvo-Marcilese L, Cusminsky GC et al (2011) Asociaciones De Foraminíferos Bentónicos En Secciones Holocenas Del Estuario De Bahía Blanca (Buenos Aires, Argentina): Caracterización Paleoambiental. Ameghiniana 48(2):210–225

Calvo-Marcilese L, Pratolongo P (2009) Foraminíferos de marismas y llanuras de marea del estuario de Bahía Blanca, Argentina: distribución e implicaciones ambientales. Revista Esp de Microp 41(3):315–332

Coimbra JC, Carreñoa AL et al (2007) Ostracodes (Crustacea) from Cananéia-Iguape estuarine/lagoon system and geographical distribution of the mixohaline assemblages in southern and southeastern Brazil. Iheringia, Sér Zool 97(3):273–279. https://doi.org/10.1590/S0073-47212007000300010

Cusminsky G, García A et al (2006) Ostrácodos (Crustacea, Ostracoda) y carófitos (Chlorophyta, Charales) de la Formación Desencuentro (Mioceno superior), provincia de La Rioja, Argentina. Ameghiniana 43:327–338

Cusminsky G, Gómez E et al (1995) Estudio sedimentológico y micropaleontológico de sedimentos superficiales del estuario de Bahía Blanca, provincia de Buenos Aires. 6° Congreso de Paleontología y Bioestratigrafía. Actas: 107–115

De Francesco GC, Tietze E et al (2013) Mollusk successions of Holocene shallow-lake deposits from the Southeastern Pampa Plain, Argentina. Palaios 28:815–862. https://doi.org/10.2110/palo.2013.100

Evangelista Nogueira A, Feijó Ramos M (2016) The genus *Perissocytheridea* Stephenson, 1938 (Crustacea: ostracoda) an evidence of brackish water facies along the Oligocene- Miocene, Pirabas Formation, eastern Amazonia, Brazil. J S Am Earth Sci 65:101–121. https://doi.org/10.1016/j.jsames.2015.11.007

Ferrero L (2009) Foraminíferos y ostrácodos Pleistoceno tardío (Mar Chiquita, provincia de Buenos Aires, Argentina). Ameghiniana 46:637–656

García A, Herbst R (1997) Charophyta del Mioceno superior de los valles de Santa María y de El Cajón, provincia de Catamarca, Argentina. Ameghiniana 34(1):61–68

Geslin E, Debenay JP et al (1998) Abnormal wall textures and test deformation in *Ammonia* (hyaline foraminifer). J Foram Res 28:148–156

Herbst R, Anzotegui LM et al (2000) Síntesis paleontológica del Mioceno de los valles Calchaquíes, noroeste argentino. In: Aceñolaza FG, Herbst R (eds) El Neógeno de Argentina. INSUGEO, Serie Correlación Geológica, vol 14, pp 263–288

Horn MY, Martínez AM et al (2016) Análisis palinofacial de la Aloformación Playa del Zorro (Mioceno Tardío Plioceno) valle del Cajón, Catamarca, Argentina. Rev Bras Paleontol 19(1): 85–94. https://doi.org/10.4072/rbp.2016.1.07

Laprida C (2006) Ostrácodos recientes de la llanura pampeana, Buenos Aires, Argentina: ecología e implicancias paleolimnológicas. Ameghiniana 43:181–204

Laprida C, Ballent S (2007) Ostracoda. In: Camacho H, Longobucco MI (eds) Los Invertebrados Fósiles. Fundación de Historia Natural Félix de Azara, Universidad Maimónides, pp 599–624

Laprida C, Orgeira MJ et al (2009) El registro de la Pequeña Edad de Hielo en las lagunas pampeanas. Revista de la Asociación Geológica Argentina 65:603–611

Laprida C, Valero-Garcés B (2009) Cambios ambientales de épocas históricas en la pampa bonaerense en base a ostrácodos: historia hidrológica de la laguna de Chascomús. Ameghiniana 46:95–111

Loeblich AR, Tappan H (1988) Foraminiferal genera and their classification. Van Nostrand Reinhold Company, New York

Maddocks RF (ed) (1983) Applications of Ostracoda. In: Proceedings of the 8th international symposium on Ostracoda, July 1982. Published by the Department of Geosciences, University of Houston, p 677

Madeira-Falcetta M (1974) Ecological distribution of the thecamoebal and foraminiferal associations in the mixohaline environment of the southern Brazilian littoral. An Acad Bras Ciênc 46:667–687

Malizzia DC, Strecker M et al (1990) La Formación El Morterito (Neógeno de Sierras Pampeanas); características sedimentológicas y paleogeográficas. In: Abstracts of the 3rd Reunión Argentina de Sedimentología, Actas: 175–180

Márquez M, Ferrero L et al (2016) Holocene palaeoenvironmental evolution of The Pampa coastal plain (Argentina) based on calcareous microfossils. Rev Bras Paleontol 19(1):25–40. https://doi.org/10.4072/rbp.2016.1.03

Martínez D (2005) Asociaciones de ostrácodos modernos del estuario de Bahía Blanca, Buenos Aires, Argentina. Ameghiniana 42(4):669–684

Meisch C (2000) Freshwater Ostracoda of Western and Central Europe. Spektrum Akademischer Verlag GmbH, Heidelberg

Moore R, Pitrat C (eds) (1961) Treatise on invertebrate palaeontology. Part Q. Arthropoda 3. Crustacea, Ostracoda. Geological Society of America and University of Kansas Press, Lawrence

Morton LS (2004) Taxonomía y paleoecología de la malacofauna (Bivalvia-Gastrópoda) dulceacuícola del Neógeno en los valles de Santa María y del Cajón, Catamarca y Tucumán, Argentina. Facultad de Cs. Ex. y Naturales y Agrimensura, UNNE. PhD Thesis, p 236

Murray JW (2006) Ecology and applications of benthic foraminifera. Cambridge University Press, Cambridge

Nicolaidis D, Coimbra J (2008) *Perissocytheridea carrenoae* sp. nov. (Crustacea, Ostracoda) and associated calcareous micro fauna from Yecua Formation (Miocene), Bolivia. Revista Brasileira de Paleontologia 11(3):179–186. https://doi.org/10.4072/rbp.2008.3.04

Parra LE, Morales JA (2003) Descripción e interpretación de la columna estratigráfica tipo de la serie vulcano-sedimentaria neógena del Valle del Cajón (Catamarca, Argentina). Geogaceta 33:55–58

Pint A, Engel M et al (2017) How to discriminate Athalassic and marginal marine microfaunas: Foraminifera and other fossils from an early Holocene continental lake in northern Saudi Arabia. J Res Foram 47(2):175–187. https://doi.org/10.2113/gsjfr.47.2.175

Ramón Mercau J, Plastini MS et al (2014) A review of the genus Limnocythere (Podocopida: Limnocytheridae) in the Pampean región (Argentina), with the description of a new specie, *Limnocythere cusminskyae* sp. Nov. Zootaxa 3821: 26–36. https://doi.org/10.11646/zootaxa.3821.1.2

Sandberg PA, Plusquellec PL (1974) Note of the anatomy and passive dispersal of *Cyprideis* (Cytheracea, Ostracoda). Geosci Man 6:1–26

Turner J (1973) Descripción Geológica de la Hoja 11d, Laguna Blanca, provincia de Catamarca. Servicio Nacional Minero y Geológico. Boletín 142:1–71

Van Morkhoven FP (1962) Ostrácodos Post - Paleozoicos. Su morfología, taxonomía y utilidad económica. Elsevier, Amsterdam, p 478 (Vol 2)

Vides M (2004) La Aloformación Playa del Zorro, Quebrada del Totoral, valle del Cajón, Provincia de Catamarca. In: Abstracts of the 10th Reunión Argentina de Sedimentología. San Luis, Argentina, 1–3 September 2004

Walton WR, Sloan BJ (1990) The genus Ammonia Brunnich, 1772: its geographic distribution and morphologic variability. J Foram Res 20:128–156

Weyrauch WK (1963) Cuatro nuevas especies de Hydrobiidae de Argentina y Perú. Acta Zoológica Lilloana 19:243–259

Whatley R, Mayburry C (1990) Ostracoda and global events. British Micropaleontology Society Publication series. Chapman and Hall, Cambridge, London

Whatley RC (1988) Population structure of Ostracods: some general principles for the recognition of palaeoenvironments. In: De Deckker P, Colin JP, Peypouquet JP (eds) Ostracoda in the earth sciences. Elsevier, Amsterdam, pp 245–256

Würdig N (1983) Applications of ostracoda. In: Maddocks RF (ed) Fresh and brackish-water ostracodes from the east coast of the state of Rio Grande do sul, Brazil. University of Houston, Houston, pp 591–604

Zabert LL (1981) Ostrácodos cuarternarios de Taco Pozo (provincia de Chaco, Argentina) con algunas consideraciones paleoecológicas. FaCENA 4:77–87

Zabert LL, Herbst R (1986) Ostrácodos Pleistocénicos del arroyo Perucho Verna, Provincia de Entre Ríos, Argentina. Ameghiniana 23:213–224

Late Pleistocene Diatoms of the Lower Basin from the Quequén Salado River, Argentina

Marcela A. Espinosa[1(✉)] and Elisa Beilinson[2]

[1] Instituto de Geología de Costas y del Cuaternario, Universidad Nacional de Mar del Plata, Instituto de Investigaciones Marinas y Costeras, CONICET-UNMDP, CC 722 7600, Mar Del Plata, Argentina
maespin@mdp.edu.ar

[2] Centro de Investigaciones Geológicas, CONICET-UNLP Diag. 113 #275, La Plata 1900, Argentina

Abstract. Diatom assemblages from a site called Cueva del Tigre were studied. It is located in the lower basin of the Quequén Salado River, Buenos Aires province (38° 50′ 2.2″ S–60° 32′ 7.1″ W). The results are part of a project that includes sedimentologic, stratigraphic and palaeontologic aspects with the aim of reconstructing the environmental history of south-eastern Buenos Aires during the Late Cenozoic. The studied sedimentary succession is 1.90 m thick and has a tabular geometry with alternating levels of coarse sandstone and clayey siltstones. Diatom analysis of the 17 samples corresponding to four levels (N1, N2, N3 and N4) was performed using conventional techniques of taxonomic identification and treatment. A total of 74 diatom taxa were recognized and grouped according to salt tolerance and life form. Cluster analysis allowed dividing the sedimentary sequence into two diatom zones. The deposit begins with a freshwater/shallow lakes with associated vegetation (N1: 7 samples) dominated by *Cyclotella meneghiniana* Kützing 1844 (plankton) accompanied by the epiphytes *Cocconeis placentula* Ehrenberg 1838 and *Cymbella cistula* (Ehrenberg) Kirchner 1878. The overlapping levels: N2 (4 samples), N3 (3 samples) dated on 29,360 ± 670 years ^{14}C BP (33,128 cal. years BP) and N4 (3 samples) represent brackish conditions in a shallow pond, where benthic epipelic diatoms dominate: *Caloneis westii* (W.Smith) Hendey 1964, *Campylodiscus clypeus* Ehrenberg 1840 and *Surirella striatula* Turpin 1828. Today, these taxa live in temperate waters with salinities of 2–10‰. The comparison between fossil and modern samples through NMDS analysis showed that modern diatom assemblages from the Quequén Salado River are distinctly dissimilar to the Pleistocene assemblages under study, but they strongly correlate with diatom assemblages from Buenos Aires shallow lakes in agreement with the autoecological interpretation.

Keywords: Diatoms · Palaeoenvironments · Pleistocene · Argentina South America

G. C. Cusminsky et al. (eds.), *Advances in South American Micropaleontology*, Springer Earth System Sciences, https://doi.org/10.1007/978-3-030-02119-1_6

1 Introduction

The Marine Isotopic Stage 3 (MIS 3; ca. 56,000–25,000 ^{14}C years BP) was an interstadial period during the Pleistocene with relatively warmer conditions than today, where climatic conditions fluctuated over millennia scales (Rabassa and Ponce 2013). The updating and reinterpretation of diatom studies that cover this period in South America showed the need and importance of selecting new sites for high-resolution interdisciplinary studies in the region (Espinosa 2016).

In the south-eastern part of the Buenos Aires province, Argentina, in a region set between the Tandilia and Ventania ranges, rivers like Quequén Salado, Indio Rico and Las Mostazas modelled banks that have been the subject of many geological studies (Schillizzi et al. 2006; Isla et al. 2014). The first studies in the Quequén Salado River correspond to Frenguelli (1928, 1945), who assigned these deposits to the *Lujanense* (upper Pleistocene) and *Platense* (Holocene) stages. More recently, Aramayo et al. (2005) recognized deposits of Late Pleistocene and Holocene ages (between 16,000 and 4800 years BP) in the south-east coast of Buenos Aires province between the localities of Pehuen-Có and Monte Hermoso, ca. 50 km south-west from the Quequén Salado inlet. Schillizzi et al. (2006) analysed profiles near Oriente (12 km upstream) assigning Late Pleistocene and Holocene ages. Isla et al. (2014) described and discussed the Neogene records outcropping at the Irene village. Beilinson et al. (2015) studied the stratigraphy and sedimentology of more than 10 km of sea cliffs of Buenos Aires province where the main macromammals sites are found and concluded that during accumulation of the Middle to Late Pleistocene succession, glacio-eustasy and/or climate controlled the balance between generation of accommodation space and sediment supply. A geological model for the Quequén Salado river valley was proposed by Beilinson et al. (2017), a case of downcutting and headward erosion that explains the spatial distribution of facies and fossil taxa: the younger in the distal sector of the Quequén Salado middle basin and the older in the lower basin.

Several biological and sedimentologic proxy records preserved in Late Quaternary sequences constitute a valuable source of information for understanding the palaeoenvironmental changes that characterized the aquatic environments. Diatoms have been used as proxy indicators to reconstruct Late Quaternary environmental and climatic changes in every continent, being in continental aquatic ecosystems much more common than in marine or coastal studies (Mackay et al. 2003). Even so, the number of diatom-based Quaternary climatic reconstructions from marine and coastal environments is increasing in recent years (Romero 2010; Cermeño et al. 2013).

Diatoms have been widely employed to reconstruct changes in salinity, depth and trophic status related to Holocene sea level changes and shoreline position in Southern South America (e.g. Espinosa et al. 2003, 2012; Garcia Rodriguez et al. 2004; Fayó and Espinosa 2014; Dos Santos et al. 2016). However, despite the high potential of diatom records for environmental reconstructions, a limited number of detailed studies have been conducted for Pleistocene age deposits in South America (Espinosa 2016). Furthermore, few studies have focused on the interaction between climate variability and aquatic ecology (Mackay et al. 2012). This highlights the need for increased knowledge of environmental responses to changes in climate in aquatic ecosystems in the southern hemisphere that can be used as analogues for palaeoclimatic interpretations. The lack of long-term, continuous or near-continuous climate records, particularly those spanning through the Late Pleistocene, limits detailed understanding of the aquatic communities' responses to climatic changes.

2 Methods

2.1 Regional Setting

The Quequén Salado River (Fig. 1) is located in the south-eastern Pampean Region, reaching to the Atlantic Ocean as a mesotidal estuary (2–3 m tidal range) and flowing across the region known as the *Humid Pampas*. It has a mean discharge of 10.76 m^3/seg (max: 28.8 m^3/seg, min: 4.6 m^3/seg, Marini and Piccolo 2000). The zone has a temperate climate (annual mean temperature of 14 °C) and an annual mean rainfall of 800 mm (Marini and Piccolo 2004). The river runs along the hilly plain (also called "Undulated Pampa") between the Tandilia and Ventania ranges. It is characterized by the presence of riffles every 1 or 2 km in the middle and lower sections, while its channel can reach 15–20 m wide. The lower valley is oriented NNW–SSE and is flanked by cliffs of 8–15 m high. In the last 5 km, the river runs across a sandy barrier composed of vegetated dunes (Marini and Piccolo 2000). A site named Cueva del Tigre was studied. It is located 12 km from the outlet (38° 50′ 2.2″ S–60° 32′ 7.1″ W). The outcrop consists of two stratigraphic units (Fig. 2). The lower one, 1.90 m thick and tabular geometry, comprises sandstones, silty sandstones and mudstones. The upper one is presented as a carbonate lenticular body *tufa* with a maximum high of 2.60 m

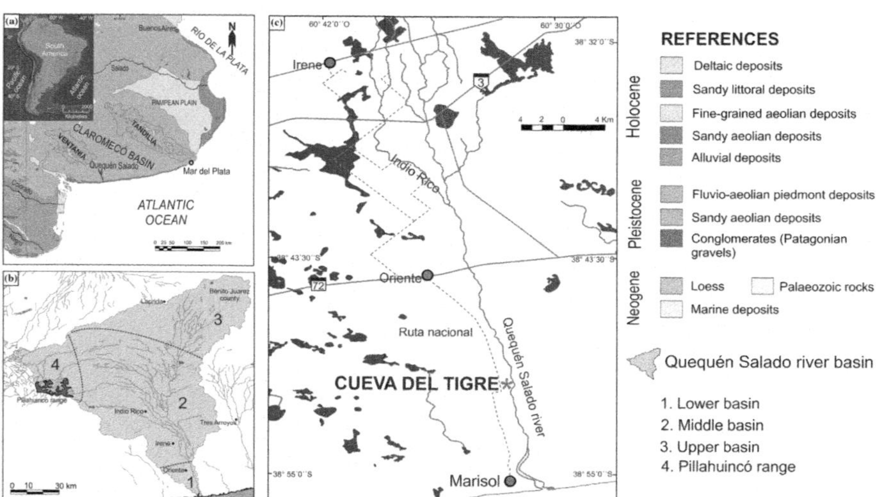

Fig. 1. Location map. **a** Geological map of the Buenos Aires province. **b** Quequén Salado River basin showing the upper, middle and lower reaches. **c** Cueva del Tigre sampling site. Modified from Beilinson et al. (2017)

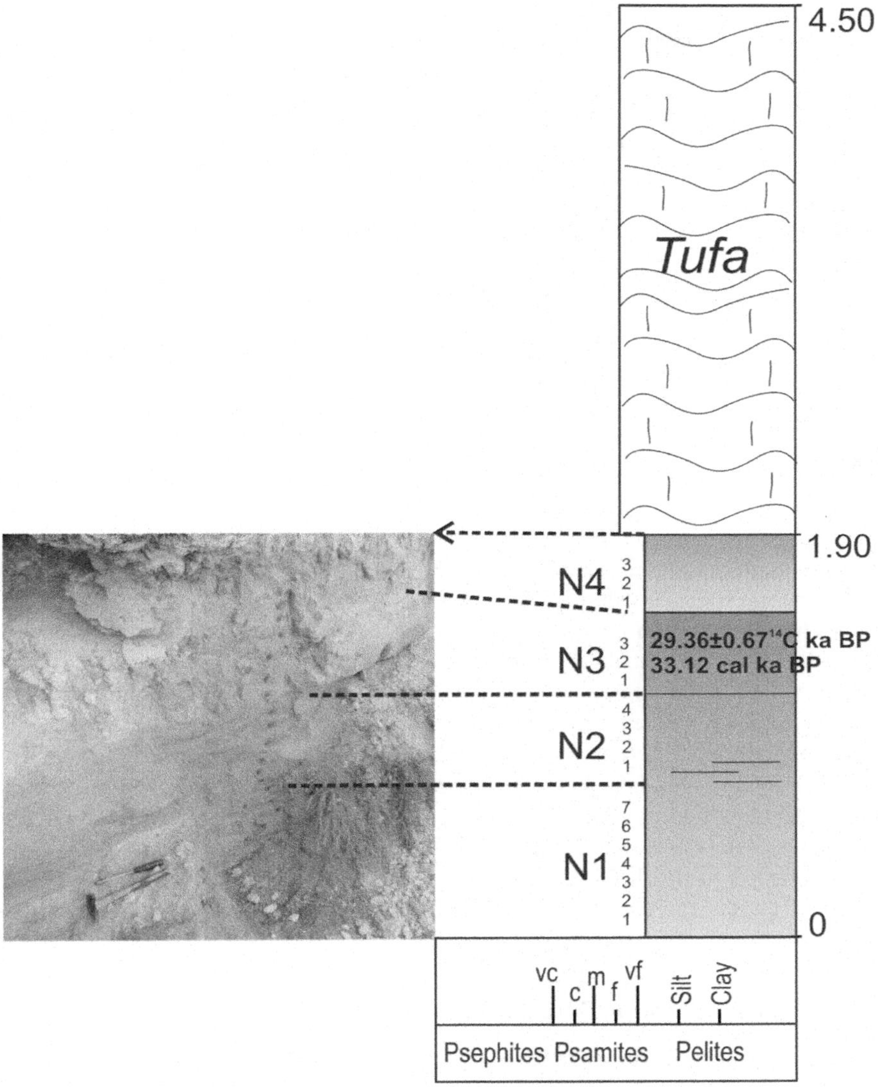

Fig. 2. Sedimentologic profile of the study site

and lateral extension of tens of metres. Its base is concave and irregular, showing an interdigitation with the non-calcareous deposits that underlie it.

Two different profiles were studied at the Cueva del Tigre site. Even though they are about 100 m from each other, correlation between them is easy to make. Taking into account the position of the levels within the Quequén Salado valley, it is observed that the tufa and diatom succession correspond to the lower terrace, and that the upstream profiles emerge at higher elevations, on the upper terrace of Quequén Salado River. Then, based on this geomorphological interpretation, the upstream profiles would be older than the levels with diatoms. Since the contact relationships between both units cannot be observed, it is difficult to discern if it is transitional or not, but there is definitely no faulting between both profiles.

Absolute chronological data for the Quequén Salado River basin is scarce. Mari et al. (2015) dated *Heleobia* sp. shells from sandy silt sediments corresponding to level 3 (N3) of the studied profile giving a radiocarbon age of 29,360 $\pm$ 670 years BP/33,128 years cal. BP (LP 3031).

2.2 Diatom Analysis

Diatoms were extracted from 17 sediment samples corresponding to four levels: N1, N2, N3 and N4 (Fig. 2). 5 g of dry sediment was oxidized with hydrogen peroxide (30%) and hydrochloric acid (10%) to remove organic matter and carbonates, washed four or five times with distilled water and diluted to a total volume of 100 ml. Diatom residues were mounted onto microscope slides using Naphrax®. On each slide at least 300 diatom valves were counted in random transects. All counts were performed under oil immersion (1000×), using a Zeiss light microscope equipped with phase contrast optics. The identification of species was based on the local and standard diatom taxonomic literature. Nomenclature and taxonomic authorities follow AlgaeBase (Guiry and Guiry 2018). Photographs of the diatom species were archived. The relative abundance of each species is presented as a proportion of the total frustule counts, and species representing less than 5% of the relative abundance were removed to increase the significance of the data. Ecological groups proposed by De Wolf (1982), Denys (1991/1992) and Vos and De Wolf (1988, 1993) were used to classify diatom taxa according to their salinity and habitat preferences. Diatom diagram and cluster analyses were performed using the TILIA and TILIAGRAPH software (Grimm 1991). The purpose of cluster analysis was to define diatom zones based on a stratigraphic constrained classification (minimum variance, Euclidean distance).

2.3 Comparison with the Modern Data set

Fossil assemblages are compared with a modern diatom data set for the region (Hassan et al. 2007, 2011) by means of *non-metric multidimensional scaling (NMDS*). NMDS is a robust ordination technique for community analysis (Minchin 1987) and was used to create similarity matrixes. A numerical measure of the closeness between the similarities in the lower dimensional space is called stress. Stress values range from 0 to 1, with 0 indicating perfect fit and 1 indicating worst possible fit. NMDS based on Bray–Curtis similarity measure was performed with the program

PAST versus 1.81, (Hammer et al. 2008).

3 Results

3.1 Diatoms

Seventy-four diatom taxa were identified. Twenty-two taxa constituted >5% of the total assemblage in at least one sample (Fig. 3). Some photographs of the dominant diatom species are presented in Fig. 4. The observed pattern of taxon occurrence and a cluster analysis on this subset permitted to define two major diatom assemblage zones, described as follows:

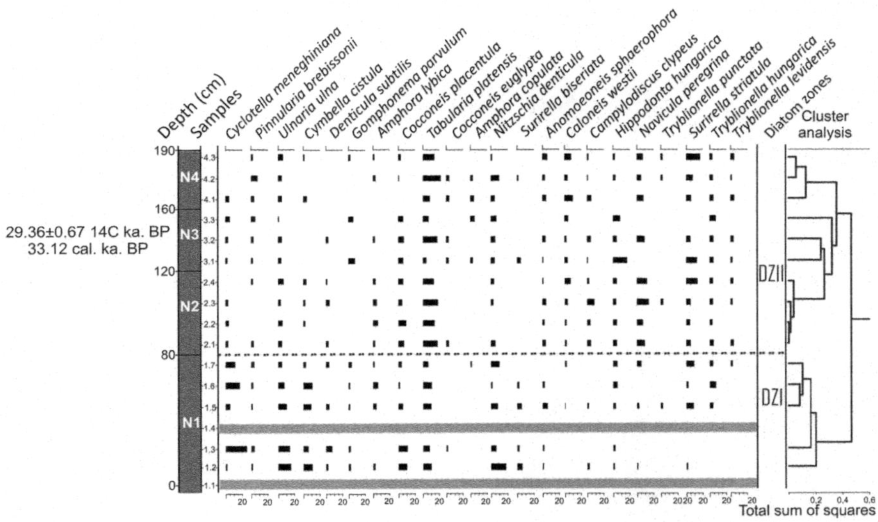

Fig. 3. Diatom diagram of Cueva del Tigre sequence. Grey shaded shows the barren levels (1.1 and 1.4)

Zone I (DZI; 0–80 cm) is characterized by the abundance of *C. meneghiniana* Kützing 1844 (plankton) accompanied by the epiphytes *Cocconeis placentula* Ehrenberg 1838, *Ulnaria ulna* (Nitzsch) Compère 2001, *Tabularia platensis* (Frenguelli) Sar and Sala 2009 and *Cymbella cistula* (Ehrenberg) O. Kirchner 1878. This assemblage is related to brackish/freshwater lagoons. Low, but sustained, abundances of benthic diatom species *Nitzschia denticula* Grunow 1880 and *Hippodonta hungarica* (Grunow) Lange-Bertalot, Metzeltin and Witkowski 1996 occur in this zone.

Zone II (DZII; 80–190 cm) is characterized by highly abundant, *T. platensis* (Frenguelli) Sar and Sala 2009. The increase of this brackish and epiphyte taxa defines the boundary between zones I and II in the same way as the appearance of the brackish, benthic and eutrophic taxa *Caloneis westii* (Smith) Hendey 1964, *Campylodiscus clypeus* Ehrenberg 1840, *Navicula peregrina* (Ehrenberg) Kützing 1844 and *Surirella striatula* Turpin 1828. Today, all of them live commonly in periodic water bodies or wet subaerial environments.

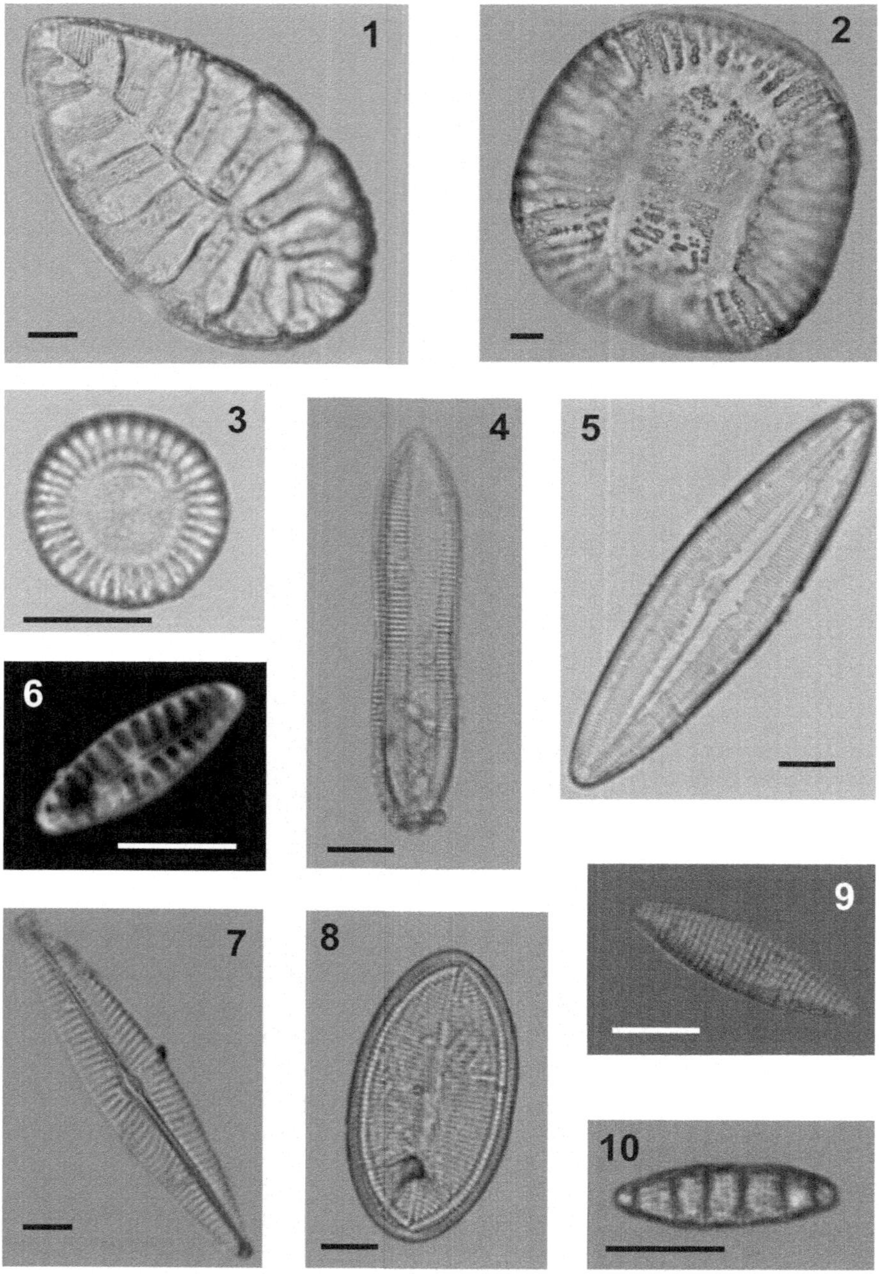

Fig. 4. Dominant diatom taxa from Cueva del Tigre sediment profile. Light microscope. **1** *Surirella striatula*; **2** *Campylodiscus clypeus*; **3** *Cyclotella meneghiniana*; **4** *Tryblionella hungarica*; **5** *Caloneis westii*; **6** *Hippodonta hungarica*; **7** *Navicula peregrina*; **8** *Cocconeis placentula*; **9** *Nitzschia denticula*; **10** *Denticula subtilis*. Scale bars = 10 μm

3.2 Non-Metric Multidimensional Scaling Analysis (NMDS)

NMDS arranged diatom samples in a specified dimensional space according to the rank order of their ecological similarities. The plot reveal differences in the structure of the assemblages. Samples showed a distributional pattern clearly related to their environment type. At first, NMDS analysis was performed between fossil samples from Cueva del Tigre profile and modern samples of the Quequén Salado River (Fig. 5a; Hassan et al. 2007) obtaining a stress value of 0.09. Figure 5b shows fossil samples (*N*) in the positive quadrant and modern samples (QS) on the negative quadrant. Quequén Salado River assemblages are distinctly dissimilar to the fossil assemblages from the Pleistocene succession under study.

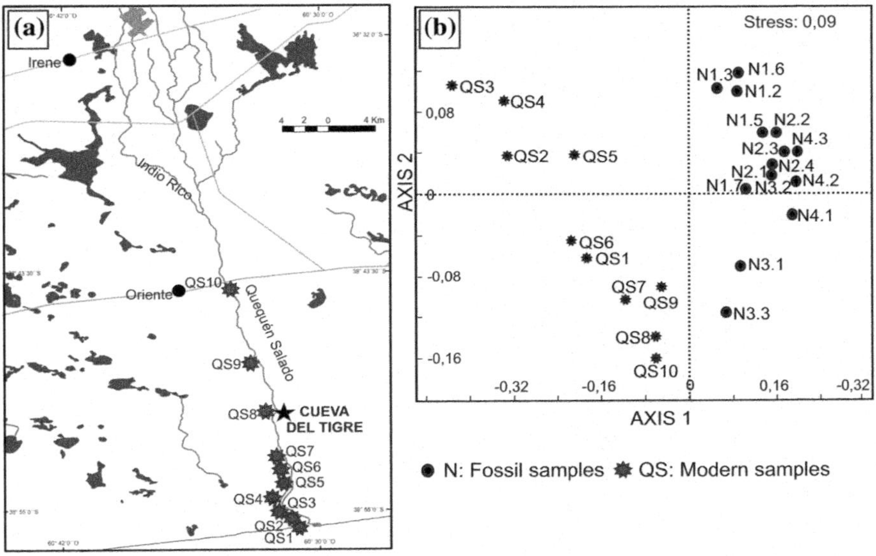

Fig. 5. a Modern sample sites along Quequén Salado River (Hassan et al. 2007). **b** NMDS analysis between Cueva del Tigre fossil samples (N) and modern samples (QS)

In order to find modern analogues that allow us to complete the characterization of the Pleistocene environments recorded at *Cueva del Tigre*, a second NMDS analysis was carried out between the fossil samples and the modern ones belonging to shallow lakes of the central zone of Argentina (Buenos Aires, La Pampa and Mendoza). Stress value obtained was 0.25. A total of 34 shallow ponds (most <2 m in depth) distributed across Central Argentina (Fig. 6a) were sampled by Hassan et al (2011). Most of them are thermally homogeneous and saturated with dissolved oxygen. Figure 6b shows that fossil assemblages belonging to Cueva del Tigre profile (*N*) remained together and close to modern samples of some shallow lakes from the Pampa Ecoregion (1, 5, 6 and 11: Buenos Aires). These samples were characterized by high percentages of organic matter (up to 85%), and pH ranges between 8.25 and 9.8 and annual precipitation of 800–1,000 mm (Hassan et al. 2011).

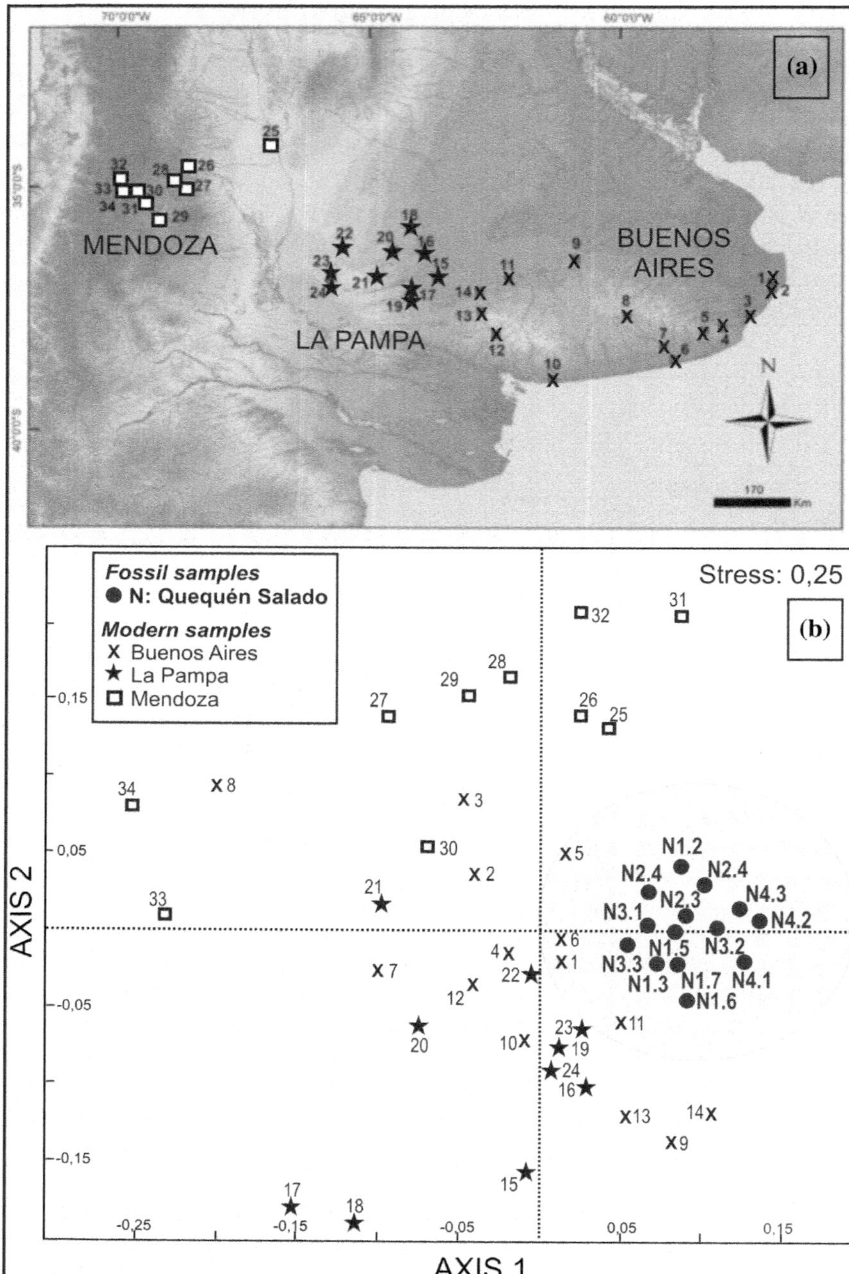

Fig. 6. **a** Modern sample sites from Central Argentina (Hassan et al. 2011). **b** NMDS analysis of modern and fossil samples (N)

4 Discussion and Final Remarks

The Quequén Salado River basin (Pampa region) was studied by means of stratigraphic, sedimentologic and palaeomagnetic approaches as well as palaeontologic analysis of vertebrate remains by Beilinson et al. (2017); Grill and Fernández (2016) analysed the palynomorphs content (spores, pollen and organic-walled marine microplankton) of two localities in the lower basin of the Quequén Salado River (Estancia Thomas Profile and QS1 Archaeological Site) to provide a palaeoenvironmental and palaeoclimatic reconstruction from the Late Pleistocene to the present. They inferred aeolian environments of extreme arid conditions during the late Pleistocene (ca. 16,000 years BP) shown by bad pollen preservation and scarce populations of ostracods and diatoms.

The diatom analysis from Cueva del Tigre succession provides new data about the palaeoenvironments of the lower basin of the Quequén Salado River ca. 30,000 years BP ago, during the Marine Isotopic Stage 3 (MIS 3). This sequence represents the second MIS 3 diatom record for Buenos Aires province: the other is a sedimentary succession located on the middle basin of the Luján River (Blasi et al. 2010). Although some Pleistocene records may have poor microfossil preservation (Grill and Fernández 2016), Cueva del Tigre profile contains well-preserved diatom frustules, but the abundance is not comparable with that found in other Holocene deposits from Buenos Aires (e.g. Espinosa et al. 2012). Both MIS 3 records from Buenos Aires province (Quequén Salado and Luján rivers) present important similarities on diatom content.

According to the cluster analysis of the diatom assemblages, it was possible to identify two zones in the evolution of the water body. The first zone (DZI) clustered seven samples of N1 (level 1). Diatom assemblages are dominated by the planktonic centric diatom *Cyclotella meneghiniana*. This species is common in swamps, shallow lakes and hot springs (Owen et al. 2004) and reflects a slightly deep freshwater/brackish environment. *C. meneghiniana* can live in different habitats, from freshwater to brackish (Tuchman et al. 1984; Hakansson and Chepurnov 1999 and is widely distributed in the modern diatom flora of the Pampean shallow lakes, dominating under alkaline (pH 8–10.5) and eutrophic conditions (Hassan et al. 2011). The accompanying flora is composed of the epiphytes *C. placentula*, *Ulnaria ulna* and *C. cistula*. All of them are freshwater taxa, but *C. placentula* is very tolerant to salinity and temperature changes, 0.2–30.4% and 4.7–33.5 °C, respectively (Trobajo Pujadas 2007), and is common in slow-flowing streams (Haworth 1976; Gasse 1986) and shallow lakes of Central Argentina (Hassan et al. 2011).

The diatom assemblage that characterizes DZII is composed by brackish taxa: the epiphyte *T. platensis* and the benthic and eutrophic *C. westii*, *C. clypeus*, *N. peregrina* and *S. striatula*. Today, all of them live commonly in periodic water bodies or wet subaerial environments.

In the Luján River basin, the lapse between 44,000 and 37,700 years B.P. is characterized by lacustrine deposits with abundant diatoms, where the planktonic and brackish *C. meneghiniana* dominate accompanied by the epiphytes *Epithemia adnata* and *C. placentula* (Blasi et al. 2010). This assemblage is very similar to DZI (the bearing level has not radiocarbon dates yet), and it is possible that a correlation exists between both levels with diatoms.

The relative abundance of the most dominant species within both assemblages (DZI and DZII) varies and might indicate differences in climatic conditions, mainly changes in water availability.

According to NMDS analysis, modern assemblages from the Quequén Salado River are distinctly dissimilar to the fossil assemblages from the Pleistocene succession of Cueva del Tigre. In this sense, most of the recorded diatom taxa in the fossil sediments are absent in the modern surface sediments of the river. Then, the environment at the time of the MIS 3 was different from the modern, with diatom assemblages that suggest fluctuations from a deep freshwater/brackish environment to shallow brackish water dominated by saline diatom taxa. Diatom analysis indicated that the interval would have had relatively humid and mild conditions in agreement with the regional patterns (Rabassa and Ponce 2013).

Acknowledgements. The authors wish to acknowledge to C. Deschamps, R. Tomassini, G. Gasparini, M. Zárate and J. Rabassa for assistance with fieldwork and sedimentologic descriptions. J. P. Lancia processed the samples. The criticism of anonymous reviewers on earlier version of this manuscript is acknowledged. Financial support was provided by Consejo Nacional de Investigaciones Científicas y Técnicas (PIP 2014-0496) and Agencia Nacional de Promoción Científica y Tecnológica (PICT 2016-1146).

References

Aramayo SA, Gutierrez Tellez B et al (2005) Sedimentologic and paleontologic study of the southeast coast of Buenos Aires province, Argentina: a late Pleistocene-Holocene paleoenvironmental reconstruction. J S Am Earth Sci 20:65–71

Beilinson E, Gasparini GM et al (2015) Insights into Pleistocene palaeoenvironments and biostratigraphy in southern Buenos Aires province (Argentina) from continental deposits. J S Am Earth Sci 60:82–91

Beilinson E, Gasparini GM et al (2017) The Quequén Salado river basin: geology and biochronostratigraphy of the Mio-Pliocene boundary in the southern Pampean plain, Argentina. J S Am Earth Sci 76:362–374

Blasi A, Castiñeira Latorre C et al (2010) Paleoambientes de la cuenca media del Río Luján (Buenos Aires, Argentina) durante el último período glacial (EIO 4-2). Lat Am J Sedimentol Basin Anal 17(2):85–111

Cermeño P, Marañón E et al (2013) Response of marine diatom communities to late quaternary abrupt climate changes. J Plankton Res 35(1):12–21

De Wolf H (1982) Method of coding of ecological data from diatoms for computer utilization. Meded Rijks Geol Dienst 36:95–99

Denys L (1991/1992) A check-list of the diatoms in the Holocene deposits of the Western Belgian coastal plain with the survey of their apparent ecological requirements. In: Introduction, ecological code and complete list. Geological Service of Belgium, Professional Paper 246:1–41

Dos Santos-Fischer C, Stalliviere Corrêa IC et al (2016) Paleoenvironmental insights into the quaternary evolution of the southern Brazilian coast based on fossil and modern diatom assemblage. Palaeogeogr Palaeoclimatol Palaeoecol 446:108–124

Espinosa MA (2016) Response of diatoms to late quaternary climate changes. In: Gasparini GM, Rabassa J, Deschamps C, Tonni EP (eds) Marine isotope stage 3 in southern South America, 60 KA B.P.–30 KA B.P., Springer Earth System Sciences, pp 299–319

Espinosa MA, De Francesco CG et al (2003) Paleoenvironmental reconstruction of Holocene coastal deposits from the southeastern Buenos Aires province, Argentina. J Paleolimnol 29:49–60

Espinosa MA, Hassan GS et al (2012) Diatom inferred salinity changes in relation to Holocene sea level fluctuations in estuarine environments from Argentina. Alcheringa Austral J Palaeontol 36:373–386

Fayó R, Espinosa MA (2014) Reconstrucción paleoambiental de la planicie costera de Mar Chiquita (provincia de Buenos Aires, Argentina) durante el Holoceno, basada en diatomeas. Ameghiniana 51(6):510–528

Frenguelli J (1928) Observaciones geológicas en la región costanera sur de la provincia de Buenos Aires. Univ. Nac. Litoral, Anales II, Sección de Historia y Geografía, Santa Fé, Argentina, p 145

Frenguelli J (1945) Las diatomeas del Platense. Revista del Museo de La Plata. Paleontología 16:77–221

Garcia Rodriguez F, Sprechmann P et al (2004) Holocene trophic state changes in relation to sea level variation in Lake Blanca, SE Uruguay. J Paleolimnol 31:99–115

Gasse F (1986) East African diatoms:taxonomy, ecological distribution. Bibliotheca Diatomologica 11. J Cramer, Stuttgart, pp 202

Grill SC, Fernández AL (2016) Multi-proxy analysis of late quaternary sediments in the lower basin of the Quequén Salado river (Buenos Aires province, Argentina): an update In: Martínez M, Olivera D (eds) Palinología del Meso-Cenozoico de Argentina. Publicación Electrónica de la Asociación Paleontológica Argentina 16(2):129–147

Grimm E (1991) Tilia software, Illinois state museum, research and collection center. Springfield, Illinois

Guiry MD, Guiry GM (2018) Algaebase. Worldwide electronic publication. National University of Ireland, Galway, http://www.algaebase.org

Hakansson H, Chepurnov V (1999) A study of variation in valve morphology of the diatom *Cyclotella meneghiniana* in monoclonal cultures: effect of auxospore formation and different salinity conditions. Diatom Res 14:251–272

Hammer O, Harper DAT et al (2008) PAST Palaeontological Statistics, ver. 1.81, http://folk.uio.no/ohammer/past

Hassan GS, Espinosa MA et al (2007) Dead diatom assemblages in surface sediments from a low impacted estuary: the Quequén Salado River, Argentina. Hydrobiologia 579:257–270

Hassan GS, Tietze E et al (2011) Problems and potentialities of using diatoms as paleoclimatic indicators in Central Argentina. In: Compton JC (ed) Diatoms: ecology and life cycle. Nova Science Publishers, New York, pp. 1–42

Haworth E (1976) Two late-glacial (late Devensian) diatom assemblage profiles from northern Scotland. New Phytol 77:227–256

Isla F, Dondas A et al (2014) Sedimentología de las formaciones Irene y La Toma del Neógeno de la Cuenca de Claromecó, Buenos Aires. Geoacta 39:1–13

Mackay AW, Jones VJ, Battarbee RW (2003) Approaches to Holocene climate reconstructions using diatoms. In: Mackay AW, Battarbee AW, Birks RW, Oldfield F (eds) Global change in the Holocene, pp 294–309

Mackay AW, Bezrukova EV et al (2012) Aquatic ecosystem responses to Holocene climate change and biome development in boreal, central Asia. Quat Sci Rev 41:119–131

Mari F, Di Lello C, Huarte R et al (2015) Dataciones radiocarbónicas en la cuenca inferior del Río Quequén Salado, Provincia de Buenos Aires. Libro de resúmenes del VI Congreso Argentino de Cuaternario y Geomorfología, 8 al 12 de abril de 2015, Ushuaia. Tierra del Fuego, Argentina, p 43
Marini MF, Piccolo MC (2000) El balance hídrico en la cuenca del río Quequén Salado, Argentina. Papeles de Geografía 31:39–53
Marini MF, Piccolo MC (2004) Water quality for supplementary irrigation in the Quequén Salado River basin (Argentina). Papeles de Geografía 39:157–172
Minchin PR (1987) Simulation of multidimensional community patterns: towards a comprehensive model. Vegetation 71:145–156
Owen RB, Renaut RW et al (2004) Swamps, springs and diatoms: wetlands of the semi-arid Bogoria-Baringo Rift, Kenya. Hydrobiologia 518(1–3):59–78
Rabassa J, Ponce JF (2013) The Heinrich and Dansgaard-Oeschger climatic events during marine isotopic stage 3: searching for appropriate times for human colonization of the Americas. Quat Internat 299:94–105
Romero OE (2010) Changes in style and intensity of production in the Southeastern Atlantic over the last 70,000 yr. Mar Micropal 74:15–28
Schillizzi R, Gutierrez Tellez B et al (2006) Reconstrucción paleoambiental del Cuaternario en las barrancas del Río Quequén Salado, provincia de Buenos Aires, Argentina. III Congreso Argentino del Cuaternario y Geomorfología, Córdoba 10–13 octubre 2006, pp 649–658
Trobajo Pujadas R (2007) Ecological analysis of periphytic diatoms in Mediterranean coastal wetlands (Empordá wetlands, NE Spain). In: Witkowski A (ed) Diatom Monographs 7:1–210
Tuchman ML, Theriot E et al (1984) Effects of low-level salinity concentrations on the growth of *Cyclotella meneghiniana* Kütz. (Bacillariophyta). Arch Protistenk 128:319–326
Vos P, De Wolf H (1988) Methodological aspects of paleoecological diatom research in coastal areas of the Netherlands. Geol Mijnbouw 67:31–40
Vos P, De Wolf H (1993) Diatoms as a tool for reconstructing sedimentary environments in coastal wetlands; methodological aspects. Hydrobiologia 269(270):285–296

New Micropaleontological Record (Foraminifera and Ostracoda) from the Late Glacial and Holocene in Salinas del Bebedero, San Luis, Argentina: A Paleoenvironmental Interpretation

Lydia Calvo Marcilese[1(✉)], Sabina D'Ambrosio[2], Leandro D. Rojo[3], and Gabriela C. Cusminsky[4]

[1] Y-TEC (YPF TECNOLOGÍA), Avenida del Petróleo s/n e 129 y 143, Código Postal 1923, Buenos Aires, Argentina
lydia.calvo@ypftecnologia.com

[2] IANIGLA-CONICET, Av. Ruiz Leal s/n Parque General San Martín, Código Postal 5500, Mendoza, Argentina
sabina.dambrosio@gmail.com

[3] Facultad de Ciencias Exactas y Naturales de la Universidad Nacional de Cuyo (FCEN-UNCuyo), Museo de Historia Natural de San Rafael (MHNSR) Ballofet S/N frente a Parque Mariano Moreno, Código Postal 5600, San Rafael, Mendoza, Argentina
ldrojo@uncu.edu.ar

[4] INIBIOMA (CONICET), Quintral 1250, Código Postal 8400, San Carlos de Bariloche, Río Negro, Argentina
gcusminsky@gmail.com

Abstract. Since the 1980s, just a few studies have been conducted related to foraminifera and ostracoda in endorheic environments of Argentina. In core SB III from Salinas del Bebedero, a preliminary report on 51 sediment samples between 986 and 506 cm (*ca.* 12,600–3600 yrs BP) recorded only 12 fertile samples. Seven samples showed typical marginal marine microfaunal assemblages with high abundance of the foraminifer *Ammonia tepida* associated with few specimens of *Elphidium excavatum*. Concerning ostracods, they are poorly represented except by six levels with *Limnocythere* aff. *staplini* and *Cyprideis salebrosa hartmanni*. It is worthy to note that ostracods and foraminifers were not recorded coexisting, with the exception of one sample (830–828 cm). The remarkably high abundance of foraminiferal assemblages, with more than 170 individuals per gram in some levels, is outstanding and deserves special attention. The *Ammonia–Elphidium* assemblage is well known in shallow marginal marine environments at global scale, whereas the two ostracod species found are common of lacustrine environments with high salinity range. Previous hypothesis indicates that birds and/or the wind as vectors were responsible for the presence of that particular microfauna in Salinas del Bebedero. These preliminary studies aim to contribute to the analysis of materials from athalassic environmental basins in order to increase the knowledge about the typical marine microfauna living in continental environments and their potential implications in paleoenvironmental interpretation.

G. C. Cusminsky et al. (eds.), *Advances in South American Micropaleontology*,
Springer Earth System Sciences, https://doi.org/10.1007/978-3-030-02119-1_7

Keywords: Saline lakes · Paleoenvironmental reconstruction · Micropaleontological analysis · San Luis · Argentina

1 Introduction

Salinas del Bebedero (33° 30′ S, 66° 40′ W) is an endorheic saline basin from an arid–semiarid region of central western Argentina. It has an accumulation of lacustrine sediments comprising the last *ca.* 20 ^{14}C ka (González 1994) that have attracted geologists and paleontologists for its potential in paleoenvironmental and paleoclimatic studies. Pioneer investigations were conducted at the beginning of the twentieth century, in which paleoshorelines were recognized as a result of ancient lacustrine water levels higher than the present ones (Doering 1906; Gerth 1914; Frenguelli 1922; Deletang 1929, among others). In 1980 and 1990, other geological and micropaleontological studies (González et al. 1980, 1981, 1998; González 1982, 1983, 1994; González and Weiler 1984; Maidana 1994; González and Maidana 1998; García 1999) also inferred lacustrine water levels with fluctuations associated with fluvial input to Salinas del Bebedero, from the Andes via Desaguadero–Salado fluvial system and Arroyo Bebedero creek. Particularly, four main lacustrine stages were documented, attributed to cold global events occurred in the past 25 ka (24,000 cal. years BP, 20,800 cal. years BP, 17,900 cal. years BP, 15,800/14,400 cal. years BP) (González 1981, 1982). At the end of 1990's, a new multiproxy research initiative (the Patagonian Lake Drilling Project—PATO, coordinated by V. Markgraf and D. Aristegui; González et al. 1998) performed four drillings in Salinas del Bebedero with the main goal to recover and analyze paleoclimatic proxies as pollen and diatoms between others, in order to comprehend interhemispheric dynamics of climatic changes in the Americas at glacial–interglacial stages. First results from that project were obtained for pollen and diatoms from core SBIV, although no numerical ages were reported and consequently inferences could not be chronologically calibrated temporally (González et al. 1998). Preliminary pollen results from other core—SBIII—indicated a permanent salt lake by inferring halophytic vegetation at the margins, but without chronologies that permit to reconstruct the past environment (Rojo 2003). More recently, new studies on core SBIII based on pollen, sedimentology, and radiocarbon dates confirmed that Salinas del Bebedero has been a saline lake at least since 12.6 cal. ka BP. No great past environmental changes have apparently affected the Monte–Espinal vegetation that surrounded the saline lake (Rojo et al. 2012).

Numerous deposits containing typical "marine" or "marginal marine" Quaternary microfossils are exposed in inland settings like saline lakes in central Argentina, constituting valuable natural archives for understanding past environmental changes. Several of these deposits take place at different hydrological settings along gradients of aridity and altitude and possess abundant foraminiferal tests and ostracod carapaces whose characteristics remain practically unknown. In fact, the only studies conducted on Quaternary microfossils from Salinas del Bebedero and surrounding areas have been carried out in the 1980s and 1990s (González et al. 1980, 1981, 1998; González 1981; Maidana 1994; González and Maidana 1998; García 1999).

This study sustained the utility of microfossils, foraminifera, and ostracoda, as tools for paleoenvironmental reconstructions and then interpreted in paleoclimatic terms in the area, by recognizing climate changes during the last 12 ka BP as well as short climatic fluctuations caused by local circumstances. Preliminary results of fossil ostracoda and foraminifera from the semiarid region of Salinas del Bebedero, in central western Argentina, are reported. The aim is to establish the paleoecological implications of these shells to serve as a basis for paleoclimatic and paleohydrologic reconstructions for the region.

At the same time, we point out the significance of having assemblages, which include species of strong marginal marine affinities and lagoonal biota in inland saline lakes, and why these species are found in continental settings. This kind of lakes needs attention in improving the understanding and the nature of saline waters.

2 Study Area and Environmental Setting

The saline-lacustrine endorheic Salinas del Bebedero located in the western part of San Luis Province (Fig. 1) has water inputs from temporary tributaries. Arroyo Bebedero gets the Salinas from the southeast, Nogolí-San Jerónimo-Balde from the north, and San Luis River from the northeast (González 1981). At present, summer precipitation coincides with the highest evapotranspiration rate favoring salt concentration, which is commercially exploited (González 1994). In the late Quaternary ($\sim$ last 20,000 yrs), the saline-lacustrine basin would have received glacifluvial seasonal meltwaters from the

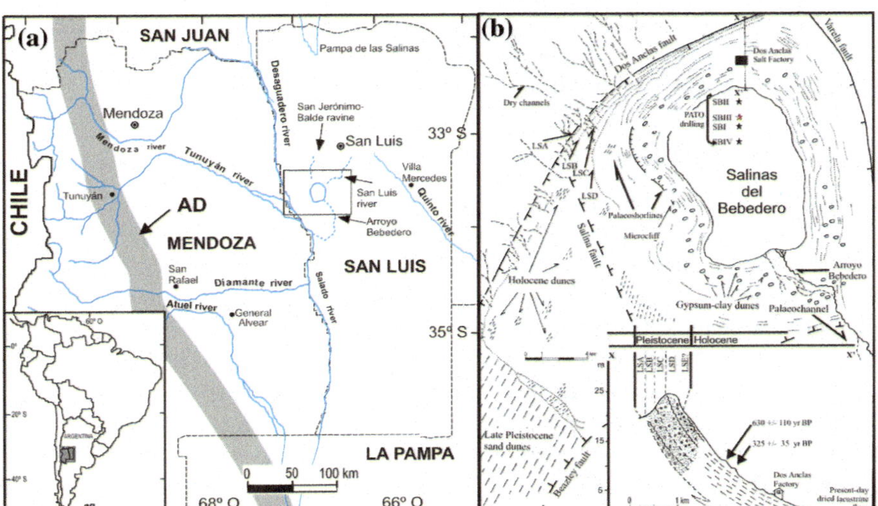

Fig. 1. **a** Study area of Salinas del Bebedero (indicated by a rectangle). Arid diagonal (AD) (Bruniard 1982). **b** Salinas del Bebedero geomorphology, modified from González (1981, 1994). Cores are indicated by black stars (core III by a red star); lacustrine stages (LS) (=paleoshorelines)

Cordillera de los Andes by means of Desaguadero–Salado fluvial system and Arroyo Bebedero (González 1994), nowadays disconnected of each other.

2.1 Geologic and Geomorphologic Framework

Salinas del Bebedero is located in the western part of the geological province of Sierras Pampeanas Orientales (Ramos 1999). Its position coincides with the Beazley basin that evolved during the Mesozoic and the Cenozoic eras. Additionally, it is situated in the southern tip of Depresión Longitudinal Central (González Díaz 1981), a basin derived from the Andean orogeny limited by four main faults (Dos Anclas, Varela, Beazley and Salina) (Fig. 1b). Neogene sedimentites are recognized to the west, Pleistocene–Holocene sand dunes to the southwest, and the Arroyo Bebedero floodplain and paleochannel to the south (González Díaz 1981).

The Salinas del Bebedero per se is circular in shape, composed by salt deposits with predominance of NaCl in the center of the basin and surrounded by aeolian loess-like sediments. The sedimentary accumulation reaches at least 60 m depth (González 1981). Abrupt high margins are recognized in the southern and western part of the saline, interpreted as ancient paleoshorelines since at least the Last Glacial Maximum (González 1994). Based on these radiocarbon-dated paleoshorelines and geomorphology, González et al. (1981) and García (1999) inferred the presence of two different lacustrine stages, named Major Lacustrine Period (MaLP) (paleoshoreelines LSA, LSB, and LSC, Fig. 1b) and Minor Lacustrine Period (MiLP) (paleoshorelines LSD, LSE, and Holocene undifferentiated ones), separated by a period of heavy rainfall (marked by dry channels, Fig. 1b). The authors showed different shoreline heights where MaLP correspond to higher water levels of 25 m above the bottom of the present shallow lake. This MaLP was dated between 15,503 ± 547 to 17,064 ± 665 years BP. The younger MiLP, corresponds with paleoshorelines around 20 m above the bottom of the present lake surface.

2.2 Environmental Context

The region is dominated by an arid subtropical climate (Prohaska 1976), with 17.5 °C mean annual temperatures and 738.5 mm annual precipitation (San Luis weather station, 2000–2009 period, Servicio Meteorológico Nacional). An east–west precipitation gradient characterizes the study area, with ~800 mm/yr to the east at Villa Mercedes city, and less than 300 mm/yr to the west at the San Luis-Mendoza provincial boundary. A north–south direction arid belt, the arid diagonal (AD) (Bruniard 1982), receives <250 mm of annual precipitation, all across South America (Fig. 1). It marks a limit between Atlantic and Pacific sources of humidity in the region. The east of AD humidity is brought by the contribution of Atlantic air masses, whereas to the west it is supplied by the influence of Pacific. Due to this atmospheric circulation, the main humidity source in Salinas del Bebedero comes from the eastern Atlantic coast during summer as seasonal precipitation.

3 Materials and Methods

The materials analyzed in this study belong to the SBIII core, one of four cores taken in 1998 from Salinas del Bebedero along a transect (Fig. 1b) through the northern part of the salt lake, reaching from near-shore (core II) toward the center of the lake (cores I, III, and IV). Coring was carried out with a percussion-drill coring system, using a closed tube containing a plastic liner of 1 m in length and 4 cm in diameter. A total of 391 contiguous sediment samples from the 10-m core SBIII were taken every 2 cm and stored in plastic bags for further analysis.

For this study, 54 samples were selected from the lowermost 5-m section comprising the late glacial—Holocene transition and the early-to-middle Holocene intervals. The uppermost 5-m section will be considered in a new future step of the investigation; here, the aim was to focus the attention in the boundary between the Pleistocene and the Holocene when main environmental/climate changes were documented (González 1994; Rojo et al. 2012). This core has been described in a previous work (Rojo et al. 2012), the grain size was equally dominated by sand (~47%) and silt (~43%), meanwhile the presence of clay was low (~10%). Three sedimentary intervals were identified based on clastic and evaporite facies (Fig. 2). SB III.1 interval (10–8.50 m) is characterized by high clastics and low evaporites (chlorides), SB III.2 interval (8.5–6 m) with similar values between clastics and evaporites (chlorides and sulfurs), and SB III.3 interval (6–4.5 m) similar to SB III.1, with higher clastics than evaporites. This information permitted to infer that water levels were higher than

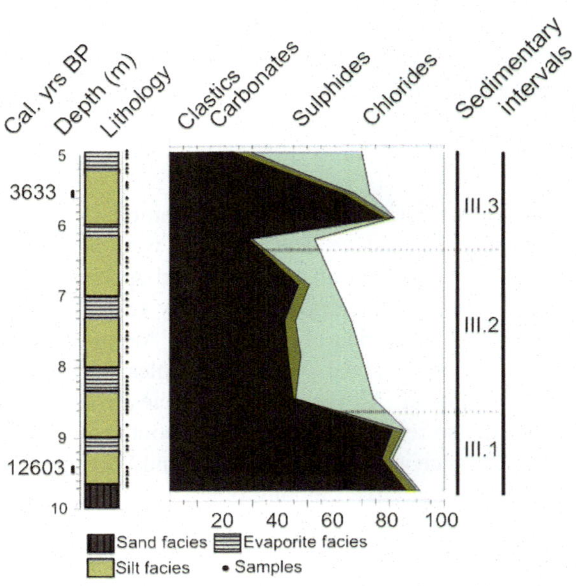

Fig. 2. Lithology of Salinas del Bebedero core SBIII. Samples are indicated by black circles. Soluble (clastics) and insoluble (salts) components and sedimentary intervals are shown on the right. Modified from Rojo et al. (2012)

present at least two peaks at the bottom and top (SB III.1 and SB III.3, respectively) of the sequence (Rojo et al. 2012).

Two radiocarbon dates from bulk sediments were obtained in previous studies (Rojo et al. 2012) and calibrated using Oxcal 4.1 (Bronk Ramsey 2001) (Table 1).

Table 1. Radiocarbon ages of SBIII core

Core	Radiocarbon ages (^{14}C yrs BP)	Calibrated ages (cal. yrs BP) with 2 sigma	Median (cal BP)	d^{13}C (‰)	Depth (m)	Material/technique	Laboratory numbers
SBIII	3444 ± 54	(3476–3826)[a]	3633	−18, 5	5, 56	Sediment/AMS	AA81414[c]
	10.700 ± 170	(12.103–13.055)[b]	12.603	−17, 6	9, 43	Sediment/AMS	AA81413[c]

[a]OxCal 4.1, SHCal 04; [b]OxCal 4.1, IntCal 09; [c]University of Arizona AMS Facility

Microfaunal collection and foraminiferal analyses were conducted following standard protocols of Boltovskoy (1965). Dry sediment (approximately 5 g) was processed and disaggregated using tap water. Samples were then washed through a sieve of 63-mm mesh and dried at room temperature. From the residue, the entire available tests were picked and studied under a binocular microscope. Samples were standardized as individuals per gram of dry sediment. All microfossils collected were plotted on a range-chart (including abundance and species diversity per sample). Systematic determinations for foraminifera were based mainly on Loeblich and Tappan (1988), Buzas-Stephens et al. (2002) and Ferrero (2006, 2009). *Elphidium* species were analyzed following Boltovskoy (1954a, b, 1957), Boltovskoy et al. (1980) and Calvo-Marcilese (2011).

For the ostracod assemblage analyses, each dried sample (approximately 5 g) was washed with tap water on a 63-μm sieve. The ostracods were picked and studied under a binocular microscope. Identification at the genus and species level was performed using local literature and general references commonly used for ostracod systematic (Cusminsky et al. 2005, 2011; Laprida et al. 2006; D'Ambrosio et al. 2015, 2016b; Palacios-Fest et al. 2016). For statistical analyses, samples were standardized as individuals per gram of dry sediment. All the ostracod species from Salinas del Bebedero have extant representatives, so reference information about ecological factors affecting their distribution and abundance is available (Ornellas and Würdig 1983; Ferrero 1996; Laprida 2006; D'Ambrosio et al. 2017, among others).

To explore relationships between species abundance and depth along the core, SBIII Tilia and Tilia Graph programs were used (TG View 2.0.2; Grimm 2004).

4 Results

From a micropaleontological point of view, this core showed only 12 fertile samples. In seven of these samples (830–828, 772–770, 756–754, 696–694, 672–670, 636–634, and 522–520 cm), typical marginal marine or lagoonal assemblages of foraminifera were found, with numerous specimens of *Ammonia tepida* and scarce individuals of *Elphidium excavatum*. The preservation was from good to moderate with specimens presenting effects of secondary alteration and reworking. Only these two species were recovered. Abundance values ranged from 1 to 175 specimens per gram of dry sediment.

The ostracods are only represented in six samples (864–862, 856–854, 852–850, 836–834, 830–828, and 578–576 cm) where the presence of *Limnocythere* aff. *staplini* Gutentag and Benson (1962) and *Cyprideis salebrosa hartmanni* Ramirez (1967) is remarkable. The ostracod assemblages, in the five samples, are well represented with males, females, and juveniles, but only one sample (864–862 cm) has numerous specimens. Abundance values ranged from 1 to 69 individuals per gram of dry sediment.

Ostracods and foraminifera were not found coexisting (Fig. 3), with the exception of one sample (830–828 cm). This sample presents one ostracod and one foraminifer, and according the ostracod valves presents effects of secondary alteration and reworking. According to the preservational features, this sample is a taphocoenosis and does not represent the original environmental conditions where the ostracods lived (Boomer et al. 2003).

From the bottom to the samples 756–770 cm of the core, the section shows a gradual change in the dominance of the microfauna, with decreasing abundance of species like *L.* aff. *staplini* and *C. salebrosa hartmanni* within ostracoda and *A. tepida* and *E. excavatum* (Foraminifera). Ostracods are relative abundant at the bottom (830–828/864–862 cm), whereas foraminifers are well represented in the top of this part of the section (756–754/772–770 cm). The upper part of the core was almost sterile, with just two samples containing scarce specimens of *A. tepida* and *Limnocythere* aff. *staplini* (522–520/750–748 cm).

5 Discussion

The presence of foraminiferal assemblages, in some samples with more than 170 individuals per gram of dry sediment, is noteworthy for this kind of environments and deserves special attention. The *Ammonia–Elphidium* association is well known for being typical of shallow marginal marine environments at global scale. The genus *Ammonia* has a wide global distribution and great morphological variety, bearing high salinity ranges. Boltovskoy (1957) defined for the Rio de la Plata estuary, Argentina, different biofacies of *A. beccarii* f. *parkinsoniana*, which would correspond to a middle river area with salinities ranging from 0.5 to 25 g L^{-1}. In this study, for the individuals included in the genus *Ammonia*, we decided to adopt the criterion proposed by Buzas-Stephens et al. (2002) and used for Calvo Marcilese et al. (2013) which suggests that the name *A. beccarii* should be reserved exclusively to designate the ornamented form

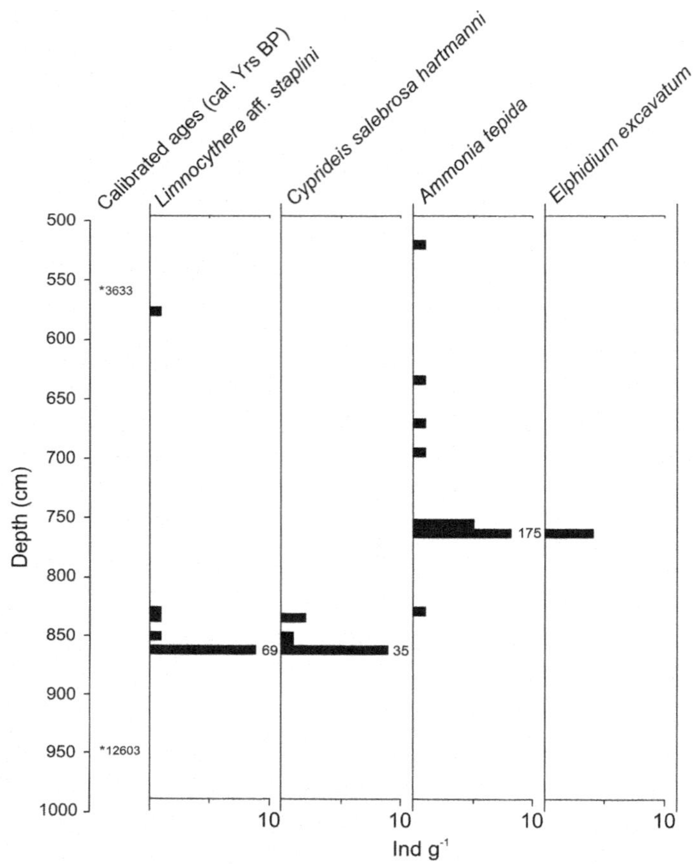

Fig. 3. Micropaleontological record at Salinas del Bebedero (core SBIII), numbers (69, 35 and 175) refer to truncated values because they are out of x axis scale

of *Ammonia*, corresponding to the original description. Here, *A. beccarii* f. *tepida* is named as *A. tepida*.

According to Murray (1991), the genus *Elphidium* is typical of coastal environments and has been found at depths ranging between 0 and 50 m in warm temperate environments. Boltovskoy (1976) indicates that this genus is widely distributed along the inner shelf of Argentina and has been recognized worldwide (Murray 2006). The species *E. excavatum* is characteristic of brackish environments in association with *Ammonia* as in Lagoa Dos Patos, Brazil, Río Quequén Grande and Bahía Blanca estuary, Mar Chiquita, Argentina (Boltovskoy et al. 1980; Cusminsky et al. 2006; Márquez et al. 2016; Márquez 2017), and Florida Bay, USA (Ishman et al. 1996).

The *Ammonia–Elphidium* assemblage was mentioned by several authors in estuarine environments worldwide as in the English Channel, North Sea, Bay of Biscay, Spain, in Nigeria, on the continental shelf in the Caribbean, Florida and in the Gulf of

Mexico (Ishman et al. 1996; Sen Gupta 2002), and in Puerto Deseado, Argentina (Boltovskoy et al. 1980) among others. Studies in cores extracted in the area of Florida, USA, suggest that the *Ammonia–Elphidium* assemblage would indicate oligohaline to mesohaline conditions with salinities between 5 and 18 g L^{-1}. A gradual increase in salinity would lead to an *Elphidium* increase and a decrease in the number of individuals of the genus *Ammonia* and vice versa (Brewster-Wingard et al. 1996). It is a common assemblage in New Zealand and its surroundings, present in intertidal mud and sand, mostly on tidal flats and beaches (Hayward and Hollis 1994). This assemblage occurs at shallow subtidal depths and in the lower half of the intertidal zone in the channels and tidal mud and sand flats at the seaward end of the Bahía Blanca estuary, Argentina (Calvo Marcilese and Pratolongo 2009).

Ostracod species identified (*Limnocythere* aff. *staplini* and *Cyprideis salebrosa hartmanni*) have extant representatives, and their ecological requirements were used to perform paleoenvironmental reconstructions (D'Ambrosio 2014). *Limnocythere* aff. *staplini* specimens were found on a salinity range of 0.4 to 42.3 g L^{-1}, but have strong affinities to shallow mesohaline lakes subject to evaporation, as Laguna Llancanelo in Mendoza, Argentina (D'Ambrosio et al. 2017). *Cyprideis salebrosa hartmanni* is a mixohaline species (1 to 29 g L^{-1}) and have preferences for a temperature range of 15 to 25 °C (Ornellas and Würdig 1983). According to Laprida (2006), this species shows affinities to hipohaline-to-oligohaline permanent lakes and lotic environments. It was found in extant lacustrine (Ramírez 1967; Laprida 2006) and estuarine settings of the Buenos Aires Province, Argentina (Martínez 2005; Cusminsky et al. 2006). *Cyprideis salebrosa hartmanni* were also found in Quaternary lacustrine sediments from Entre Ríos (Zabert and Herbst 1986) and Buenos Aires Province (Bertels and Martínez 1990; Ferrero 1996; Bertels and Martínez 1997; Laprida et al. 2009).

The location of the sampling site of the core SBIII is crucial to better interpret the microfaunal record obtained in this study from Salinas del Bebedero. In this regard, particularly, it is important to consider that core SBIII belongs from the center of the lake, and clearly separated from the shoreline (and paleoshorelines), because previous works from this saline have recovered microfossils from MaLP and MiLP paleoshorelines (García 1999; González et al. 1981). It is also important to highlight that different lithological facies and fluctuations in clastics vs evaporites proportions of core SBIII permitted to infer two higher water levels in the lake (SB III.1 and SB III.3) between a relative lower one (SB III.2); all these water levels would have been higher than the present one (Rojo et al. 2012).

In Quaternary shorelines of Salinas de Bebedero, González et al. (1981) identified six different genera, two of which are the same (*Limnocythere* and *Cyprideis*) recorded in this study. Other two (*Darwinula* and *Ilyocypris*) are typically from freshwater and *Ilyocypris* is related to lotic environment (Schwalb et al. 2002; D'Ambrosio et al. 2015; Coviaga et al. 2017), but this could be associated with freshwater inputs of paleostreams near to the sample site as Arroyo Bebedero. The last two genera (*Cypridopsis* and *Heterocypris*) are found in a wide range of environments (Meisch 2000). García (1999), analyzed four Salinas del Bebedero samples from extant water bodies and from a Quaternary fossil site (shoreline samples analyzed by González et al. 1981), registered a charophyte assemblage containing six species and provided useful data for complementary paleoecological approaches. However, the charophyte was absent in core samples used in this study, perhaps, this would be related to a difference in depth of the sampling site.

Several examples have been described in the literature concerning modern and Quaternary lagoonal microfossil biota recovered in athalassic saline lakes (Cann and De Deckker 1981; González et al. 1981; García 1993; González 1994; Perthuisot 1995; D'Ambrosio et al. 2015, 2016a, b, 2017). In Argentina, various reports (Gonzalez et al. 1981; González 1994) mentioned the presence of euryhaline species of foraminifera (*Ammonia parkinsoniana*, *Elphidium gunteri*) and non-marine Ostracods (*Limnocythere* aff. *staplini*, *L. rionegroensis*, *Cypridopsis vidua*, *Cyprideis salebrosa*) (Cusminsky et al. 2005, 2011; Laprida 2006; D'Ambrosio et al. 2017) in inland saline lakes from Mendoza, La Pampa, Río Negro, and Buenos Aires Provinces. All these sites are particularly situated near the most arid region of Argentina, as Diagonal Árida and they were not found in more humid regions (except saline lakes with marine origin sediments), so these assemblages of ostracods and foraminifers found together are indicative of this kind of extreme environments.

Concerning the Salinas del Bebedero, González et al. (1981) analyzed the presence of well-preserved microfaunal assemblages in Holocene levels which are absent in older lacustrine deposits. It is known for the Salinas that there was no previous marine connection to the sea during the Quaternary (González et al. 1981) that could explain the presence of this particular type of biota. In fact, geochemical analysis of the salt beds in Salinas del Bebedero discarded their marine origin (Delelang 1929). Previous hypothesis (González et al. 1981) sustained the presence of marginal marine microfauna in this kind of inland saline lakes because of migratory bird's activity, like flamingos, which visit different lake environments from a transect from the sea to arid regions with mud on their feet, possibly transporting foraminifera with them. This is also considered the most reliable hypothesis in order to explain the presence of foraminifera in Salinas del Bebedero. It is also important to point out that in the *Ammonia–Elphidium* assemblage, both genus are known for their ability to thrive under a wide range of salinities and to support environmental distress (Murray 2006). They also constitute the dominant forms in estuarine environments (Calvo-Marcilese 2011).

The paleoenvironmental implication of the microfaunal record helps us to a better comprehension of the evolution of Salinas del Bebedero during the last $\sim$13 ka. The ^{14}C dating shows that core SBIII corresponds to the Minor Lacustrine Period (MiLP) of González et al. (1981) and García (1999), but this core presents scarce microfossil content in general. The beginning of MiLP was defined as a freshwater environment, related with collected meltwaters from the Andes glaciers by García (1999). Then, the rest of the MiLP period was defined based on charophytes as a period with a high rate of desiccation and aridity as a saline lake, (García 1999). In agreement with García (1999), the ostracod assemblages found in the core SBIII are typically from saline and shallow environments and this could indicate high evaporation related to arid environments and high temperatures.

Around $\sim$13 to $\leq$12.6 ka, sterile levels of microfauna are evident, but there is no presence of evaporite sediments, and correspond partially with the highest clastic values (sedimentary interval III.1) as well as the highest abundance of family

Chenopodiaceae (pollen zone 1) (Fig. 4), registered in the core SBIII (Rojo et al. 2012). For that period (late Pleistocene and early Holocene), it is inferred an interval characterized by relative higher than present water levels at Salinas del Bebedero. Pollen evidence suggests a vegetation adapted to more arid conditions (due to low precipitation and/or high temperature) than the present (Rojo et al. 2012). Although apparently contradictory, that could have been possible if Salinas del Bebedero was connected via Desaguadero and Arroyo Bebedero fluvial systems to water sources from the Andes cordillera, which would have facilitated water inputs into the basin meanwhile the humidity via local precipitation was minor (Rojo et al. 2012). The absence of microfauna could be related to high sediment load by high-energy inflows into the lake. Meanwhile microfauna reflect water level of the lake (connected or not with the cordillera fluvial system), pollen evidence is related with the local environment independently of the hydrology of the lake.

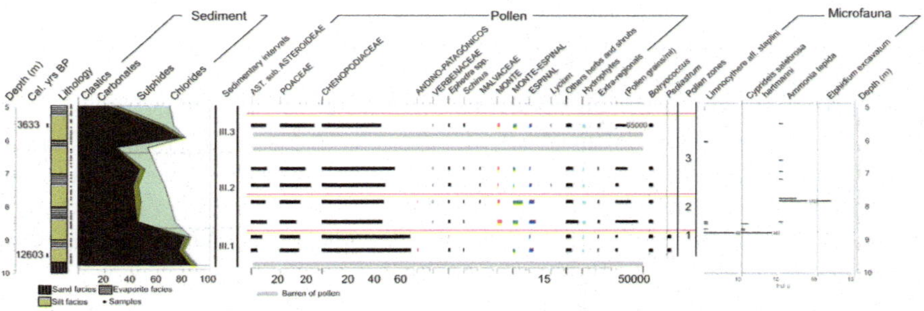

Fig. 4. Lithology, soluble, and insoluble components, pollen, and microfauna of Salinas del Bebedero core SBIII. Modified from Rojo et al. (2012)

At the beginning of the Holocene (~11.5 ka, beginning of the sedimentary interval III.2), a shallow saline lake is inferred, by the presence of *Limnocythere* aff. *staplini*, and *Cyprideis salebrosa hartmanni*. After that (early-to-middle Holocene, upper part of sedimentary interval III.2 and III.3), the presence of evaporitic sediments and sterile microfaunal levels denotes high levels of evaporation as a result of a warmer climate that led to the decrease of the hydrometric level of the lake. This arid period was interrupted in two moments by the foraminifera presence, one around the 8 ka (?), which was a less arid period showed by a great abundance of *A. tepida* and *E. excavatum*, and the second at ~4.5 ka (?) where the evaporites sediments decreasing and the scarce abundance of *A. tepida* indicates the presence of temporary water levels. The early-to-middle Holocene was an arid period with a regional aridization showed by the evaporites sediments and coincides with Páez et al. (2010) and D'Ambrosio (2014).

The fact that the Holocene microfaunal record differs from the Pleistocene record probably suggests very different environmental conditions for both periods—the Pleistocene with deep waters (cooler environment) and the Holocene with shallow waters and high salinity (warmer environment).

6 Conclusions

From this preliminary study of the microfauna of SBIII core, one foraminiferal and one ostracod assemblages could be defined. The information obtained after analyzing these microfossils, jointly with the sedimentological information, allowed the identification of the environmental characteristics during the late Quaternary. Particularly, it was interesting to highlight the correspondence between different proxies, namely sedimentology, pollen, and microfauna, which support the connection that Salinas del Bebedero would have maintained with fluvial drainage system of cordillera from central Argentina. In the context of arid conditions for that time (late Pleistocene), a reinforcement of westerlies producing high precipitation in the Andean Cordillera and consequently more snow melting water inputs to Salinas del Bebedero by Desaguadero–Salado fluvial system, would explain the high water levels for that period at Salinas del Bebedero. Alternatively, easterly polar outbreaks have been postulated as other possible sources of humidity bringing moisture to the saline (Bradbury et al. 2001), so more efforts are necessary to better comprehend the hydrology of Salinas del Bebedero. In this regard, oxygen stable isotopes could be a tool that can contribute to decipher humidity sources.

Microfaunal record contributes complementary to other proxies about past reconstruction of the water body of the lake. For example, the absence of microfauna, as recorded from the middle to the upper part of the core, would be as a consequence of warmer climate that carries to the diminution of the hydrometric level of the lake. In the bottom of the core SBIII, sterile samples associated with clastics and sandy facies could be attributed to high water (and/or freshwater) energy.

It is important to point out the need to conducting studies considering the possible occurrence of living assemblages, in order to measure all the environmental parameters that could affect the presence of foraminifera and ostracods in the Argentinian athalassic saline lakes and their potential paleoenvironmental implications.

Acknowledgements. Financial support for this study was provided by Secretaría de Ciencia, Técnica y Posgrado, and Universidad Nacional de Cuyo (M016 SeCTyP-UNCuyo 2013–2015). The authors wish to thank Jorge O. Chiesa for providing access to SBIII core and helping in sediment sampling. This paper is a contribution to the projects IB 619. Análisis de Análogos Modernos (Y-TEC S.A); PIP 2013-00021 and PICT 2014-1271.

References

Bertels A, Martínez D (1990) Quaternary ostracodes of continental and transitional littoral-shallow marine environments. Courier Forschungsinstitut Senckenberg 123:141–159

Bertels A, Martínez D (1997) Ostrácodos holocenos de la desembocadura del Arroyo Napostá Grande, sur de la provincia de Buenos Aires, Argentina. Rev Esp Micropaleontol 29:29–69

Boltovskoy E (1954a) Foraminíferos del golfo San Jorge. Revista del Instituto Nacional de Investigaciones y Museo Argentino de Ciencias Naturales Bernardino Rivadavia. Ciencias Geológicas 3:85–246

Boltovskoy E (1954b) Foraminíferos de la Bahía San Blas. Revista del Instituto Nacional de Investigaciones y Museo Argentino de Ciencias Naturales Bernardino Rivadavia. Ciencias Geológicas 3:247–300

Boltovskoy E (1957) Los foraminíferos del estuario del Río de La Plata y su zona de influencia. Revista del Instituto Nacional de Investigaciones y Museo Argentino de Ciencias Naturales Bernardino Rivadavia, Ciencias Geológicas 6:1–77

Boltovskoy E (1965) Los foraminíferos recientes. Eudeba pp. 510

Boltovskoy E (1976) Distribution of recent foraminifera of the South American region. Foraminifera 2:171–236

Boltovskoy E, Giussani G et al (1980) Atlas of benthic shelf foraminifera to the Southwest Atlantic. M. Junk Pub, The Hague, pp. 174

Boomer I, Horne DJ, Slipper IJ (2003) The use of ostracods in palaeoenvironmental studies, or what can you do with an Ostracod shell?. Paleontol Soc Pap 9:153–180

Bradbury JP et al (2001) Full and late glacial lake records along the PEP 1 transect: their role in developing interhemispheric paleoclimate interactions. In: Markgraf V (ed) Interhemispheric climate linkages. Academic Press, New York, pp 265–292

Brewster-Wingard GL et al (1996) A preliminary report on the distribution of modern fauna and flora at selected sites in North-central and North-eastern Florida Bay. Open-File Report USGS, pp 96–732

Bronk Ramsey C (2001) Development of the radiocarbon calibration program OxCal. Radiocarbon 43:355–363

Bruniard ED (1982) La diagonal árida Argentina: un límite climático real. Revista Geográfica 95:5–19

Buzas-Stephens P, Pessagno EA et al (2002) A review of species names for *Ammonia* and *Elphidium*, common foraminifera along the Texas Gulf Coast. Texas J Sci 54:3–16

Calvo Marcilese L, Pérez Panera JP et al (2013) Micropaleontological record of Holocene estuarine stages in the Bahía Blanca estuary, Argentina. J S Am Earth Sci 45:147–159

Calvo-Marcilese L (2011) Sistemática y paleoecología de los foraminifera (Protista) del holoceno del área del estuario de Bahía Blanca, Argentina. Ph.D. thesis, Facultad de Ciencias Naturales y Museo, Universidad Nacional de La Plata, pp 253, http://sedici.unlp.edu.ar/handle/10915/21866

Calvo-Marcilese L, Pratolongo P (2009) Foraminíferos de marismas y llanuras de marea del estuario de Bahía Blanca, Argentina: Distribución e implicaciones ambientales. Rev Esp Micropaleontol 41:315–332

Cann JH, De Deckker P (1981) Fossil quaternary and living foraminifera from athalassic (non-marine) saline lakes, Southern Australia. J Paleontol 55:660–670

Coviaga C, Cusminsky G, Pérez P (2017) Ecology of freshwater ostracods from Northern Patagonia and their potential application in paleoenvironmental reconstructions. Hydrobiologia https://doi.org/10.007/s10750-017-3127-1

Cusminsky GC, Pérez PA et al (2005) Recent lacustrine ostracods from Patagonia, Argentina. Rev Esp Micropaleontol 37:431–450

Cusminsky GC, Martínez D et al (2006) Foraminíferos y ostrácodos de sedimentos recientes del estuario de Bahía Blanca, Argentina. Rev Esp Micropaleontol 38:395–410

Cusminsky GC, Schwalb A et al (2011) Late quaternary environmental changes in Patagonia as inferred from lacustrine fossil and extant ostracods. Biol J Linnean Soc 103:397–408. https://doi.org/10.1111/j.1095-8312.2011.01650.x

D'Ambrosio S (2014) Reconstrucción paleolimnológica de la laguna Llancanelo (Mendoza, Argentina) a través del estudio de ostrácodos del Cuaternario. Ph.D. Thesis, Universidad Nacional de La Plata-Facultad de Ciencias Naturales y Museo, La Plata, pp 181

D'Ambrosio DS, Díaz AR, García A, Claps MC (2015) First description of the soft part anatomy of *Ilyocypris ramirezi* Cusminsky and Whatley (Crustacea, Ostracoda) from Argentina, South America. Zootaxa 3957(1):59–68

D'Ambrosio DS, Claps MC et al (2016a) Zooplankton of a protected and vulnerable wetland system in southern South America (Llancanelo area, Argentina). Int Aquat Res 8:65–80. https://doi.org/10.1007/s40071-016-0125-2

D'Ambrosio DS, Díaz AR et al (2016b) First record and revised description of *Herpetocypris helenae* G. W. Müller, 1908 (Crustacea, Ostracoda) from the Neotropics (central-west Argentina). Spixiana 39:29–38

D'Ambrosio DS, García A et al (2017) Distribution of ostracods in Westcentral Argentina related to host-water chemistry and climate: implications for paleolimnology. J Paleolimn 58(2): 101–117

Delelang LF (1929) Contribución al estudio de las salinas argentinas. La Salina del Bebedero, y sus relaciones con el sistema hidrográfico Andino o del Desaguadero. Min. Agric. Nac., Dir. Gral. de Minas Geol e Hidrol Publ 47:1–69

Doering A (1906) La formation pampéene de Córdoba. Revista del Museo de La Plata XIV: 172–190

Ferrero L (1996) Paleoecología de ostrácodos holocenos del estuario del río Quequén Grande (provincia de Buenos Aires). Ameghiniana 32:209–222

Ferrero L (2006) Micropaleontología y Paleoecología del Cuaternario del sudeste dela provincia de Buenos Aires. Ph.D. thesis, Facultad de Ciencias Exactas y Naturales, Universidad Nacional de Mar del Plata, pp 373

Ferrero L (2009) Foraminíferos y ostrácodos del Pleistoceno tardío (Mar Chiquita, provincia de Buenos Aires, Argentina). Ameghiniana 46:637–656

Frenguelli J (1922) Edad de Loess pampeano dentro del ciclo Glacial-Interglacial Prometeo. Órgano del Centro de Estudiantes Universitarios de Paraná. Año 1(16):1–8

García A (1993) *Chara halina* nov. sp. (Charophyta) de ambientes salinos de Argentina. Cryptogamie, Algal 14(1):65–75

García A (1999) Quaternary charophytes from Salina del Bebedero, Argentina: their relation with extant taxones and palaeolimnological significance. J Paleolimnol 21:307–323

Gerth E (1914) Constitución Geológica, hidrogeológica y minerales de aplicación de la provincia de San Luis. An Min Agric Sec. Geol., X (2):64

González MA (1981) Evidencias paleoclimáticas en la Salina del Bebedero (San Luis). 8° Congreso Geológico Argentino (San Luis). Actas 3:411–438

González MA (1982) Oscilaciones pleistocénicas del nivel lacustre en la actual Salina del Bebedero (San Luis, Argentina). Su relación con la última glaciación en la Cordillera de Mendoza y San Juan. Abstracts Commission Genesis and Lithology of Quaternary Deposits, South American Regional Meeting, INQUA y Universidad Nacional del Comahue, Neuquén. Resúmenes, pp. 41–43

González MA (1983) Pleistocene and Holocene lake levels in the actual Salina del Bebedero, Argentina. ^{14}C dates. Relations with the latest Pleistocenic glaciation. In: Symposium on Desert Encroachment, Fast Tropical Erosion, and Coastal Subsidence and Submergence, INQUA-IGCP (Hamburgo), Abstracts p 88

González MA (1994) Salinas del Bebedero Basin (República Argentina). In: Global Geological Records of Lake Basins, pp 381–386. Cambridge Univ Press, Cambridge

González Díaz EF (1981) Geomorfología. Geología de la Provincia de San Luis. In: Relatorio del VIII Congreso Geológico Argentino, pp 345–353

González MA, Maidana NI (1998) Post-Wisconsinian paleoenvironments at Salinas del Bebedero basin, San Luis, Argentina. J Paleolimnol 20:353–368

González MA, Musacchio E et al (1980) Sobre la presencia de foraminíferos en sedimentos holocenos de la Salina del Bebedero (San Luis, Argentina). In: 1° Primer Simposio sobre Problemas del Litoral Atlántico Bonaerense (Mar del Plata), Resúmenes, pp 253–269

González M, Weiler N (1984) Cambios climáticos del Holoceno medio en latitudes medias de la República Argentina. 2° Reunión del Grupo Periglacial Argentino. IANIGLA (Mendoza) Anales 6:87–105

González MA, Musacchio E et al (1981) Las líneas de costa Pleistocenas de la Salina del Bebedero, San Luis, Argentina. Implicancias paleoambientales de sus microfósiles. 8° Congreso Geológico Argentino (San Luis), Actas 3:617–628

González MA, Maidana NI et al (1998) Patagonian lake drilling transect: first results from Salinas del Bebedero cores. AGU 79(45):F479

Grimm E (2004) Tilia software. Research and Collections Center, Illinois State Museum, Springfield

Gutentag ED, Benson RH (1962) Neogene (Plio-Pleistocene) fresh water ostracodes from the central high plains. Kansas Geol Surv Bull 157(4):60

Hayward BW, Hollis CJ (1994) Brackish foraminifera in New Zealand: a taxonomic and ecologic review. Micropaleontology 40:185–221

Ishman SE, Brewster-Wingard GL et al (1996) Preliminary Paleontologic Report on Core T-24, Little Madeira Bay, Florida. Open-File Report USGS, pp 96–43

Laprida C (2006) Ostrácodos recientes de la llanura pampeana, Buenos Aires, Argentina: ecología e implicancias paleolimnológicas. Ameghiniana 43:181–204

Laprida C, Díaz A et al (2006) Ostracods (Crustacea) from thermal waters, southern Altiplano, Argentina. Micropaleontol 52:177–188. https://doi.org/10.2113/gsmicropal.52.2.177

Laprida C, Orgeira MJ et al (2009) El registro de la pequeña edad de hielo en lagunas pampeanas. Revista de la Asociación Geológica Argentina 65:603–611

Loeblich A, Tappan H (1988) Foraminiferal Genera and Their Classifications, vol 1e 2. In: Van Nostrand Reinhold, p 1182. New York

Maidana NI (1994) Fossil diatoms from Salinas del Bebedero (San Luis, Argentina). Diatom Research 9:99–119 Ecología y paleoecología de foraminíferos en ambientes litorales del sudeste de la provincia de Buenos Aires (partidos de Mar Chiquita y Gral Pueyrredón), Argentina (2017). Ph.D. Thesis, Universidad Nacional de Mar del Plata Facultad de Ciencias Exactas y Naturales, p 259

Márquez M, Ferrero L et al (2016) Holocene Paleoenvironmental evolution of the Pampa Coastal plain (Argentina) based on calcareous microfossils. Revista Brasileira de Paleontología 19(1): 25–40

Márquez M (2017) Ecología y paleoecología de foraminíferos en ambientes litorales del sudeste de la provincia de Buenos Aires (partidos de Mar Chiquita y Gral Pueyrredón), Argentina. Tesis Doctoral Universidad Nacional de Mar del Plata Facultad de Ciencias Exactas y Naturales, pp 259

Martínez D (2005) Asociaciones de ostrácodos modernos del estuario de Bahía Blanca, Buenos Aires, Argentina. Ameghiniana 42:669–684

Meisch C (2000) Freshwater Ostracoda of Western and Central Europe. In: Schwoerbel J, Zwick P (eds) Süsswasserfauna von Mitteleuropa, 8/3. Spektrum Akademischer Verlag, Heidelberg, p 522

Murray JW (1991) Ecology and paleoecology of benthic foraminifera. Longman, p 397. Wiley, New York

Murray JW (2006) Ecology and applications of benthic foraminifera. Cambridge University Press, Cambridge, p 426

Ornellas LP, Würdig N (1983) *Cyprideis salebrosa hartmanni* Ramírez, F. 1967. A new subspecies from Brazil and Argentina. Pesquisas 15:94–112

Paez MM, Navarro D et al (2010) Vegetación y paleoambientes durante el Holoceno. In: Zárate M, Gil A, Neme G (eds), Paleoambientes y ocupaciones humanas del centro-oeste de Argentina durante la transición Pleistoceno-Holoceno y Holoceno. Sociedad Argentina de Antropología, pp 175–212

Palacios-Fest MR, Cusminsky GC et al (2016) Late Quaternary lacustrine ostracods (Ostracoda, Crustacea) and charophytes (Charophyta, Charales) from the Puna Plateau, Argentina. J Micropalaeontol 35:66–78

Perthuisot JP (1995) Inland saline lakes with lagoonal biota: some reflections on the concept and nature of athalassic (non-marine), paralic and marine saline waters. Int J Salt Lake Res 4: 79–94

Prohaska F (1976) The climate of Argentina, Paraguay and Uruguay. In: Schwerdtfeger W (ed), Elsevier, Climate of Central and South America 12:13–112

Ramírez FC (1967) Ostrácodos de lagunas de la provincia de Buenos Aires. Revista del Museo de La Plata. Nueva Serie, Zoología 73:5–79

Ramos V (1999) Las provincias geológicas argentinas. In: Caminos R (ed) Geología Argentina. SEGEMAR, Buenos Aires, pp 41–96

Rojo LD (2003) Análisis palinológico-sedimentológico del Holoceno Superior de Salinas del Bebedero (San Luis, Argentina): primeros resultados. Revista del Museo de Ciencias Naturales Bernardino Rivadavia Nueva Serie 5(2):285–289

Rojo LD, Mehl A et al (2012) Mid- to late holocene pollen and alluvial record of the arid Andean piedmont between 33° and 34° S, Mendoza, Argentina: inferences about floodplain evolution. J Arid Environ 77:110–122

Schwalb A, Burns SJ et al (2002) Assemblage diversity and isotopic signals of modern ostracodes and host waters from Patagonia, Argentina. Palaeogeogr Palaeoclimatol Palaeoecol 187:323–339

Sen Gupta BK (2002) Foraminifera in marginal marine environments. In: Gupta BKS (ed), Modern foraminifera. Kluwer Academic Publ, pp 151–159

Zabert L, Herbst R (1986) Ostrácodos Pleistocénicos del arroyo Perucho Verna, provincia de Entre Ríos, Argentina. Ameghiniana 23:213–224

Calcareous Microfossils (Ostracoda and Foraminifera) from the Holocene (MIS 1) at the Salado Basin (Arroyo San Miguel Site, Pila, Buenos Aires Province, Argentina)

Nicolás A. Ramos[1,2(✉)], Ana P. Carignano[1,3], Gabriela C. Cusminsky[1,4], and Enrique Fucks[2]

[1] CONICET–Consejo Nacional de Investigaciones Científicas y Técnicas, Buenos Aires, Argentina
nicolasramos@fcnym.unlp.edu.ar, {anapcarignano, gcusminsky}@gmail.com
[2] Facultad de Ciencias Naturales y Museo, Universidad Nacional de La Plata, Calle 64 no 3, 1900 La Plata, Argentina
{nicolasramos, efucks}@fcnym.unlp.edu.ar
[3] División Paleozoología Invertebrados, Facultad de Ciencias Naturales y Museo, Universidad Nacional de La Plata, Paseo del Bosque s/n, 1900 La Plata, Argentina
[4] Instituto de Biodiversidad y Medio Ambiente, Delegación Departamental de Geología y Petróleo, Universidad Nacional del Comahue, Quintral 1250, 8400 San Carlos de Bariloche, Argentina

Abstract. A unit of 30 cm thick attributable to the Marine Isotopic Stage 1 (MIS 1) (Canal de Las Escobas Formation) was identified at the study site, between the Members Gorch and Puente Las Gaviotas of the Luján Formation. The ostracods *Cyprideis salebrosa hartmanni* Ramírez, *Leptocythere* sp., *Cytherura dimorphica* Bertels and Martínez, *Limnocythere cusminskyae* Ramón-Mercau, Plastani and Laprida, *L. solum* (Whatley and Cholich), *L. multiperforata* (Whatley and Cholich), and the foraminifers *Ammonia tepida* (Cushman), *A. parkinsoniana* (d'Orbigny) and *Elphidium gunteri* (Cole) where recognized. The ostracods recovered include common species of estuarine environments and with wide salinity tolerance. The *Ammonia-Elphidium* association is also typical of oligo-mesohaline conditions. This low diversity association, with clear dominance of *C. salebrosa hartmanni*, plus the presence of *C. dimorphica*, *Leptocythere* sp. and the Foraminifera mentioned above, is indicative of environments with marine influence. Moreover, a wedge of sediments containing shells of the bivalve *Anomalocardia brasiliana* (Gmelin), was dated obtained a radiocarbon age of 5580 ± 110 years BP (6278–6511 cal. years BP). This information suggests that Arroyo San Miguel site represents the most internal locality assignable to the MIS 1 marine transgression.

Keywords: Marine transgressions · Mid-Holocene · Canal de las Escobas formation · Radiocarbon age · Cytheroidea · *Ammonia-Elphidium*

G. C. Cusminsky et al. (eds.), *Advances in South American Micropaleontology*,
Springer Earth System Sciences, https://doi.org/10.1007/978-3-030-02119-1_8

1 Introduction

The Quaternary Period was a time of intense climatic changes, implying glacial and interglacial stages. These periodic changes are named as Marine Isotopic Stage (MIS) numbered from 1 to 100, (Shackleton 1987; Shackleton et al. 1990), where odd numbers indicate warm events and even numbers cold ones. The MIS 1 represents the last warm episode (beginning at 11,700 years BP), that flooded flat nearshore areas, racing its maximum sea level (3–4 m a.s.l.) at 6000 years BP, and with storm deposits reaching heights near 6 m a.s.l. (Schnack et al. 2005; Fucks et al. 2010; Prieto et al. 2016). This is an event recorded over large areas in the Argentinean coastal territory, and its extension has been mapped over Buenos Aires Province (Fig. 1b). Deposits of the MIS 1 can be followed along the Salado River lower section and Canal 15 ravines that cut across strata of the "Pampeano" (González Bonorino 1965) and Luján (Fidalgo et al. 1973) Formations (Pleistocene–Holocene). The Salado basin is located in the Buenos Aires Province, running in a northwest–southeast direction toward the Atlantic Ocean, in a subregion known as "Pampa Deprimida" (Frenguelli 1950). Ostracods and foraminifers studies from the Salado basin were mainly focused in the littoral deposits produced by the Holocene marine transgressions in the lower section of the Salado River. These researches were carried on by Bertels-Psotka and Laprida (1998), Laprida (1998), Laprida and Bertels-Psotka (2003), who analyzed outcrops from several sites of the basin, including samples from Las Escobas Formation (Canal 18 and Cerro de la Gloria Members, middle Holocene), and from the Destacamento Río Salado Formation (lower Holocene). Based on the paleoecological requirements of ostracod and foraminifers, these authors (Bertels-Psotka and Laprida 1998; Laprida 1998; Laprida and Bertels-Psotka 2003) suggested marine to brackish water conditions for the aforementioned formations. Other studies in the area, related to Holocene calcareous microfossils (charophytes), were performed by García (1987, 1993), who described gyrogonites and oospores of *Lamprothamnium haesseliae* Dont, 1984 and *L. succinctum* (A. Br. in Asch.) Wood, 1962 from several shallow lakes in the studied area (Fig. 1).

The aim of this work is to analyze the marine transgression extension at the Salado basin, with special focus in its microfossils content, and to record the paleoenvironmental consequences of this last transgression. We propose that the marine influence was more extended than previous studies suggests.

2 Geological Settings

The "Pampa Deprimida" is a flat plain slighting sloping to the east, with poor runoff surface, and elevations below 7 m a.s.l., with exceptions linked to lunettes up to 18 m a.s.l. The basin genesis is associated with the opening of the Atlantic Ocean during Mesozoic times, with deposits that reach 7000 m in the depocenter (Introcaso and Ramos 1984). The top sequence is mainly of Pleistocene and Holocene ages and eolian origin, which were later reworked by fluvial action under dry and humid climate alternation. These eolian deposits are represented by the "Pampeano" and La Postrera Formations, and the latter is overlain by the Luján Formation. The stratigraphy of the

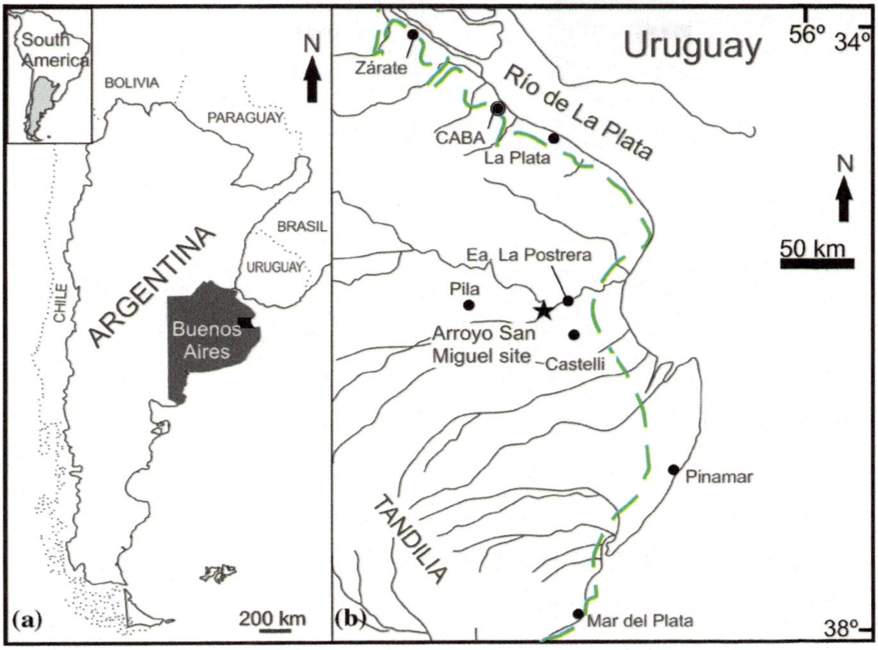

Fig. 1. Location of the studied site. **a** Argentina, gray shaded Buenos Aires Province, **b** Buenos Aires coast, the green line represents the mid-Holocene coastline. Modified from Schnack et al. (2005)

Luján Formation has been subject of debate. Ameghino (1880–1881) and Doering (1882) mentioned the "Pampeano Lacustre" for the upper deposits of the "Pampeano", later redefined by Ameghino (1884) as "Lujanense" stage. Fidalgo and De Francesco (1973) differentiated the "Pampeano" deposits, of eolian origin, from the Luján Formation and the "Aluvio" of fluvial genesis, and at the same time, divided the Luján Formation into the Guerrero (lower) and Río Salado (upper) members. Subsequently, the Luján Formation was further subdivided by Dillon and Rabassa (1985) and Dangvas and Blasi (1992, 2003). Fucks et al. (2015) recognized for the Luján Formation three members, from base to top: La Chumbiada (Dillon and Rabassa 1985), Gorchs and Puente Las Gaviotas (Fig. 2a). Toward the east, at Estancia La Postrera, these authors identify a wedge of marine sediments assigned to the Canal de Las Escobas Formation, interposed between the fluvial sediments of Gorch and Puente Las Gaviotas Members. The deposits of the "Querandina" transgression (Frenguelli 1950) were formerly named as Destacamento Río Salado Formation and Canal Las Escobas Formation by Fidalgo et al. (1973). These authors characterized the Destacamento Río Salado Formation by its dark gray or black sands to silty sands. At the same time, Las Escobas Formation was divided into Cerro de la Gloria Member (with shelly banks with sandy matrix) and Canal 18 Member (of sands with silt and clay lents). Recently, Fucks et al. (2010) reunited both Destacamento Río Salado and Las Escobas

Formations into Canal de Las Escobas Formation, that represents the MIS 1 transgression at the Salado basin. These authors further divided the latter into Destacamento Río Salado, Canal 18, Cerro de la Gloria and Canal 15 Members (Fig. 2b).

(a)

Age	Ameghino 1884	Fidalgo et al. 1973		Dillon and Rabassa 1985		Dangavs and Blasi 2003		Fucks et al. 2015	
Holocene	Ariano stage Platense stage	Luján Formation	Aluvio Río Salado Mb.	Luján Formation	Río Salado Mb.	Luján Formation	Aluvio Río Salado Mb.	Luján Formation	Puente Las Gaviotas Mb. Gorch Mb.
Late Pleistocene	Lujanense stage		Guerrero Mb.		Guerrero Mb. La Chumbiada Mb.		Lobos Mb. La Chumbiada Mb.		La Chumbiada Mb.

(b)

Age	Frenguelli 1957	Fidalgo et al.1973		Fucks et al. 2010	
Holocene	Platense Querandinense	Las Escobas Fm.	Cerro de la Gloria Mb. Canal 18 Mb.	Canal de Las Escobas Fm.	Canal 15 Mb. Cerro de la Gloria Mb. Canal 18 Mb. Destacamento Río Salado Mb.
Late Pleistocene	Belgranense	Destacamento Río Salado Fm. Pascua Fm.		Pascua Fm.	

Fig. 2. Summary of the different stratigraphic proposals **a** for the Luján Formation. (Modified from Fucks et al. 2015), **b** for the Canal de Las Escobas Formation

This proposal is considered to be more appropriate with our field observations, for further details about sedimentology, paleoenvironmental interpretations, and discussion with previous authors see Fucks et al. (2010, 2015).

The samples studied here comes from the ravines of the Salado River, at Arroyo San Miguel site (36° 02′ 36.3″ S–58° 00′ 25.3″ W), between the towns of Pila and Castelli, (Fig. 1a, b), 60 km west from the present coast. The Luján Formation crops out totally or partially at the study area; between Gorchs and Puente Las Gaviotas Members a lense of sediments was assigned to the Canal de Las Escobas Formation (Fig. 3).

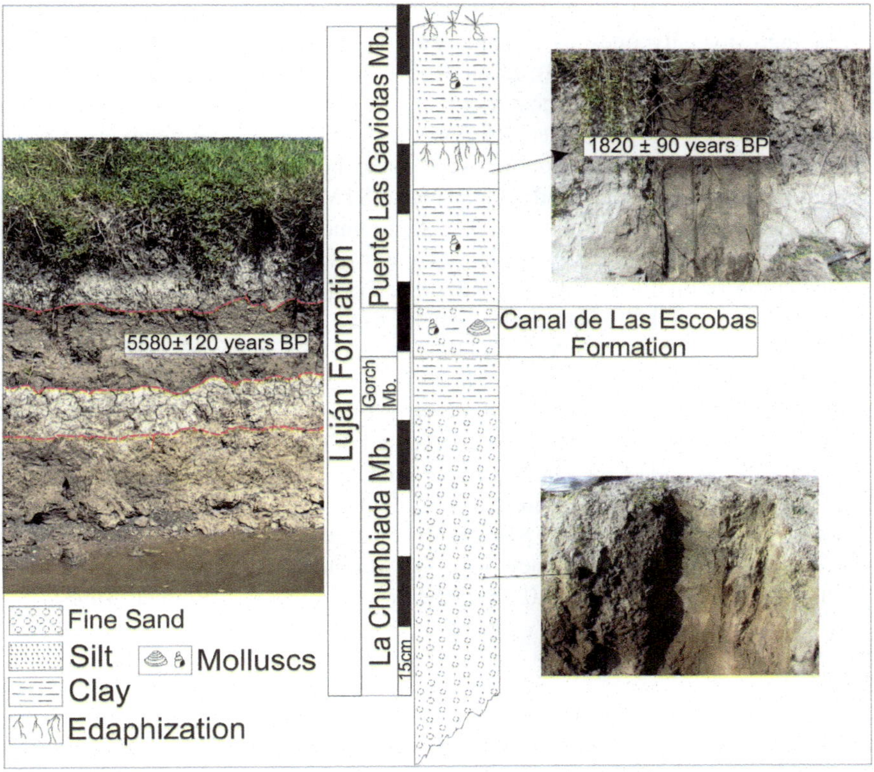

Fig. 3. Integrated sedimentary profile of the Salado River ravine at the Arroyo San Miguel site

3 Materials and Methods

A 34 cm core was extracted using a half PCV tube, of 5 cm of diameter, and was sampled at 1 cm interval, obtaining a total of 11 samples of 2 cm^3. Following standard methodology, samples were left in a water solution for 24 hrs, washed through a Tyler 230 sieve under tap water, and dried at room temperature. From the total weight of the dry samples, only 1 gr was separated for the analysis, and every organic remain was picked out. Absolute abundance was counted with the greater number of left or right valves, plus the number of carapaces to calculate the MNI (minimum number of individuals). The classification followed for suprageneric levels in Ostracoda is that proposed by Martin and Davis (2001) and Karanovic (2012), Loeblich and Tappan (1988), with modifications by Sen Gupta (1999) were followed for suprageneric levels of Foraminifera taxonomy. The figures specimens were stored in the repository of Museo de La Plata Micropaleontogía under the numbers MLP–Mi 1974–1984. The abundance and the species richness were analyzed. Also, the diversity index of Shannon–Weaver H(S) was used for each sample, calculated according to the expression $H = -\sum pi.\ln pi$, in which pi is the relative abundance of the ith species.

Values lower than 2 implies low diversity assemblages, and higher values than 3 represent high diversity ones. Dominant ostracod species molt stages were counted according to Boomer et al. (2003), to recognize if the assemblages were in situ or if they have been subjected to postmortem transport. Also, the male/female and valve/carapace ratios were considered for paleoenvironmental interpretations (Boomer et al. 2003). The textural descriptions were made both in field and the laboratory, separating sand fractions from mud with a sieve ASTM N° 230. Grain size recognition followed Folk (1966), grouping clays and silts into mud. For radiocarbon dating, valves of the bivalve *Anomalocardia brasiliana* (Gmelin) were obtained from a wedge of sediments at the studied site and analyzed at the Tritium and Radiocarbon Laboratory (LATYR-CIG, CONICET). The mollusk shells were separated from the sediment by washing with tap water under a ASTM N20 sieve and then subjected to ultrasonic bath to remove traces of adhered material. Next, the specimens were treated with a HCl solution to remove the 20% in weight of the surface carbonate. Lastly, the dried sample was transformed into benzene. The age is expressed in radiocarbon years before present (year 1950 BP), corrected by isotopic fractioning, with table δ13C estimated values (Stuiver and Polach 1977). The calibration of continental samples from the Southern Hemisphere is performed with the SHCal13.14c curve (Hogg et al. 2013), for marine samples the calibration is performed with the Marine 13. 14c curve (Reimer et al. 2013), and the program CALIB 7.0.0 together with Stuiver and Reimer (1993).

4 Results

The sediments are represented by fine muddy sands, with a mud percentage higher than 40% in the first 15 cm. This percentage decreases toward the top of the core, with no more than 30% of mud content.

A paucispecific (Kidwell et al. 1986) accumulation of *A. brasiliana* (Gmeling) and *Heleobia australis* (D'Orbigny) was observed on the field. The *A. brasiliana* specimens were mostly disarticulated but not fragmented, in a concordant biofabric with black muddy-sandy sediment, like the ones described for transgressive facies (Schnack et al. 2005). The shells did not show evidence of reworking, and they were suitable for radiocarbon dating, giving 5580 ± 110 years BP (6278–6511 cal. years BP) (LP-3466).

From the eleven samples studied, microfossils were scarce in the first five, and the abundance and diversity of them increased significantly in the remaining six samples (Fig. 4). In these six samples, a moderately diverse microfauna was recovered, with Shannon–Weaver H(S) index value around 2. Eight ostracod species were obtained, represented mostly by members of *Cytherocopina*, from which adults and juveniles valves were preserved. Three species of Foraminifera were identified, *Ammonia tepida* (Cushman 1926), *A. parkinsoniana* (d'Orbigny 1839) and *Elphidium gunteri* (Cole 1931) (Fig. 5).

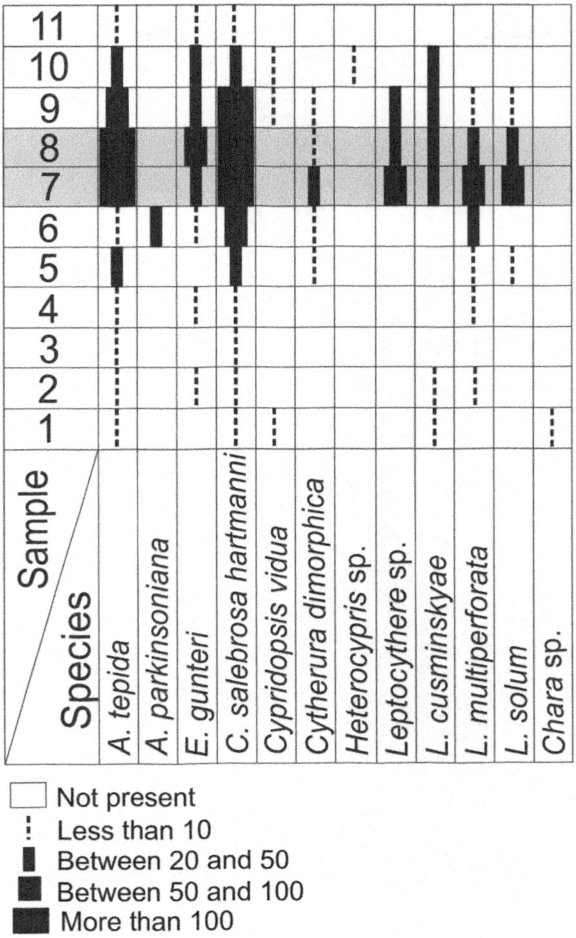

Fig. 4. Distribution and relative abundances of foraminifers and ostracods recovered from the studied core. Radiocarbon dated level shaded gray

The dominant species in the associations was *Cyprideis salebrosa hartmanni* Ramírez 1967. To a lesser extent, along the profile samples, the accompanying species varied in abundance (*Limnocythere cusminskyae* Ramón-Mercau, Plastani and Laprida 2014; *Leptocythere* sp., *Cytherura dimorphica* (Bertels and Martínez 1997), *L. multiperforata* (Whatley and Cholich 1974) and *L. solum* (Whatley and Cholich 1974). Other species were scarce, rare or absent in all the profile [e.g., *Cypridopsis vidua* (Muller 1776)].

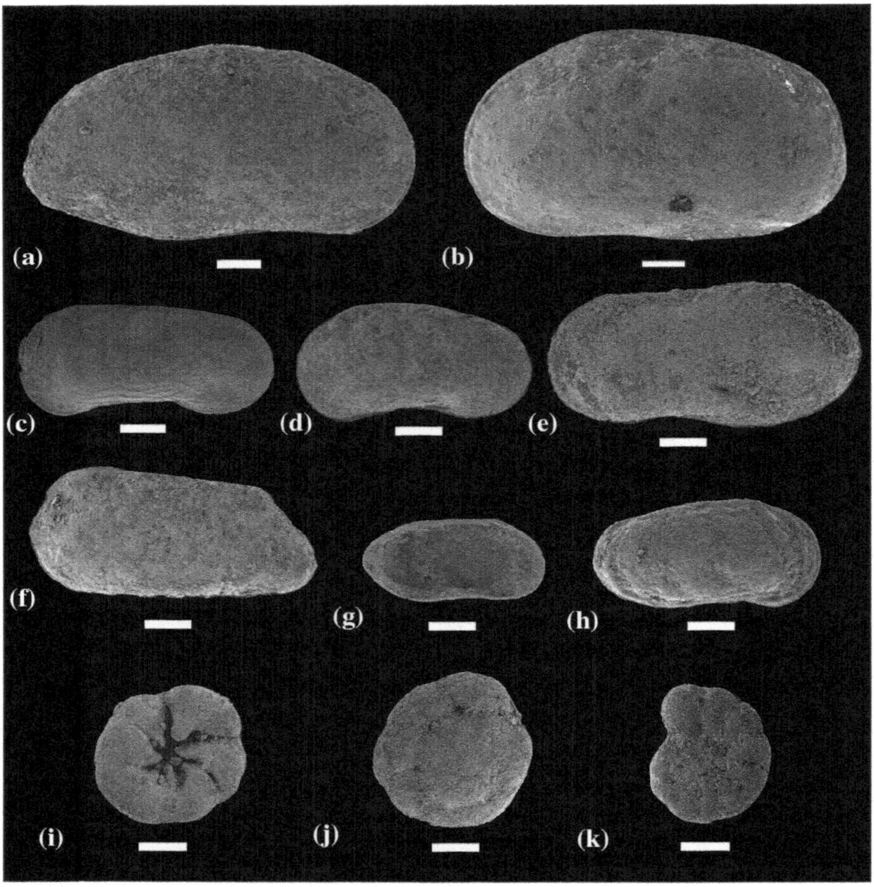

Fig. 5. Calcareous microfossils. **a–b** *Cyprideis salebrosa hartmanni* (Ramírez) **a** male right valve, external view, MLP–Mi 1974, **b** female left valve, external view, MLP–Mi 1975; **c–d** *Limnocythere cusminskyae* Ramón Mercau et al., **c** male left valve, external view, MLP–Mi 1976, **d** female left valve, external view, MLP–Mi 1977; **e** *Limnocythere solum* (Whatley and Cholich) male left valve, external view, MLP–Mi 1978; **f** *Limnocythere multiperforata* (Whatley and Cholich), male left valve, external view, MLP–Mi 1979; **g** *Cytherura dimorphica* Bertels and Martínez, left valve, internal view, MLP–Mi 1980; **h** *Leptocythere* sp., right valve, external view, MLP–Mi 1981; **i–j** *Ammonia tepida* (Cushman), **i**, umbilical view, MLP–Mi 1982, **j** spiral view, MLP–Mi 1983; **k** *Elphidium gunteri* (Cole), lateral view, MLP–Mi 1984. Scale bar = 0.01 mm

In samples seven and eight the number of individuals increases significantly and the shells presented recrystallization of calcium carbonate from diagenetic origin, giving them a cretozoid aspect, which at the scanning electronic microscope is easily differentiable from the bioestratinomic correlation that can occur on shells and carapaces that have undergone transport.

5 Discussion

Different proxies had been used to assess the depositional paleoenvironments of the Holocene, and the calcareous microfossils have proved to be an essential tool. The abundance of specimens of *C. salebrosa hartmanni* allowed assessing the population age structure of this species. The population graphics obtained (Fig. 6) suggest an autochthonous thanatocoenosis, deposited under moderate to low energy (at least for samples 6–10). Thereby, the association can provide good paleoenvironmental information (Whatley1983, 1988; Boomer et al. 2003). Moreover, the male/female ratio was lower than 1, suggesting r-strategy reproduction under relative environmental instability (Boomer et al. 2003).

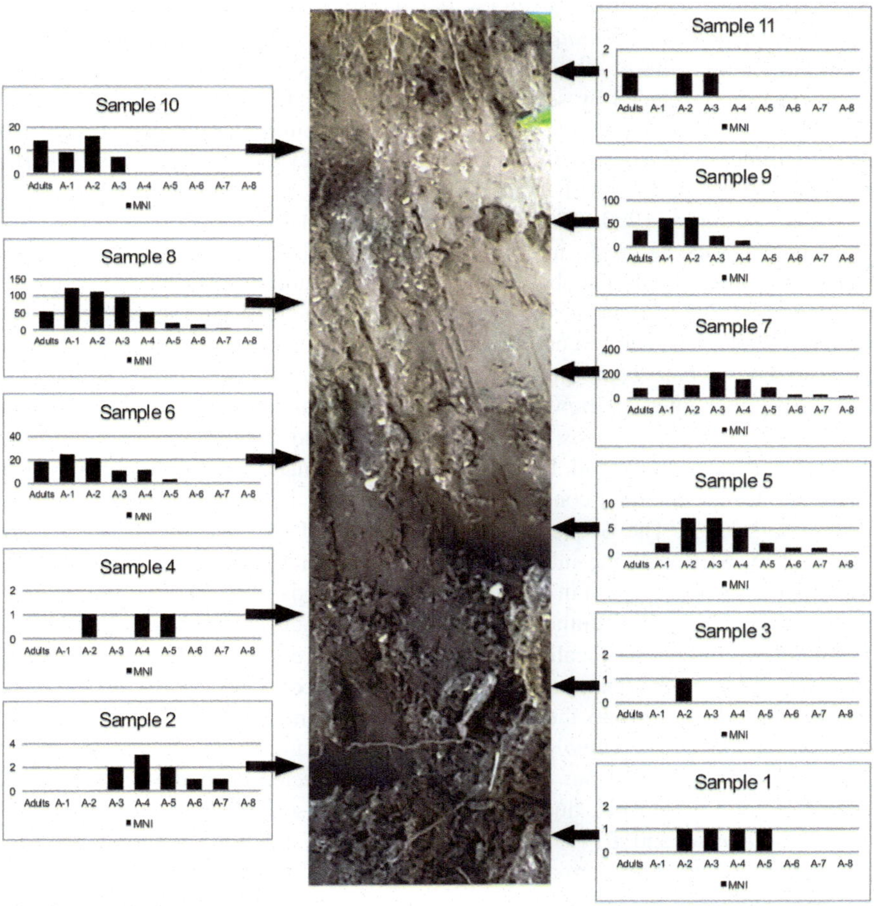

Fig. 6. *Cyprideis salebrosa hartmanni* population age structures along the core

The components of the microfossils association recovered in this study were recorded at several sites in the Buenos Aires Province, allowing comparison and evaluation of our results, and thus to propose a depositional paleoenvironment for the Arroyo San Miguel site. In this sense, Whatley and Cholich (1974) studied Pleistocene?–Holocene core and outcrop samples from the Laguna Salada Grande, and described two new ostracods species, *Pampacythere solum* and *P. multiperforata,* later reassigned by Laprida (1998) to *Limnocythere*. Taking into account the associated macro (gastropods and bivalves) and microfauna (foraminifers from the genera *Ammonia* and *Elphidium*), Whatley and Cholich (1974) considered the depositional paleoenvironment as a lagoon, within the oligo-mesohaline range, and with fresh and marine water influence. A similar association was recovered from a core drill at the Laguna Hinojales, were Prieto et al. (1998) recorded dinoflagellate cysts, marine-brackish and freshwater diatoms, together with the ostracods *Limnocythere staplini* (= *L. cusminskyae*), *L. multiperforata* (as *Pampacythere multiperforata*), *L. solum* (as *P. solum*), *Cyprideis multidentata*, and *Cypridopsis* sp., and the foraminifers *A. tepida* and *Elphidium* sp. These authors proposed a brackish paleoenvironment for the Hinojales lagoon, produced by sea ingression prior to 4500 years BP.

The genus *Leptocythere* Sars, 1925 is known from littoral, continental shelf, and estuarine sediments of Argentina (Whatley and Moguilevsky 1975); while *Cytherura* Sars 1866 is considered as a mesohaline and littoral taxon (van Morkhoven 1963); Bertels and Martínez (1990, 1997) recovered a rich association of ostracods from outcrops of the Napostá creek, near its outlet to the Atlantic Ocean, from levels correlated with Canal de Las Escobas Formation. These authors divided the outcrop in three sequences; two of them of marine origin, the sequence I was characterized by a lower interval of total marine conditions, suggested by a low diversity association of marine ostracods where *Leptocythere* sp. and *C. dimorphica* were some of the components. An upper interval showed a shift from polyhaline conditions (*Leptocythere* sp., *Callistocythere litoralensis* (Rossi de García) and *Papillosacythere parallela* Whatley, Chadwick, Coxil and Toy, 1987 as the dominant species, together with *C. dimorphica*) to mesohaline conditions suggested by the dominance of *Cyprideis salebrosa hartmanni*. The Sequence III was interpreted by the authors as a shallow marine with marked estuarine influence, based on the presence mixed microfauna.

A comprehensive study on the ecology of the ostracods from the Pampean plains is that of Laprida (2006). This author sampled 15 water bodies at continental and littoral environments from several localities in the Buenos Aires province and recorded the ostracods species and the temperature, pH, and oxygen concentration of the sites. As a result, she was able to group the different taxa regarding their paleoecological preferences. Assemblage of *C. dimorphica, C. multidentata* and *Perissocytheridea* sp. was considered as exclusively from estuarine waters connected with the sea, subject to variations of salinity and pH, due to tidal influence. A second group, comprising the species *C. salebrosa* and *L. cusminskyae* (as *L.* aff. *L. staplini*) among others, was considered as typical from permanent lagoons and oligohaline streams, environments that suffered slow and gradual salinity fluctuations (Laprida 2006). Laprida and Valero-Garcés (2009) studied sediment cores from the Laguna de Chascomús, analyzing its paleoenvironmental evolution during the last 500 years, based on ostracod assemblages together with sedimentological and geochemical analyzes. They observed that

Cyprideis salebrosa hartmanni was dominant in low diversity ostracods assemblages, during periods of lower water availability, in oligo to mesohaline waters. It is important to note that Würdig (1983), studying the living ostracods from a coastal lagoonal system (coast of Rio Grande do Sul, Brazil), found that the assemblages of salt marsh environments were dominated by the species *Cyprideis salebrosa.*

Cusminsky et al. (2006) studied the recent foraminifer and ostracod fauna of the Tres Brazas channel at the Bahía Blanca estuary. The tidal flats environment was characterized by the dominance of *Ammonia parkinsoniana* (sensu Calvo Marcilese 2011) and *E. gunteri* (sensu Calvo Marcilese 2011) and *Buccella peruviana* f. *campsi,* with minor components represented by bulliminids and lagenids between the Foraminifera. Ostracods fauna was represented by the species *Loxocythere variasculpa, Neocythereis ruidis and Cyprideis salebrosa hartmanni* as the most abundant, together with *Leptocythere darwini*? and *C. dimorphica* among other species.

Recently, Márquez et al. (2016) sampled an outcrop at Las Brusquitas creek, close to its outlet at the Atlantic Ocean, and recovered a rich association of ostracods, foraminifers, and charophytes. Between 6300–6200 and 6000–5700 cal years BP, an association composed by the ostracods *C. salebrosa hartmanni, L. cusminskyae, L. solum, L. multiperforata, C. dimorphica, Leptocythere* sp. (among the most represented), together with the foraminifers *Buccella peruviana, A. tepida, A. parkinsoniana* and *E. gunteri* and *Elphidium margaritaceum* as minor components. These authors registered between 6300–5700 and 2300–2000 cal years BP, a dominance of *B. peruviana* and *C. salebrosa hartmanni* (brackish estuarine paleoenvironment); toward the end of this stage, an increase in the abundance of *Ammonia* and *Elphidium* was recognized, and a more restricted paleoenvironment was suggested.

The Foraminifera genera *Ammonia* and *Elphidium* inhabit marine (inner shelf), brackish, and hypersaline waters (lagoons) (Murray 2006). This association is typical of shallow marginal marine environments worldwide, although records in inner waters are known. For example, Calvo Marcilese et al. (2016) recovered an association of *A. tepida, Elphidium excavatum* together with the ostracods *L.* cf. *cusminskyae* and *C. salebrosa hartmanni,* from a core drilled at the laguna Salinas del Bebedero (San Luis Province), and proposed the dispersion to this paralic environment through wind or birds. Brewster-Wingard et al. (1999) studied the benthic foraminifer assemblages from the Florida Bay (United States of America) and concluded that the *Ammonia-Elphidium* association would indicate oligohaline to mesohaline conditions with salinities between 5–18%.

5.1 Associated Macrofauna

At the studied site, between Gorchs and Puente Las Gaviotas Members, a wedge of black muddy sands was distinguished by its mollusk shells content, dominated by *A. brasiliana* (Gmelin), a sublittoral marine/estuarine superficial infaunal species that lives in soft substrata (Aguirre and Farinati 2000). Also, the species *H. australis* (d'Orbigny) was registered, the most abundant species in Holocene sediments of estuarine origin, that inhabits wave-protected, tidally influenced environments such as coastal lakes, and that becomes abundant at salinity ranges of 1–40% (De Francesco and Isla 2003; Aguirre and Farinati 2000).

5.2 MIS 1 Transgression

The warmer phase of the MIS 1 took place at mid-Holocene (circa 6000 years BP), with temperatures higher than present days, producing a marine transgression that displaced the Buenos Aires coastal line at least 30 km from its actual position, flooding tidal flats and sometimes river mouths (Shacketon 1987; Isla 1998; Schnack et al. 2005; Prieto et al. 2016).

Fucks et al. (2015) described a wedge of marine sediments at Estancia La Postrera (20 km northeast from our study site), assigned to the Canal de Las Escobas Formation, where shells of the marine bivalve *Tagelus plebeius* with articulated valves and in vertical position, were registered. The radiocarbon age obtained in valves of *T. plebeius* placed the deposits at 5920 ± 90 years BP (6556–6788 cal years BP). This wedge is interposed between the fluvial sediments (Gorchs and Puente Las Gaviotas Members) and was assigned to the MIS 1 transgression. These authors suggested that the marine transgression entered exclusively following the Salado River depression, and advanced toward the continent while the topography enabled it. Considering the radiocarbon ages obtained by Fucks et al. (2015) at different sites of the Salado River, we suggest that the Arroyo San Miguel site deposits were formed during the Hypsithermal at mid-Holocene, simultaneously with the last part of the Gorchs Member, and before a hiatus, gave base to Puente Las Gaviotas Member deposition.

Prado et al. (2013) described a phase of dryer conditions during mid-Holocene. These authors suggested that the decrease of austral summer insolation weakened the South American monsoon system circulation, producing a hydrological reversal to a super-dry phase in the Southern Hemisphere. According to these authors, it generated a decrease in precipitation, saltier conditions along the South American continental margin, and lower lake levels, with an extreme magnitude around 4700 cal year BP (Piovano et al. 2009). We consider that these particular climatic conditions allowed the

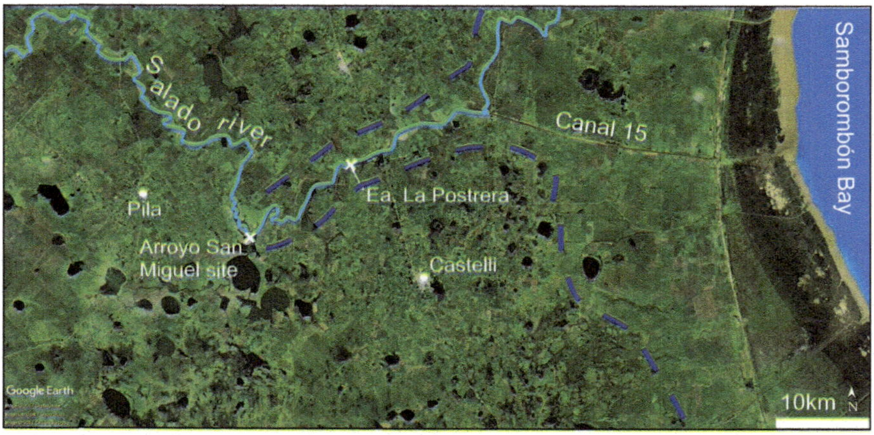

Fig. 7. Estimated MIS 1 transgression following the depression of the Salado River (blue dotted line), proposed in this work (Image from Google Earth, between 57° 47′ 54.93″ W, 57° 15′ 26.29″ W and 35° 43′ 28.29″ S, 35° 58′ 25.26″ S)

transgression of the sea through the Salado River 60 km from the present coastline, taking advantage of a probably lower flow rate (Fig. 7).

6 Conclusions

A calcareous microfossils assemblage and mollusk fauna associated were recovered from a wedge of sediments assigned to Canal de Las Escobas Formation, interposed between the Gorchs and Puente Las Gaviotas members of the Luján Formation. The mixed characteristic of the microfauna, together with the presence of the bivalve *A. brasiliana* suggested deposition under brackish conditions, extending the marine influence further west in the Buenos Aires Province, at 5580 $\pm$ 110 years BP (6278–6511 cal years BP). This age is assumed for the transgressive maximum occurred in the Salado River basin, which was favored by dryer climatic conditions and a lower flow rate.

Acknowledgements. Authors wish to thank Lic. Patricia Sarmiento from the Servicio de Microscopía de La Plata, and Lic. Florencia Mari from the LATYR; many thanks to Dr. Lydia Calvo-Marcilese for her help with the foraminifers identifications. Special thanks are to L. Ferrero and an anonymous reviewer, whose useful comments and correction greatly improved the manuscript. Founds were provided by PICT 2014-1271 and 2014-1964; PIP 2013-00021.

References

Aguirre ML, Farinati EA (2000) Aspectos sistemáticos, de distribución y paleoambientales de *Littoridina australis* (D'Orbigny 1835) (Mesogastropoda) en el Cuaternario marino de Argentina (Sudamérica). Geobios 33:569–597

Ameghino F (1884) Excursiones geológicas y paleontológicas en la provincia de Buenos Aires. Boletín Academia Nacional de Ciencias Córdoba 6:61–257

Bertels A, Martínez DE (1990) Quaternary ostracodes of continental and transitional littoral-shallow marine environments. Courier Forschchungsinstitut Senckenb. 123:141–160

Bertels A, Martínez DE (1997) Ostrácodos holocenos de la desembocadura del arroyo Napostá Grande, sur de la provincia de Buenos Aires, Argentina. Revista Española de Micropaleontología 29:20–69

Bertels-Psotka A, Laprida C (1998) Ostrácodos (Arthropoda, Crustacea) del Miembro Cerro de la Gloria, Formación Las Escobas (Holoceno), provincia de Buenos Aires, República Argentina. Revista Española de Micropaleontología 30:103–128

Boomer I, Horne DJ et al (2003) The use of Ostracodes in paleoenvironmental studies, or what can you do with an ostracod shell? Paleontol Soc Pap 9:153–180

Brewster-Wingard GL, Ishman SE (1999) Historical trends in salinity and substrate in central Florida Bay: paleoecological reconstruction using modern analogue data. Estuaries 22:12–15

Calvo Marcilese L (2011) Sistemática y Paleoecología de los Foraminifera (Protista) del Holoceno el área del estuario de Bahía Blanca, Argentina. Ph.D. Thesis, Facultad de Ciencias Naturales y Museo, Universidad Nacional de La Plata, pp 253

Calvo-Marcilese L, D'Ambrosio L, et al (2016) Nuevo registro microfaunístico (foraminíferos y ostrácodos) del Cuaternario tardío en Salinas del Bebedero, San Luis, Argentina. In: Actas del 11° Congreso de la Asociación Paleontológica Argentina, General Roca, p 123, 17–21 Oct 2016

Cusminsky GC, Martínez D et al (2006) Foraminíferos y ostrácodos de sedimentos recientes del estuario de Bahía Blanca, Argentina. Revista Española de Micropaleontología 38:395–410

Dangavs N, Blasi A (1992) Formación lobos, nueva unidad estratigráficas en la cuenca del río Salado, Provincia De Buenos Aires. In: Actas de las 3° Jornadas Geológicas Bonaerenses, La Plata, pp 17–32, 9–13 Mar 1992

Dangavs N, Blasi A (2003) El miembro lobos de la formación luján: cambio de rango de la Fm. Lobos, unidad estratigráfica del Pleistoceno superior de la cuenca del río Salado de la provincia de Buenos Aires. In: Actas del 2° Congreso Argentino de Cuaternario y Geomorfología, San Miguel de Tucumán, pp 67–74, 22–24 Sept 2003

De Francesco CG, Isla FI (2003) Distribution and abundance of hydrobiid snails in a mixed estuary and a coastal lagoon, Argentina. Estuaries 26:790–797

Dillon A, Rabassa J (1985) Miembro la chumbiada, formación luján (pleistoceno, provincia de buenos aires): una nueva unidad estratigráfica del valle del río Salado. In: Actas de las 1°Jornadas Geológicas Bonaerenses, Tandil, p 27, 11–15 Nov 1985

Doering A (1882) Informe oficial de la Comisión Científica agregada al Estado Mayor General de la Expedición al Río Negro (Patagonia) realizada en los meses de Abril, Mayo y Junio de 1879, bajo las órdenes del general d. Julio A. Roca. Comisión Científica de la Exploración al Río Negro, Congreso de la Nación, Buenos Aires, pp 259–340

Fidalgo F, De Francesco F et al (1973) Geología superficial en las hojas castelli, J. M. Cobo y monasterio (Provincia de Bs. As.). In: Actas del 5° Congreso Geológico Argentino, Carlos Paz, pp 27–39, 22–28 Oct 1973

Folk RL (1966) A review of grain-size parameters. Sedimentology 6:73–93

Frenguelli J (1950) Serie II. rasgos generales de la morfología y la geología de la provincia de Buenos Aires. Publicaciones LEMIT 33:3–72

Fucks E, Schnack E et al (2010) Nuevo ordenamiento estratigráfico de las secuencias marinas del sector continental de la Bahía Samborombón, provincia de Buenos Aires. Revista de la Asociación Geológica Argentina 67:27–39

Fucks E, Pisano MF et al (2015) Stratigraphy of the fluvial deposits of the Salado river basin, Buenos Aires province: lithology, chronology and paleoclimate. J S Am Earth Sci 60:129–139

García A (1987) Estudio del gametangio femenino de Charophyta actuales de Argentina. Análisis comparado con el registro fósil correspondiente. Ph.D. Facultad de Ciencias Naturales y Museo, Universidad Nacional de la Plata, La Plata, pp 312

García A (1993) Quaternary and extant *Lamprothamnium* Groves (Charophyta) from Argentina. Hydrobiologia 267:143–154

González Bonorino F (1965) Mineralogía de las fracciones arcilla y limo del pampeano en el área de la Ciudad de Buenos Aires y su significado estratigráfico y sedimentológico. Revista de la Asociación Geológica Argentina 20(1):67–148

Hogg AG, Hua Q, Blackwell PG et al (2013) SHCal13 southern hemisphere calibration, 0–50,000 Years cal BP. Radiocarbon 55:1889–1903

Introcaso A, Ramos V (1984) La cuenca del Salado. un modelo de evolución aulacogénica. In: Actas del 9° Congreso geológico Argentino, San Carlos de Bariloche, Río Negro, pp 27–46, 5–9 Nov 1984

Isla FI (1998) Holocene coastal evolution in Buenos Aires Province, Argentina. Quat S Am Antarct Peninsula 16:297–321, A.A. Balkema, Rotterdam

Karanovic I (2012) Recent freshwater Ostracods of the World. Springer, London, p 608

Kidwell SM, Fursich FT, Aigner T (1986) Conceptual framework for the analysis and classification of fossil concentrations. Palaios 1:228–238

Laprida C (1998) Micropaleontological assemblages (Foraminiferida and Ostracoda) from late quaternary marginal marine environments (Destacamento Río Salado Formation), Salado Basin, Argentina. Rev Paléobiol 17:461–478

Laprida C (2006) Ostrácodos recientes de la llanura pampeana, Buenos Aires, Argentina: ecología e implicancias paleolimnológicas. Ameghiniana 43:181–204

Laprida C, Bertels-Psotka A (2003) Benthic foraminifers and paleoecology of a Holocene shelly concentration, Salado Basin, Argentina. Geobios 36:559–572

Laprida C, Valero-Garcés B (2009) Cambios ambientales de épocas históricas en la pampa bonaerense en base a ostrácodos: historia hidrológica de la laguna de Chascomús. Ameghiniana 46:95–111

Loeblich AR Jr, Tappan H (1988) Foraminiferal Genera and Classification. Van Nostrand Reinhold Company, New York, p 970

Márquez M, Ferrero L et al (2016) Holocene palaeoenvironmental evolution of the Pampa coastal plain (Argentina) based on calcareous microfossils. Revista Brasilera de Paleontologia 19:25–40

Martin JW, Davis GE (2001) An update classification of the recent Crustacea. Nat Hist Los Angeles County Contrib Sci 39:1–123

Murray J (2006) Ecology and applications of Benthic Foraminfera. Cambridge University Press, Cambridge, p 426

Piovano E, Ariztegui D et al (2009) Hydrological variability in South America below the Tropic of Capricorn (Pampas and eastern Patagonia, Argentina) during the last 13.0 ka. In: Vimeux F, Sylvestre F, Khodri M (eds) Past climate variability from the last glacial maximum to the Holocene in South America and surrounding regions: from the last glacial maximum to the Holocene, developments in Paleoenvironmental research series. Springer, Berlin, pp 323–351

Prado LF, Wainer I, Chiessi CM et al (2013) A mid-holocene climate reconstruction for eastern South America. Clim Past 9:2117–2133

Prieto AR, Stutz S et al (1998) Evidencias de la transgresión holocénica en la laguna Hinojales (37° 34′ S, 57° 27′ O). In: Actas de las 5º Jornadas Geológicas y Geofísicas Bonaerenses, Mar del Plata, pp 257–258, 9–11 Dec 1998

Prieto AR, Mourelle D et al (2016) Relative sea-level changes during the Holocene in the Rio de la Plata, Argentina and Uruguay: a review. Quatern Int 442:35–49

Reimer PJ, Bard E et al (2013) IntCal13 and Marine13 radiocarbon age calibration curves 0–50,000 years cal BP. Radiocarbon 55:1869–1887

Schnack E, Isla E et al (2005) Estratigrafía del Cuaternario marino tardío en la provincia de Buenos Aires. In: De Barrio R, Etcheverry R, Caballe M, Llambías E (eds) Relatorio del 16º Congreso Geológico Argentino, Geología y Recursos Minerales de la provincia de Buenos Aires, Asociación Geológica Argentina, Buenos Aires, pp 159–182

Sen Gupta BK (1999) Modern Foraminifera. Louisiana State University, Kluwer Academic Publishers, p 371

Shackleton NJ (1987) Oxygen isotopes, ice volume and sea level. Quatern Sci Rev 6:183–190

Shackleton NJ, Berger A, Peltier WR (1990) An alternative astronomical calibration of the lower Pleistocene timescale based on ODP Site 677. R Soc Edinburgh Trans Earth Sci 81:251–261

Stuiver M, Polach HA (1977) Discussion reporting of 14C data. Radiocarbon 19:355–363

Stuiver M, Reimer P (1993) Extended 14C data base and revised Calib 3.0 14C age calibration program. Radiocarbon 35:215–230

Van Morkhoven FP (1963) Post paleozoic Ostracoda their morphology, taxonomic and economic use, vol 2, generic descriptions. Elsevier, New York, USA, p 478

Whatley RC (1983) Some simple procedures for enhancing the use of Ostracoda in palaeoenvironmental analysis. Bull Nowegian Pet Direct 2:129–146

Whatley RC (1988) Population structure of ostracods: some general principles for the recognition of palaeoenvironments. In: De Deckker P, Colin J-P, Peypouquet JP (eds) Ostracoda in the earth sciences. Elsevier, New York, pp 245–256

Whatley RC, Cholich TC (1974) A New quaternary ostracod genus from Argentina. Paleontol 17:669–684

Whatley RC, Moguilevsky A (1975) The family Leptocytheridae in Argentinean waters. Bull Am Paleontol 65:501–527

Würdig N (1983) Fresh and brackish-water ostracodes from the east coast of the state of Río Grande do Sul, Brazil. In: Maddocks RF (ed) Appl of Ostracoda, Houston Geosciences, pp 591–604

Modern Benthic Foraminifera "Phylum Foraminifera (D'Orbigny 1826)" of the Panama Bight: A Census Report Based on Thanatocoenoses from the Continental Slope

Angélica Ballesteros-Prada(✉)

Instituto de Investigaciones en Biodiversidad y Medioambiente, Consejo Nacional de Investigaciones Científicas y Técnicas-Universidad Nacional del Comahue, Buenos Aires, Argentina
amballesterosprada@comahue-conicet.gob.ar

Abstract. This contribution compiles the published reports of foraminifera assemblages in Panama Bight (Eastern Equatorial Pacific), and new information obtained from the analysis of a sample set located on the continental margin of Colombian Pacific with the purpose to generate a taxonomic list, and to explore the presence and richness of the species throughout the region. A total of 417 species and 149 genera have been reported so far, mostly hyaline organisms. The species richness suggests a latitudinal variation, with the highest values in the continental margin of Ecuador and Colombia and the lowest values in the continental slope of Panama.

Keywords: Foraminifera · Continental margin · Taxonomy · Micropaleontology · Species description

1 Introduction

Foraminifera are a group of unicellular organisms that belong to the phylum Foraminifera, this group contains fossil organism as recent organisms and comprises more than 60,000 species, 2455 genera and 296 families (Loeblich and Tappan 1988). Its exoskeleton protects the cytoplasm from physical, chemical and biological stress and after the death of individuals gives them a greater capacity to be preserved in the fossil record (Armstrong and Brasier 2008; Murray 2006). For instance, in some modern and fossil sedimentological scenarios, Foraminifera shells may be the dominant component of the sediments.

Ecologically, benthic Foraminifera are characterized by living adhered or buried in the substrate and by their wide range of geographical and bathymetric distribution. Their lifestyle and type of reproduction allow large populations to appear (Armstrong and

Electronic supplementary material The online version of this chapter (https://doi.org/10.1007/978-3-030-02119-1_9) contains supplementary material, which is available to authorized users.

G. C. Cusminsky et al. (eds.), *Advances in South American Micropaleontology*,
Springer Earth System Sciences, https://doi.org/10.1007/978-3-030-02119-1_9

Brasier 2008). The density of the populations and the ease of sampling, together with the characteristics mentioned above, influence that in small samples, we find a sufficiently robust statistical base for environmental, paleoecological, paleoclimatic, and paleoceanographic analyses (Murray 1991; Murray 2006; Armstrong and Brasier 2008).

Even though molecular analyses now provide new relationships between species, it may affect the way the current classification is made (e.g., Pawlowski 2000; Pawlowski and Lecroq 2010). The taxonomic identification of these organisms is still carried out under microscopes, based mainly on the characteristics, composition and structure of the shell, the shape and arrangement of both the chambers, and the openings, and the ornamentation. However, this information is difficult to standardize, which generates problems in the identification and the appearance of a large number of synonyms. This does not allow a real understanding and synthesis of the information, being relegated exclusively to a local use. That is why the publication of taxonomic listings and taxonomic revisions still are of great importance, since they help to group the dispersed information.

Several works based on associations of benthic and planktonic foraminifers have been published for the Panama Basin, mostly with the objective of reconstructing the paleoceanography of the area from the early Holocene and analyzing the factors that influence the distribution and abundance of the benthic Foraminifera (e.g., Betancur and Martínez 2001, 2003; Martínez et al. 2006). Pioneer workers such as Bandy and Arnal (1957), Bandy and Rodolfo (19564), Golik and Phleger (1977) studied surface samples from the Panamanian Gulf, Central America (including the Panamanian Gulf) and the Perú–Chile trench area of the west coast of South America (covering the northern zone of the Panama basin), respectively. Each of these authors described the associations of benthic Foraminifera characteristic for zones at specific depths. More studies such as Parada and Tchegliakova (1990) described the coastal fauna of benthic holes in Gorgona Island in the Panama basin. Betancur and Martínez (2003) subdivided the basin from the microfaunal content of the surface samples into three regions: (1) north eutrophic region ($\sim$ 8° and 5.5° N), (2) mean mesotrophic region ($\sim$ 5.5 and 2° N), and (3) southern eutrophic region (south of 2 N). More recently, Patarroyo and Martínez (2013a, b) explored deep environments in the Colombian Pacific basin (500–4000 m). These authors propose the homogeneous presence throughout the basin of species of the families Bolivinidae and Buliminidae, as opposed to Uvigerinidae, which present their highest percentage abundances in the western margin of the Cocos ridge and in the northeast flank of the Carnegie ridge. Moreover, they propose *Cibicides wuellerstorfi* (Schwager) as a potential marker species (proxy) in the reconstruction of intense bottom currents in the ridges of Carnegie and Cocos for the Holocene. Moreover, the continental margin of Ecuador was of particular interest in the south of the Panama basin (Cruz et al. 1980; Gualancañay 1983, 1987, 1989, 2000, 2007, 2009, 2013). In these works, they analyzed the composition of the microfauna of benthic Foraminifera and its relationship with environmental variables.

As it can be seen, there are different works based on benthic Foraminifera for the Panama Bight. However, this information is dispersed in different publications and in some case, the taxonomy used is outdated. With the exception of the taxonomic and paleoecological work of the Caribbean and the equatorial Pacific, published by Bornmalm (1997), very few studies summarize the taxonomy for this zone. Therefore, the present work intends to integrate the previous information for the study area and thus to generate a taxonomic listing of the species of recent benthic Foraminifera in the Panama Bight. This inventory compiles the taxonomic information of the benthic foraminifers of the Panama Bight. This compilation will be useful to micropaleontologists, researchers and students interested in the biogeography and biodiversity of this group of organisms.

2 Study Area

The Panama Bight is an oceanographic area located in the Equatorial Pacific, to the north limits with the isthmus of Panama (9° N) and to the south with Puntilla Santa Elena (2° S). Longitudinally, it extends westward from the coasts of Panama, Colombia and Ecuador around 81°W (Rodríguez-Rubio and Schneider 2003) (Fig. 1). Geomorphologically, a short continental shelf (less than 50 km wide) between 0 and 200 m deep characterizes it, followed by a slope of variable width and an acute inclination (López 2009; García-Hansen and Cortés-Altamirano 2004). This region is characterized by highly atmospheric and oceanographic dynamics (Mitchell and Wallace 1992). It presents drastic climatic variations resulting from the influence of the Intertropical Convergence Zone, which controls the properties of seawater and the upper ocean circulation (Wyrtki 1966; Forsbergh 1969; Mitchell and Wallace 1992).

In addition, this area is oceanographically characterized by the presence of a cold-water tongue located south of the equatorial front, zonal currents (South Equatorial Current, Equatorial Countercurrent, Colombia Current), highly dynamic upwelling systems (Legeckis 1988; Tomczak and Godfrey 1994; Chávez et al. 1999; Rodríguez-Rubio and Stuardo 2002) and large river discharges (e.g., Restrepo and Kjerfve 2000).

3 Materials and Methods

This work was based on the information reported by Bandy and Arnal (1957); Bandy and Rodolfo (1964); Golik and Phleger (1977); Culver and Buzas (1987); Pascual and Pujos (2001); Betancur and Martínez (2003); Gualancañay (2013) and Patarroyo and Martínez (2013a, b). Added to the data obtained from the analysis of a new set of 31 samples in the continental margin of the Colombian Pacific (Table 1, Fig. 1).

This new data set was collected during 2009 by the company GEMS-HRT in the areas of technical evaluation (TEA's) of Egoró and Merayana, Colombian Pacific Coast. The samples were disintegrated and washed over a 63-μm sieve using only water. Between 100 and 300 benthic foraminifers per sample were picked and arranged in microslides. The information of the species reported in each work was compiled to create a presence–absence matrix. The taxonomy was homogenized, and the reworked species were eliminated. The suprageneric classification of this study was based on Loeblich and Tappan (1988) with modification following Pawlowski et al. (2013). The identification of the species was based mainly on Bornmalm (1997), and additional literature on Foraminifera from the Pacific, the Caribbean and deep settings worldwide (Lankford and Phleger 1973; Coulbourn 1980; Resig 1981; Hermelin 1989; Finger 1990, 1992; Burke et al. 1993; Ujie 1995; Zapata and Gutierrez 1995; Hess and Kunt

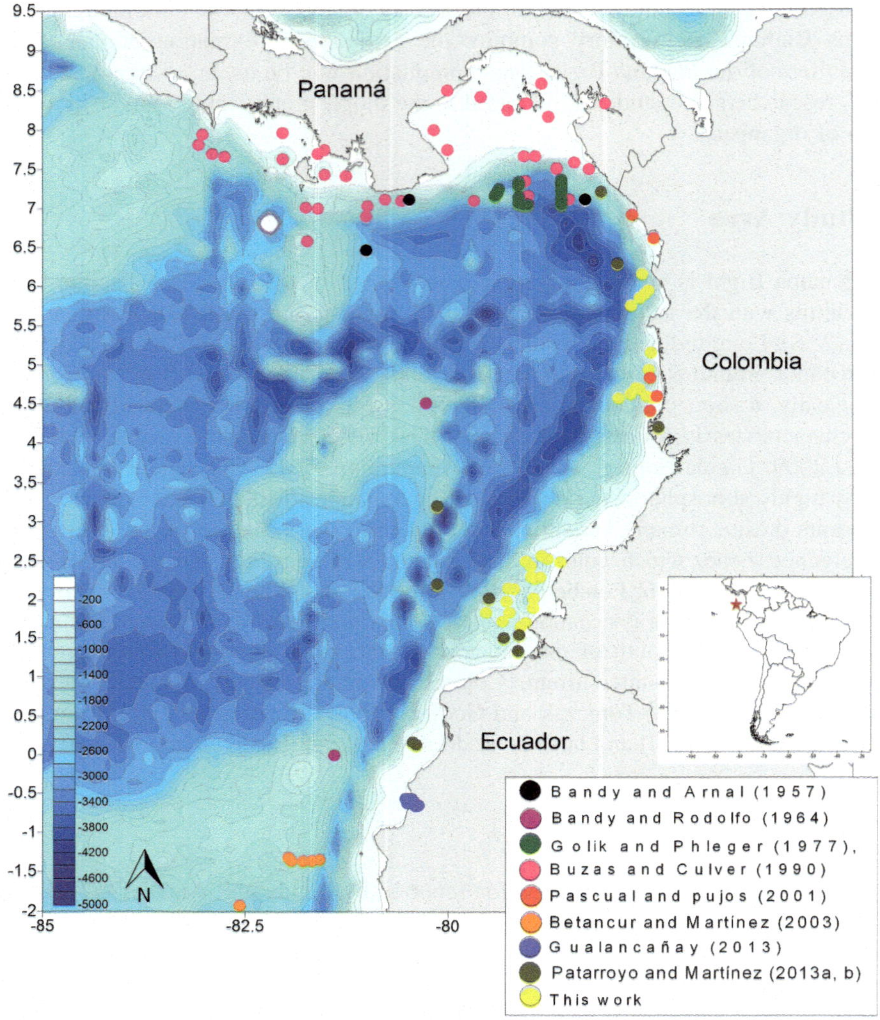

Fig. 1. Location map of the study area. The scale of depth is in meters. The star indicates the location of the Panama Bight in South America, and the different points of color indicate the location of the samples according to the different authors

1996; Scott et al. 2000; Holbourn and Henderson 2002; Betancur and Martínez 2003; Figueroa et al. 2005; Abu-Zied et al. 2011).

Optical microscope photographs were made using an AixoCam ICc3 mounted on a ZEISS Stereo Discovery. V8 stereomicroscope. Optical microscopic photographs were processed using the focus stacking software Helicon Focus to produce an image with a greater depth of field. SEM images were taken with FEI—Inspect S50 SEM at the Characterization of Materials Department, Bariloche Atomic Center (CAB), San Carlos de Bariloche, Argentina. Foraminifer specimens were stored in faunal slides and deposited in the collections of Foraminiferal Regional of the Micropaleontology Laboratory—Biostratigraphy Group, Colombian Petroleum Institute—Ecopetrol (Archivador N°09).

4 Results and Discussion

A total of 417 species were reported and grouped into 149 genera. Calcareous forms predominated, mainly represented by hyaline Foraminifera, followed by agglutinated forms and a very low proportion of porcelaneous forms (Table 2). The number of species reported by the authors for sampling point ranges from 1 to 75 per sample. The lowest species richness was observed in the continental margin of Panama, toward the northern zone of the Panama Bight and increases toward the south (Ecuador–Colombia) (Fig. 2). This increase may be related to the flow of allochthonous species transported by colder currents from the South, such as the Humboldt Current (Boltovskoy and Gualancañay 1975; Boltovskoy and Wright 1976; Murray 1991).

The faunal composition reported in these papers corresponds zoogeographically to the Panamanian province (Boltovskoy and Wright 1976; Crouch and Poag 1987), which was characterized by the presence of *Bifarina hancocki* (*Rectobolivina hancocki*), *Bolivina spissa*, *Bulimina pagoda*, *Cancris panamensis*, *Uvigerina incilis* and *Valvulineria araucana* (Boltovskoy and Gualancañay 1975). In addition, there were warm water species such as *Bolivina plicata*, *Bolivina pseudoplicata*, *Globobulimina pacifica* and *Nonion pizarrense* (Boltovskoy and Gualancañay 1975; Boltovskoy and Wright 1976; Murray 1991; Pascual and Pujos 2001).

The species with the wider distribution in the Panama Bight were *Globobulimina pacifica*, *Chilostomella oolina*, *Siphouvigerina proboscidea*, *Epistominella smithi*, *Uvigerina peregrina* and some specimens of the genus *Hoeglundina*. These cosmopolitan species have been reported as indicators of environments with high flow of organic matter and high productivity (Jorissen et al. 1995; de Mello e Sousa et al. 2006). *Uvigerina peregrina* presents a high range of bathymetric distribution (De Gupta 2010); however; it is abundant in sediments with high organic matter supplies (Altenbach and Sarnthein 1989; Altenbach et al. 1999; Fontanier et al. 2002) and areas of high productivity and low oxygenation (Eichler et al. 2014). The species *Cassidulina carinata* and *U. peregrina* have been proposed as opportunistic species that thrive in conditions of high flow of organic carbon, which makes them a good indicator of biological productivity (Hayward et al. 2006; Ovsepyan and Ivanova 2009). *Globobulimina pacifica* has been reported for low oxygenation sediments (Bernhard et al. 1997). Shepherd et al. (2007) report that *Epistominella smithi* increases in flowering seasons (Blooms) on the surface of the water column. This increase in percentage abundance is associated with the richness of degraded organic matter in Holocene sediments of the Coast Dome Rica (Ovsepyan and Ivanova 2009).

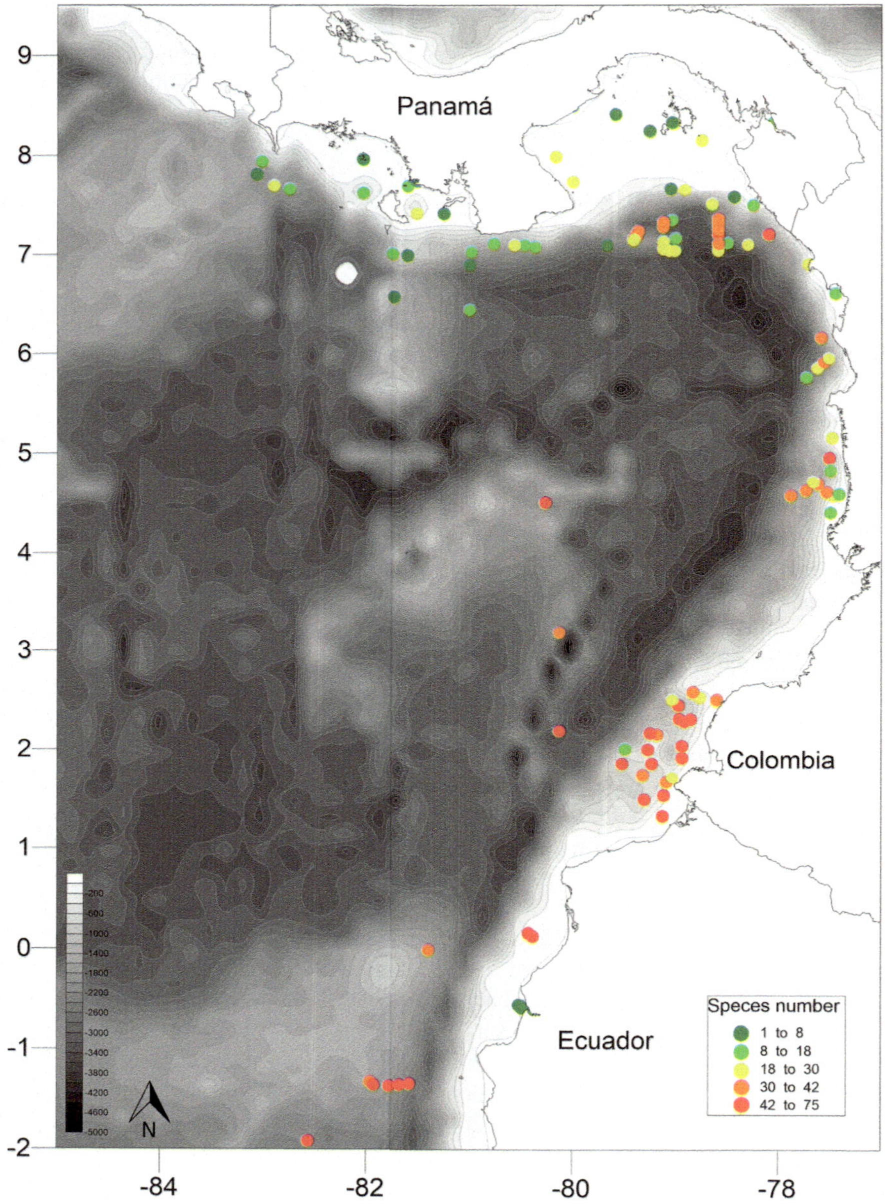

Fig. 2. Distribution of benthic foraminiferal species richness for sampling area

4.1 Taxonomic Reference List

Taxonomic list of species is reported by the different authors in the Panama Bight. The taxa illustrated in the figures come from study done by the author in the margin of the Colombian Pacific.

Phylum FORAMINIFERA (d'Orbigny 1826)
Class MONOTHALAMEA Haeckel 1862
Order ASTRORHIZIDA Lankester 1885
Suborder SACCAMMININA Lankester 1885
Superfamily PSAMMOSPHAEROIDEA Haeckel 1894
Family PSAMMOSPHAERIDAE Haeckel 1894
Subfamily PSAMMOSPHAERINAE Haeckel 1894
Genus *Psammosphaera* Schulze 1875
Psammosphaera fusca Schultz 1875
Figure 3, 1
Psammosphaera fusca Schultze 1875

Description

Test free, spherical or subspherical unilocular in cross section, large, up to 6 mm in diameter. Wall composed of sand grains on an inner organic coating; the grains may be of different sizes and may include calcareous Foraminifera shells. The primary opening is simple, rounded, elongated and absent to lip.

Remarks

The shells collected from the samples are mostly fractioned. The wall is composed of coarse grains of sand and minerals, and some shells of planktonic and benthic Foraminifera, an important character in the description of the species.

Genus *Lagenammina* Rhumbler 1911
Lagenammina atlantica (Cushman 1944)
Figure 8, 1
Proteonina atlantica Cushman 1944

Description

Test flask-shaped and symmetrical with an elongate neck, aperture terminal and wall covered by large quartz grain.

Class GLOBOTHALAMEA Pawlowski, Holzmann and Tyszka 2013
Subclasses TEXTULARIIA
Order TEXTULARIIDA
Suborder TEXTULARIINA Delage and Hérouard 1896
Superfamily EGGERELLOIDEA Cushman 1937
Family EGGERELLIDAE Cushman 1937
Subfamily EGGERELLINAE Cushman 1937
Genus *Martinottiella* Cushman 1933
Martinottiella communis (D'Orbigny 1826)

Figure 3, 2
Clavulina communis D'Orbigny 1846

Description

Test free, elongate; initial portion trochospiral, reduced to triserial and later uniserial; indistinct chambers in the early portion, more notorious in the uniserial portion. Sutures slightly depressed in uniserial stage. Wall finely agglutinated, smooth, finely

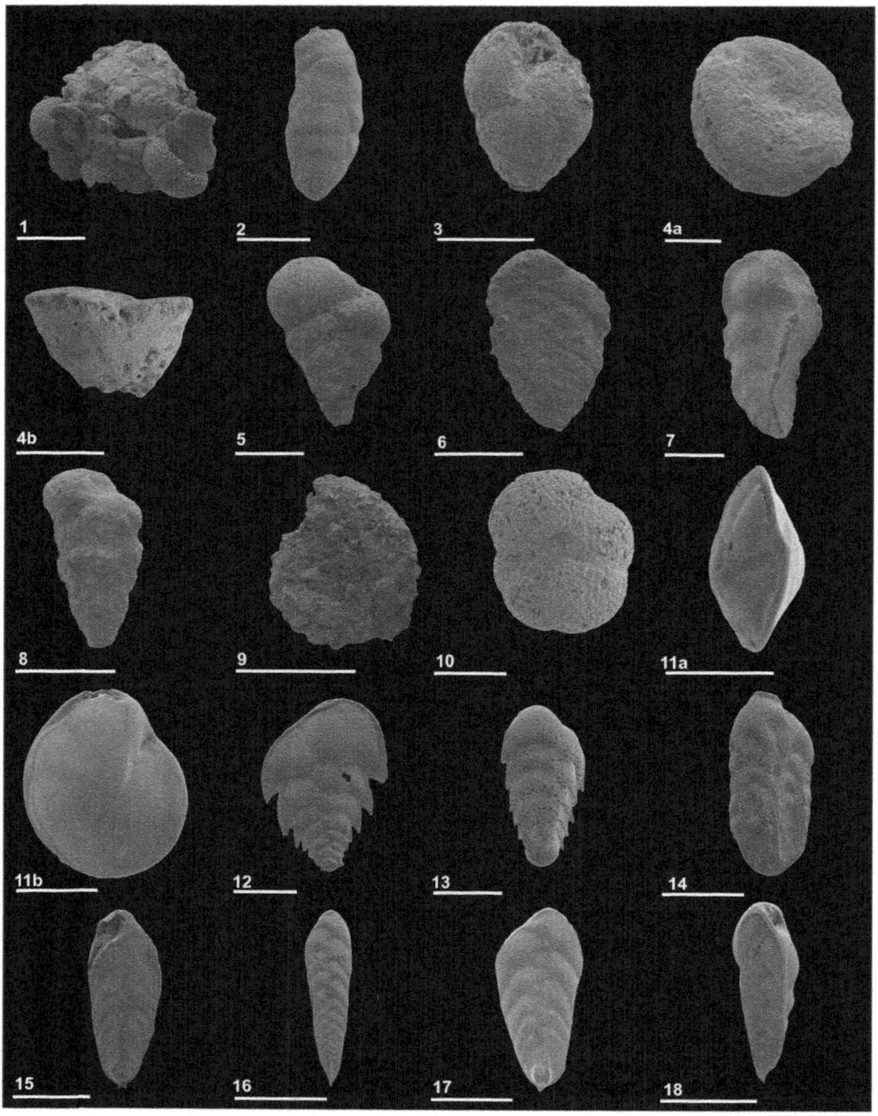

◄ **Fig. 3.** **1** *Psammosphaera fusca* (Schultze 1875), lateral view, 300 μm. **2** *Martinottiella communis* (d'Orbigny 1826), lateral view, 200 μm. **3** *Eggerella bradyi* (Cushman 1911), lateral view, 300 μm. **4** *Dorothia goesi* (Cushman 1911), a. lateral view, 500 μm; b. apertural view, 500 μm. **5** *Textularia catenata* Cushman 1911, lateral view, 300 μm. **6** *Textularia foliacea* Heron-Allen and Earland 1915, lateral view, 400 μm. **7** *Pseudogaudryina atlantica* (Bailey 1851), lateral view, 500 μm. **8** *Gaudryina* cf. *G. rudis,* lateral view, 500 μm. **9** *Discammina compressa* (Goës, 1882), lateral view, 400 μm. **10** *Portatrochammina pacifica* (Cushman 1925), latera view, 200 μm. **11** *Hoeglundina elegans* (d'Orbigny, 1826) a. edge view, 500 μm; b. ventral view, 500 μm. **12** *Bolivina alata* (Seguenza 1862), lateral view, 200 μm. **13** *Bolivina cochei* Cushman and Adams 1935, lateral view, 100 μm. **14** *Bolivina plicata* d'Orbigny 1839, lateral view, 200 μm. **15** *Bolivina spissa* Cushman 1926, lateral view, 200 μm. **16–17** *Bolivina subaenariensis* Cushman 1922, *a.* lateral view 400 μm, b lateral view, 300 μm. **18** *Bolivinita quadrilateral* (Schwager 1866), lateral view, 400 μm

perforate. Aperture a small round opening or slightly elliptical at the center of the ultimate chamber.

Remarks

Most of the specimens in the samples are divided into fractions; however, the diagnostic characters may be identified.

Genus *Eggerella* Cushman 1933
Eggerella bradyi (Cushman 1911)
Figure 3, 3
Verneuilina bradyi Cushman 1911

Description

Test free, conical, early stages trochospiral, later stages triserial; chambers inflated, increasing rapidly in size, so that the last whole one forms more than one-half of the test. Wall smooth, finely agglutinated. Sutures distinct and depressed. Aperture a low thin slit at the base of the ultimate chamber.

Genus *Dorothia* Plummer, 1931
Dorothia goesi (Cushman 1911)
Figure 3, 4a, 4b
Textularia goesii Cushman 1911

Description

Test broadly conical in outline, tapering abruptly to the apex with margins slightly convex, subcircular in end view; chambers low and broad without internal division; sutures limbate, distinct; wall more or less coarsely agglutinated, but usually smoothly finished; aperture linear in a depression at the base of the apertural chamber.

Family PSEUDOGAUDRYINIDAE Loeblich and Tappan 1985
Subfamily PSEUDOGAUDRYININAE Loeblich and Tappan 1985
Genus Pseudogaudryina Cushman 1936
Pseudogaudryina atlantica (Bailey 1851)

Figure 3, 7
Textularia atlantica Bailey 1851

Description

Test free, elongate, early stage triserial, later biserial, but test triangular throughout, so that the two series of angular biserial chambers are dissimilar, one series being roughly triangular in section and the other quadrangular in section, maintaining the triangular test shape. Wall agglutinated, canaliculate. Aperture an interiomarginal arch.

Superfamily: TEXTULARIOIDEA Ehrenberg 1838
Family: TEXTULARIIDAE Ehrenberg 1838
Subfamily TEXTULARIINAE Ehrenberg 1838
Genus *Textularia* Defrance 1824
Textularia catenata Cushman 1911
Figure 3, 5
Textularia catenata Cushman 1911

Description

Test free, elongate, biserial, lobate periphery. Chambers globular, increasing in size rapidly. Sutures depressed. Agglutinated wall, finely perforate. Aperture subterminal, rounded, in the lower base of the ultimate chamber.

Textularia foliacea Heron-Allen and Earland, 1915
Figure 3, 6
Textularia foliacea Heron-Allen and Earland 1915

Description

Test elongate, biserial throughout, narrow, laterally compressed, oval in end view; periphery rounded. Seven to nine pairs of chambers increasing regularly in width so as to give a leaf-shaped outline to the test; sutures depressed but at time very obscure, straight and oblique; wall coarsely agglutinated with material highly heterogeneous in size, surface rough; aperture small, a low arch at the base of the apertural face.

Order LITUOLIDA
Suborder VERNEUILININA Mikhalevich and Kaminski 2000
Family VERNEUILINIDAE Cushman 1911
Subfamily VERNEUILININAE Cushman 1911
Genus *Gaudryina* D'Orbigny 1839
Gaudryina cf. *G. rudis*
Figure 3, 8
Gaudryina rudis Wright 1900

Description

Test free, conical, biserial but early stage triserial; aperture interiomarginal. Wall coarsely agglutinates.

Remarks

This specimen is longer than the one reported for the Chinese sea (Lei and Li 2016). However, it is morphologically similar to that reported by Spezzaferri et al. (2014)

Suborder LITUOLINA Lankester 1885
Superfamily LITUOLOIDEA
Family DISCAMMINIDAE Mikhalevich 1980
Genus *Discammina* Lacroix 1932
Discammina compressa (Goës 1882)
Figure 3, 9
Lituola irregularis var. *compressa* Goës 1882

Description

Test planispiral, compressed, lenticular, somewhat involute, depressed in the umbilicus; peripheral margin acute or somewhat rounded; chambers and sutures not distinct; wall coarsely agglutinated, surface rough, color brown; aperture crescentic, at the base of the last-formed chamber.

Suborder TROCHAMMININA Saidova 1981
Superfamily TROCHAMMINOIDEA Schwager 1877
Family TROCHAMMINIDAE Schwager 1877
Subfamily TROCHAMMININAE Schwager 1877
Genus *Portatrochammina* Echols 1971
Portatrochammina pacifica (Cushman 1925)
Figure 3, 10
Trochammina pacifica Cushman 1925

Description

Test small, composed of several coils, 4–5 chambers in the last-formed one, all the chambers visible from the dorsal side, only those of the last-formed coil from the ventral side, which is umbilicate, periphery rounded. Sutures distinct but only slightly depressed, nearly radial; wall finely arenaceous, smoothly finished, of a yellowish-brown color, with a few larger fragments; aperture a narrow slit on the ventral side, at the base of the last-formed chamber.

Order LOFTUSIIDA Kaminski and Mikhalevich 2000
SuborderLOFTUSIINA Kaminski and Mikhalevich 2000
SuperfamilyLOFTUSIOIDEA Brady 1884
Family CYCLAMMINIDAE Marie 1941
Subfamily CYCLAMMININAE Marie 1941
Genus *Cyclammina* Brady 1879
Cyclammina cancellata Brady 1879
Figure 8, 2
Cyclammina cancellata Brady 1879

Description

Test large, biconvex, planispirally coiled, round to somewhat lobate in outline. Coiling is involute, with 10–16 chambers in the last whorl. Periphery rounded to subacute. Sutures are sigmoidal and slightly depressed, curved anteriorly. Apertural face is convex, not alveolar, and contains coarse agglutinated grains. Above the apertural zone of the convex apertural face is a flat, alveolar. Primary aperture a basal slit, with supplementary apertures in the apertural face and septae.

Order ROBERTINIDA Loeblich and Tappan 1984
Suborder ROBERTININA Loeblich and Tappan 1984
Superfamily CERATOBULIMINOIDEA Cushman 1927
Family EPISTOMINIDAE Wedekind 1937
Genus *Hoeglundina* Brotzen 1948
Hoeglundina elegans (d'Orbigny 1826)
Figure 3, 11a, 11b
Rotalia (Turbinuline) elegans d'Orbigny 1826

Description

Trochospiral, bioconvex test, rounded outline, medium size. Up to 10 chambers in the last whorl, with flush or slightly depressed sutures, curved on the spiral side, straight and oblique on the umbilical side, all of which is showed under the surface when the specimen is wet. Smooth calcareous hyaline walls. The primary aperture is small, interiomarginal and the supplementary aperture is an elongate lateromarginal slit, parallel to the peripheral keel and opening on the umbilical side. The primary aperture may be absent in some specimens.

Hoeglundina sp.1
Figure 8, 3a, 3b

Description

Trochospiral, bioconvex test, rounded outline, medium size, smooth calcareous hyaline walls. Up to 10 chambers in the last whorl, with flush or slightly depressed sutures, curved on the spiral side and unlike *Hoeglundina elegans* presents depressed sutures on the umbilical side. The aperture is an elongate lateromarginal slit, parallel to the peripheral keel and opening on the umbilical side.

Order ROTALIIDA Delage and Hérouard 1896
Superfamily SERIOIDEA Holzmann and Pawlowski 2017
Family BOLIVINITIDAE Cushman, 1927
Subfamily BOLIVINITINAE Cushman 1927
Genus BOLIVINA d'Orbigny 1839
Bolivina alata (Seguenza 1862)
Figure 3, 12
Vulvulina alata Seguenza 1862

Description

Test forms an elongate, strongly compressed, biserial series; lobulate in outline and elliptical in cross section with a sharply keeled periphery. The slightly inflated chambers increase rapidly in height and are separated by strongly curved, slightly depressed sutures. Chambers are prolonged by sharp, downward-pointing spines that merge into the peripheral keel. Chamber walls are calcareous and finely perforate. The primary aperture is a loop-shaped opening with a lip and an internal toothplate.

Bolivina cochei Cushman and Adams 1935
Figure 3, 13
Bolivina cochei Cushman and Adams 1935

Description

Test elongate, compressed, biserial series; lobulate in outline and elliptical in cross section. Sutures strongly curved, slightly depressed. Chambers are prolonged by little spines. Walls are calcareous and coarsely perforate. The primary aperture is a loop-shaped opening.

Bolivina plicata d'Orbigny 1839
Figure 3, 14
Bolivina plicata d'Orbigny, 1839

Description

Test elongate, broadly rounded, slightly tapering, periphery slightly lobulated, broadly oval in end view, biserial. Some 20 chambers in the adult test, chambers low, about twice as wide as high, regularly increasing in size, frontal overlap in the later chambers considerable, up to 2/3 of the test width, numerous crenulations present. Sutures undulate, aperture loop-shaped, often somewhat twisted, surrounded by a highly raised everted apertural lip, lip asymmetrical, upper part of the toothplate not always apparent in the aperture, a single ridge commonly running from the apex of the aperture toward and somewhat onto the previous chamber. Hyaline wall, radial, finely perforate; ornamented by the raising of the crenulations due to the deposition of secondary lamellae, coalescing to form a few sinuous costae, most prominent in the lowest third of the test, the remnants of the crenulations and the intervening arches of the sutures.

Bolivina spissa Cushman, 1926
Figure 3, 15
Bolivina spissa Cushman, 1926

Description

Test elongate, compressed, biserial series; lobulate in outline and test free, biserial, compressed, widest near apertural end; periphery acute, seven to nine pairs of chambers in the adult, increasing gradually in size. Sutures strongly curved, slightly depressed. Walls calcareous, hyaline, coarsely perforate. Aperture elongate, narrow, surrounded by a rim, with narrow toothplate.

Bolivina subaenariensis Cushman, 1922
Figure 3, 16, 17
Bolivina subaenariensis Cushman, 1922

Description

Test free, biserial, compressed, widest near apertural end, twice as long as broad; periphery acute, carinate, seven to nine pairs of chambers in the adult, increasing gradually in size. Sutures strongly curved, distinct, slightly depressed. Wall calcareous, hyaline, finely perforate, ornamented with two or more costae, the center two extending from the initial, end two-thirds the length of test; additional costae shorter, on either side of the other two. Aperture elongate, narrow, surrounded by a rim, with narrow toothplate.

Family BOLIVINITIDAE Cushman 1927
Subfamily BOLIVINITINAE Cushman 1927
Genus *Bolivinita* Cushman 1927
Bolivinita quadrilatera (Schwager 1866)
Figure 3, 18
Textilaria quadrilatera Schwager 1866

Description

Test forms an elongate, tapered compressed, biserial series, quadrangular in cross section with concave side, and a keeled periphery. The low, broad chamber increased gradually in side and distinct, oblique and raised sutures separate it. Frontal overlap of the chamber occupies the middle third of the test width. Chamber walls are calcareous, finely perforate, and smooth or with a few longitudinal striate in the early stage, and ornamented by lateral keels on the outer edges. A spine is often present on the proloculus. The oval primary aperture starts slightly above the basal suture, extends up the apertural face and is surrounded by curving lip that merges into the upper edge of the toothplate, thus delimiting a distinct sulcus.

Family CASSIDULINIDAE d'Orbigny 1839
Subfamily CASSIDULININAE d'Orbigny 1839
Genus *Islandiella* Nørvang 1958
Islandiella californica (Cushman and Hughes 1925)
Figure 4, 1
Cassidulina californica Cushman and Hughes 1925

Description

Test free, subglobular in side view, coiled biserially, ovate in transverse section. Chambers 4–5 pairs in last whorl, increasing gradually in size as added. Sutures narrow, smooth, curved. Wall calcareous, hyaline, smooth, finely perforate. Aperture narrow, in a depression of the apertural face, varying in shape from a straight, narrow slit to a chevron-shaped opening.

Genus *Cassidulina* d'Orbigny 1826

Cassidulina delicata Cushman 1927
Figure 4, 2
Cassidulina delicata Cushman 1927

Description

Test free, compressed, small; chambers biserially arranged and enrolled, tightly coiled. Chambers elongate, about four pairs of chambers in the final whorl, slightly inflated; periphery rounded in side view, acute with a narrow keel. Sutures distinct, slightly depressed, strongly curved. Wall calcareous, hyaline, smooth, slightly perforate. Aperture an elongate broad slit parallel to the periphery along the suture of the last chamber.

Cassidulina carinata Silvestri 1896
Figure 4, 3
Cassidulina carinata Silvestri 1896

Description

Test free, compressed, small; chambers biserially arranged and enrolled, tightly coiled. Chambers elongate, about four pairs of chambers in the final whorl, slightly inflated; periphery slightly lobulated in side view, acute with a narrow keel. Sutures distinct, slightly depressed. Wall calcareous, hyaline, smooth, coarsely perforate. Aperture an elongate broad slit parallel to the periphery along the suture of the last chamber, with a tooth.

Cassidulina auka Boltovskoy and Theyer 1970
Figure 4, 4
Cassidulina auka Boltovskoy and Theyer 1970

Description

Test free, compressed, chambers biserially arranged and enrolled, tightly coiled. Chambers elongate, slightly inflated; periphery rounded in side view. Sutures distinct, slightly depressed. Wall calcareous, hyaline, smooth, slightly perforate. Aperture an elongate broad slit parallel to the periphery along the suture of the last chamber.

Genus *Evolvocassidulina* Eade 1967
Evolvocassidulina tenuis (Phleger and Parker 1951)
Figure 4, 6
Cassidulinoides tenuis Phleger and Parker 1951

Description

Test free, elongate; length 2.5 times the width, early portion coiled; chambers biserially enrolled, later portion uncoiled, and circular to ovate in transverse section. Chambers distinct slightly inflated, increasing gradually in size. Sutures distinct, slightly depressed. Wall calcareous, hyaline, smooth, perforated with small, narrow pores. Aperture an elongate opening in a depression, extending from base of ultimate chamber to a point near the apex of test.

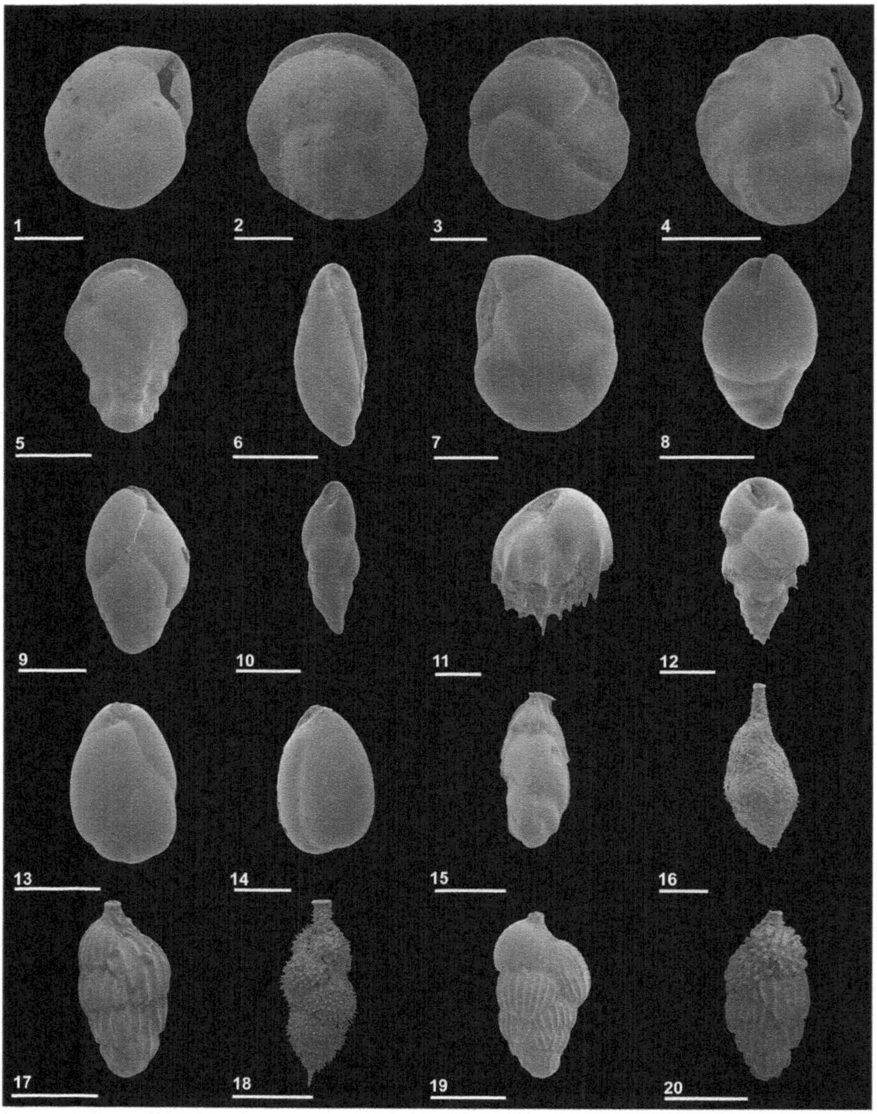
1
2
3
4
5
6
7
8
9
10
11
12
13
14
15
16
17
18
19
20

◄ **Fig. 4. 1** *Islandiella californica* (Cushman and Hughes 1925), lateral view, 200 μm. **2** *Cassidulina delicata* Cushman 1927, lateral view, 100 μm. **3** *Cassidulina carinata* Silvestri 1896, lateral view, 100 μm. **4** *Cassidulina auka* Boltovskoy and Theyer 1970, lateral view, 300 μm. **5** *Ehrenbergina pacifica* Cushman 1927, lateral view, 200 μm. **6** *Evolvocassidulina tenuis* (Phleger and Parker 1951), lateral view, 200 μm. **7** *Globocassidulina subglobosa* (Brady 1881), lateral view, 200 μm. **8** *Fursenkoina cornuta* (Cushman 1913), lateral view, 300 μm. **9** *Fursenkoina glabra* (Cushman and Wickenden 1929), lateral view, 500 μm. **10** *Fursenkoina complanata* (Egger 1893), lateral view, 200 μm. **11** *Bulimina inflata* Seguenza 1862, lateral view, 100 μm. **12** *Bulimina pagoda* Cushman 1927, lateral view, 100 μm. **13** *Globobulimina affinis* (d'Orbigny 1839), lateral view, 500 μm. **14** *Globobulimina pacifica* Cushman 1927, lateral view, 200 μm. **15** *Trifarina angulosa* (Williamson 1858) lateral view, 400 μm. **16** *Uvigerina auberiana* d'Orbigny 1839 lateral view, 100 μm. **17** *Uvigerina hollicki* Thalmann 1950, lateral view, 200 μm. **18** *Uvigerina hispida* Schwager 1866, lateral view 300 μm. **19** *Uvigerina peregrine* Cushman 1923, lateral view, 400 μm. **20** *Uvigerina hispidocostata* Cushman and Todd 1945, lateral view, 400 μm

Genus *Globocassidulina* Voloshinova 1960
Globocassidulina subglobosa (Brady 1881)
Figure 4, 7
Cassidulina subglobosa Brady 1881

Description

Test small, subglobular, biserial series, with a subcircular cross section and a rounded periphery. Slightly depressed, curved sutures separate the inflated chambers. Chamber walls are calcareous, finely perforate, and smooth. The primary aperture is a straight or curved slit extending up the apertural face, with a toothplate on the posterior side.

Subfamily EHRENBERGININAE Cushman 1927
Genus *Ehrenbergina* Reuss, 1850
Ehrenbergina pacifica Cushman 1927
Figure 4, 5
Ehrenbergina pacifica Cushman 1927

Description

Test triangular in front view, chambers numerous, low and broad, dorsal side convex, ventral side with a narrow median furrow which may be entirely closed; sutures distinct, on the dorsal side flush with the surface, on the ventral side depressed; periphery with long

Subfamily FURSENKOININAE Loeblich and Tappan 1961
Genus *Fursenkoina* Loeblich and Tappan 1961
Fursenkoina cornuta (Cushman 1913)
Figure 4, 8
Virgulina cornuta Cushman 1913

Description

Test pointed at the initial end, broadly rounded at the apertural end, whole test curved, irregularly biserial, last-formed chambers very tumid, aperture a comma-shaped, opening with a surrounding raised portion; wall thin and translucent, smooth.

Fursenkoina glabra (Cushman and Wickenden 1929)
Figure 4, 9
Bulimina patagonica d'Orbigny var. *glabra* Cushman and Wickenden 1929

Description

Test elongate, spire with five evolutions; chamber inflated, wall shiny, white, translucent or opaque, finely perforate; sutures strongly depressed; aperture ovate of medium size.

Fursenkoina complanata (Egger 1893)
Figure 4, 10
Virgulina complanata Egger 1893

Description

Test exhibits an elongate, biserial, slightly compressed, slightly twisted, fusiform shape with a pointed initial portion (often bearing a spine), a rounded apertural end, and a subcircular cross section. The moderately inflated chambers increase rapidly in height and are separated by distinct, curved, depressed sutures. Chamber walls are calcareous, smooth, and finely perforate. The primary aperture is an elongate oval slit extending up the face of the final chamber, with a toothplate.

Superfamily BULIMINOIDEA Jones 1875
Family BULIMINIDAE Jones 1875
Genus *Bulimina* d'Orbigny 1826
Bulimina inflata Seguenza 1862
Figure 4, 11
Bulimina inflata Seguenza 1862

Description

Inflated cameras that rapidly increase in size, the last spire covers about half the chambers, the ribs are strong and can become short spines; these are limited to the bottom of the chambers and are not continued in the previous round.

Bulimina pagoda Cushman 1927
Figure 4, 12
Bulimina pagoda Cushman, 1927

Description

Test fusiform, greatest breadth somewhat below the middle, nearly twice as long as broad, initial end acute, about five whorls in the adult. Chambers distinct, somewhat inflated, the last three in the adult making up much the larger part of the test, increasing rapidly in size as added, much overlapping; sutures distinct, only slightly depressed; wall mostly smooth, finely perforate, with a small number of short spines. At the basal margin of the chambers; aperture elongate, with a distinct, raised lip.

Genus *Globobulimina* Cushman 1927
Globobulimina affinis (d'Orbigny 1839)
Figure 4, 13
Bulimina affinis d'Orbigny 1839

Description

Test free, triserial, pyriform in side view, usually broadest at the middle of the test, ovate in cross section. Chambers distinct, the last three making up about three-fourths of the exterior. Sutures distinct, very weakly depressed. Wall calcareous, smooth, finely perforate, no spines or denticles. Aperture open or nearly so, the tip of the tongue fan-shaped the free border of the aperture with a weakly marked collar.

Globobulimina pacifica Cushman 1927
Figure 4, 14
Globobulimina pacifica Cushman 1927

Description

Test free, triserial, pyriform in side view, broadest are base of the test, decreasing in width toward the apertural end, ovate in cross section. Chambers inflated, strongly overlapping, last whorl only visible in side view, earlier chambers seen in basal view. Sutures distinct flush with surface. Wall calcareous, smooth, finely perforate. Aperture loop-shaped, terminal, with a broad projecting toothplate.

Genus *Protoglobobulimina* Hofker 1951
Protoglobobulimina pupoides (D'Orbigny 1846)
Figure 8, 4
Bulimina pupoides D'Orbigny 1846

Description

Test oval or ovate, with short segment but slight inflated, twice as long as board, last whorl forming one-half or more of the test, circular in transversal section; chambers somewhat inflated, strongly overlapping previous one; sutures distinct, slightly depressed, oblique; wall calcareous, hyaline, smooth, finely perforate; aperture loop-shaped, with a lip and distinct toothplate.

Family UVIGERINIDAE Haeckel 1894
Subfamily ANGULOGERININAE Galloway 1933
Genus *Trifarina* Cushman 1923
Trifarina angulosa (Williamson 1858)
Figure 4, 15
Uvigerina angulosa Williamson 1858

Description

Test elongate, sharply triangular in section, initial stage triserial, later with tendency to an irregularly uniserial arrangement; apertural end pointed; angles carinate, sutures curved, oblique, and slightly depressed; wall with fine perforations at the end of raised pustules. Surface with a few longitudinal costae around the keel; aperture terminal, ovate, produced on a neck and bordered by a narrow lip, provided with an internal toothplate.

Subfamily SIPHOUVIGERINA Parr 1950
Genus *Siphouvigerina* Parr 1950
Siphouvigerina hispida (Schwager 1866) Figs. 8, 5; 4, 18
Uvigerina hispida Schwager 1866

Description

Test forms an elongate, fusiform, triserial series, approximately two and a half times as long as broad, widest in the middle, tapered in outline in microspheric forms, subcylindrical in outline in megalospheric forms, and with a circular cross section in both. A basal spine is present in some well-preserved specimens. The inflated chambers increase markedly in height and are separated by distinctly depressed sutures. Chamber walls are calcareous, perforate, and ornamented by dense, acicular spines (microspheric forms) or blunt spines to coarse papillae (megalospheric forms). The primary aperture is a terminal, round opening at the end of a short neck, bordered by a phialine lip, and with an internal toothplate.

Genus *Uvigerina* D'Orbigny 1826
Uvigerina auberiana D'Orbigny 1839
Figure 4, 16
Uvigerina auberiana D'Orbigny 1839

Description

Test free, initial portion triserial, later twisted biserial, elongate, circular in cross section; initial end blunt and sometimes with small spine. Chambers inflated, narrowing toward the apertural neck. Wall calcareous, hyaline, finely perforate, ornamented with small, sharp spines, randomly arranged over the test.

Uvigerina hollicki Thalmann 1950
Figure 4, 17
Uvigerina hollicki Thalmann 1950

Description

Test free, triserial, 5–2 times as long as broad, widest near middle of test. Chambers inflated, embracing. Sutures depressed, indistinct. Wall calcareous, perforate, ornamented by numerous blade-like costae on each chamber. Aperture terminal, on a short neck, often with a phialine lip.

Uvigerina hootsi Rankin 1934
Figure 8, 6
Uvigerina hootsi Rankin 1934

Description

Test about twice as long as board, greatest width above the middle; chambers very distinct, strongly inflated, of uniform shaped, gradually increasing in size as added; sutures very distinct, depressed; wall smooth, matte or the earliest portion slightly ovate; aperture very small, with a very short cylindrical neck with a thin flaring lip.

Uvigerina peregrina Cushman 1923
Figure 4, 19
Uvigerina peregrina Cushman 1923

Description

Test free, triserial, elongate, about 2–5 times, as long as broad, widest near middle of test. Chambers inflated, embracing. Sutures indistinct, depressed. Wall calcareous, perforate, ornamented by numerous high blade-like costae on each chamber. Aperture circular at end of short neck, with a phialine lip.

Uvigerina hispidocostata Cushman and Todd 1945
Figure 4, 20
Uvigerina hispidocostata Cushman and Todd 1945

Description

Test free, triserial, elongate, about 2–5 times, as long as broad, widest near middle of test. Chambers inflated, embracing. Sutures indistinct, depressed. Wall calcareous, perforate, ornamented by numerous high blade-like costae on each chamber; costae break up into spinose or irregular short segments toward the aperture lend. Aperture circular at end of short neck, with a phialine lip.

Superfamily DISCORBOIDEA Ehrenberg 1838
Family CANCRISIDAE Chapman, Parr and Collins 1934
Genus *Cancris* Montfort 1808
Cancris auricula (Fichtel and Moll 1798)
Figure 5, 1
Nautilus auricula Fichtel and Moll 1798

Description

Test auriculate, lenticular in section, chambers increasing rapidly in breadth as added a flaring trochospiral oil, about six to seven in final whorl. Periphery smooth, with feathered margin. Sutures depressed, arched on the spiral side, nearly radial around the umbilicus on the opposite side. Wall calcareous, perforate, but with an imperforate area above the aperture continuing into the imperforate umbilical apertural flap, surface smooth. Aperture a low interiomarginal opening on the umbilical side.

Cancris panamensis Natland 1938
Figure 5, 2a, 2b, 2c
Cancris panamensis Natland 1938

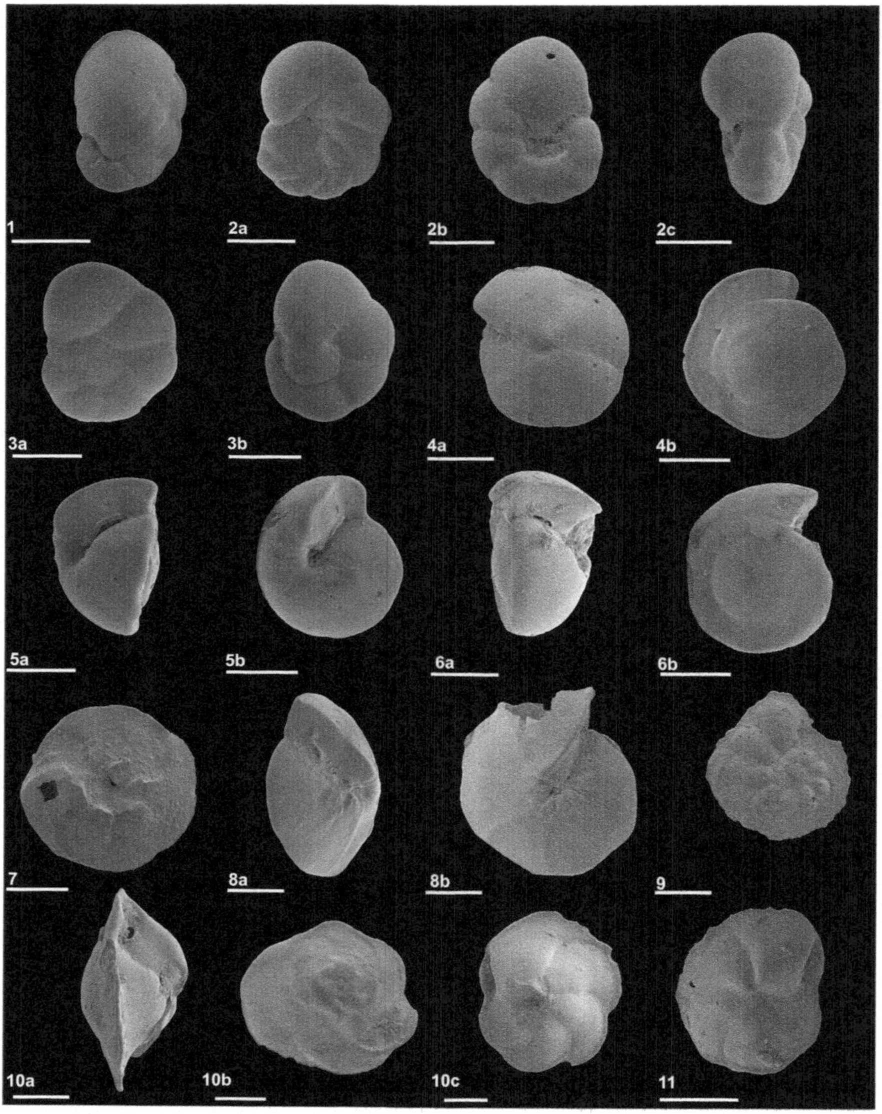

Fig. 5. **1** *Cancris auricula* (Fichtel and Moll 1798) ventral view, 400 µm. **2** *Cancris panamensis* Natland 1938 a. dorsal view, 300 µm b. ventral view, 300 µm c. edge view, 300 µm. **3** *Valvulineria glabra* Cushman 1927, a. dorsal view, 300 µm b. ventral view, 300 µm. **4** *Oridorsalis umbonatus* (Reuss 1851), 200 µm a ventral view, b dorsal view. **5** *Gyroidina altiformis* Stewart and Stewart 1930, 200 µm a. edge view, b. ventral view. **6** *Gyroidina neosoldanii* Brotzen 1936, 200 µm a. edge view, b. dorsal view. **7** *Gyroidina rothwelli* Natland 1950, 200 µm ventral view. **8** *Gyroidina soldanii* d'Orbigny 1839, 100 µm a. edge view, b. ventral view. **9** *Laticarinina pauperata* (Parker and Jones 1865), lateral view, 500 µm. **10** *Epistominella smithi* (Stewart and Stewart 1930) a. edge view, 100 µm b. dorsal view, 100 µm c. ventral view. **11** *Epistominella* sp.1, ventral view, 100 µm

Description

Test trochospiral, periphery ovate, planoconvex and sharp edges from the apertural view; wall smooth, more than six chambers in the last whorl; sutures depressed and curved on spiral view, last chamber elongate and inflated.

Genus *Valvulineria* Cushman 1926
Valvulineria glabra Cushman 1927
Figure 3, 3a, b
Valvulineria glabra Cushman 1927

Description

Test free, biconvex, periphery broadly rounded, longer than broad inside view, slightly lobate. Chambers distinct, 7–8 in the last whorl, increasing rapidly in size as added, last chambers few inflated. Wall calcareous, smooth, hyaline, coarsely perforate except for the face of the ultimate chamber and the umbilical flap, which are imperforate. Aperture interiomarginal, umbilical–extraumbilical, partially covered by a large projecting umbilical lip.

Family ROSALINIDAE Reiss 1963
Subfamily RUPERTININAE Loeblich and Tappan 1961
Genus RUPERTINA Loeblich and Tappan 1961
Rupertina stabilis (Wallich 1877)
Figure 7, 2a, 2b
Rupertia stabilis Wallich 1877

Description

Test attached by a prominent basal disk; chambers trochospiral in the early stage, later loosely coiled, growing spirally around a solid central column; chambers inflated, sutures depressed; wall thick, coarsely perforate; aperture a low umbilical slit bordered by a distinct lip.

Superfamily CHILOSTOMELLOIDEA Brady 1881
Family ALABAMINIDAE Hofker 1951
Genus *Oridorsalis* Andersen 1961
Oridorsalis umbonatus (Reuss 1851)
Figure 5, 4a, 4b
Rotalina umbonata Reuss 1851

Description

Test free, trochospiral, biconvex, peripheral margin carinate with a distinct keel, circular in side view. All chambers visible on spiral side, increasing gradually as added. Sutures radial and slightly curved on spiral side, strongly sinusoidal on umbilical side. Wall calcareous, hyaline, smooth, finely perforate. Aperture an interiomarginal slit extending from the umbilical area to the periphery, supplementary apertures on the spiral side at proximal end of the sutures of the last-formed chambers, similar apertures on the umbilical side are located along the suture between ultimate and penultimate chambers.

Family GAVELINELLIDAE Hofker 1956
Subfamily GAVELINELLINAE Hofker 1956
Genus *Gyroidina* D'Orbigny 1826
Gyroidina altiformis Stewart and Stewart 1930
Figure 5, 5a, 5b
Gyroidina altiformis Stewart and Stewart 1930

Description

Test free, trochospiral, planoconvex, spiral side flat to slightly convex, umbilical side convex with deep umbilicus. Circular in outline, peripheral edge subacute. Seven to ten chambers in the final whorl. Sutures distinct, limbate, slightly depressed and radial on the umbilical side, raised and oblique on the spiral side. Wall calcareous, hyaline, smooth, finely perforate. Aperture allows interiomarginal slit extending from periphery to umbilicus.

Gyroidina neosoldanii Brotzen 1936
Figure 5, 6a, 6b
Gyroidina neosoldanii Brotzen 1936

Description

Test trochospiral; spiral side evolute and flattened, umbilical side involute and highly convex with subangular umbilical shoulder bordering the open umbilicus; sutures radial, straight to slightly curved, flush, becoming depressed toward the umbilicus. The periphery is broadly truncate; wall finely perforate, surface smooth; aperture a short equatorial and interiomarginal slit, bordered by a narrow lip, an umbilical flap extending into the umbilicus from each chamber partially covers a small secondary aperture.

Gyroidina rothwelli Natland 1950
Figure 5, 7
Gyroidina rothwelli Natland 1950

Description

Test small, trochoid, involute spire, planoconvex, with dorsal side flat, ventral side arched, with broad deep umbilical pit; 7–9 chambers in last whorl; chambers slightly peaked to form a row of knobs around the umbilicus. Sutures oblique on both sides, distinct, often with a sharp crenulation near top ends on ventral side; slightly limbate on dorsal side, moderately depressed on ventral side; periphery with narrow carina, wall calcareous, finely perforated; aperture near the periphery along the lower margin of the septal face of the last chamber.

Gyroidina soldanii D'Orbigny 1839
Figure 5, 8a, 8b
Gyroidina soldanii D'Orbigny 1839
Family DISCORBINELLIDAE Sigal 1952

Subfamily DISCORBINELLINAE Sigal 1952
Genus *Laticarinina* Galloway and Wissler 1927
Laticarinina pauperata (Parker and Jones 1865)
Figure 5, 9
Pulvinulina repanda var. *menardii pauperata* Parker and Jones 1865

Description

Test free, slightly trochoid, compressed, biconvex to planoconvex in edge view, spiral side flattened and evolute, ventral side convex and involute. Surrounded with a broad, thin, transparent keel. Sutures radial, depressed on ventral side, flush on spiral side. Wall calcareous, hyaline, smooth, finely perforate. Aperture on the dorsal side at the inner margin of the last-formed chamber.

Family PSEUDOPARRELLIDAE Voloshinova 1952
Subfamily PSEUDOPARRELLINAE Voloshinova 1952
Genus *Epistominella* Husezima and Maruhasi 1944
Epistominella smithi (Stewart and Stewart 1930)
Figure 5, 10a, 10b, 10c
Pulvinulinella smithi Stewart and Stewart 1930

Description

Test trochoid, planoconvex, with flattened dorsal side and moderately convex ventral side, thinly keeled; chambers distinct, slightly inflated (on ventral side) with five to seven chambers visible in final whorl; sutures oblique, curved and limbate on dorsal side, ventral sutures curved and slightly depressed. Walls calcareous, hyaline and finely perforate; aperture narrow, elongate, lipped slit, almost parallel to peripheral keel, on umbilical (ventral) side.

Epistominella sp.1
Figure 5, 11

Description

Test trochoid, planoconvex, with flattened dorsal side and highly convex ventral side, thinly keeled; chambers distinct, with five to seven chambers visible in final whorl; sutures oblique, curved and limbate on dorsal side, ventral sutures curved and slightly depressed; walls calcareous, hyaline and finely perforate; aperture narrow, elongate, lipped slit, almost parallel to peripheral keel, on umbilical (ventral) side.

Superfamily NONIONOIDEA Schultze 1854
Family NONIONIDAE Schultze 1854
Subfamily NONIONINAE Schultze 1854
Genus *Nonion* Montfort 1808
Nonion pizarrensis Berry 1928
Figure 6, 2a, 2b
Nonion pizarrensis Berry 1928

Description

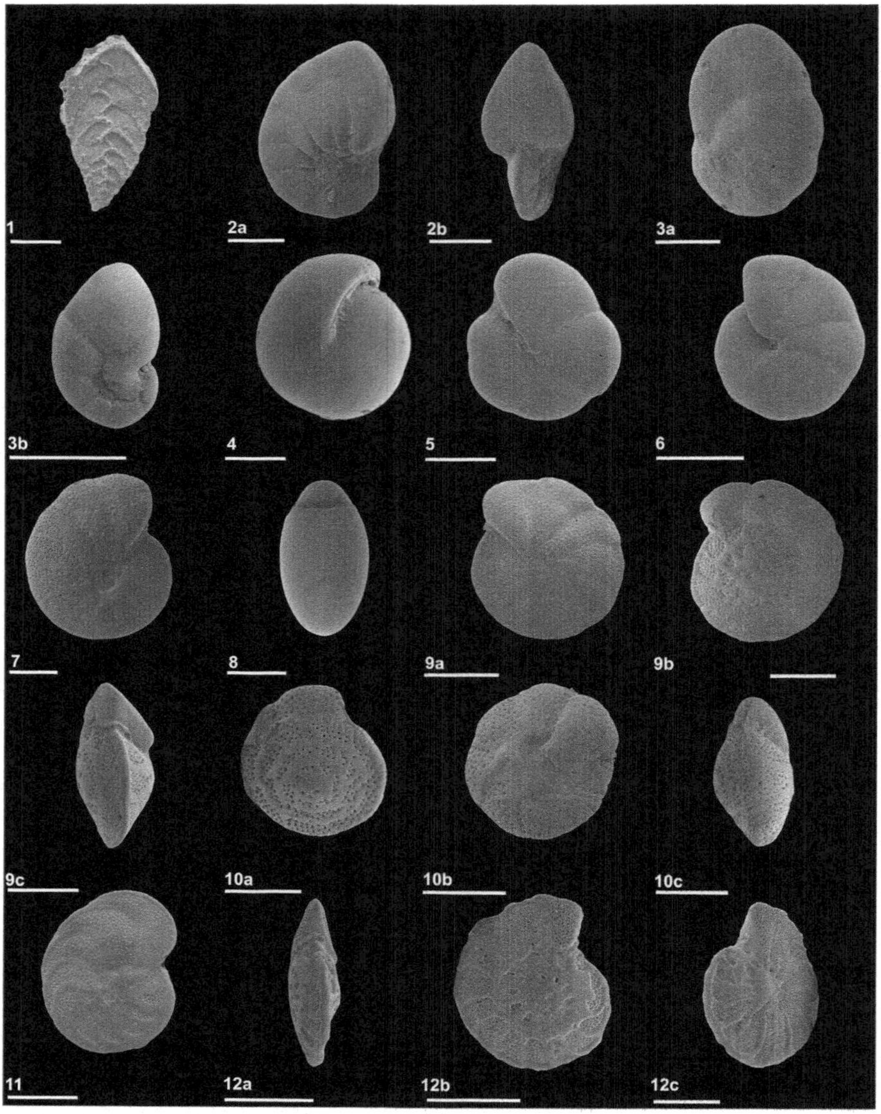

Fig. 6. **1** *Loxostomum minuta* (Natland 1938), lateral view, 100 µm. **2** *Nonion pizarrensis* Berry 1928, a. lateral view, 200 µm b. Apertural view, 200 µm. **3** *Nonionella auris*(D'Orbigny 1839), a. dorsal view, 200 µm, b. ventral view, 200 µm. **4** *Pullenia bulloides* (D'Orbigny 1846), lateral view, 400 µm. **5** *Pullenia quinqueloba* (Reuss 1851), lateral view, 200 µm. **6** *Pullenia subcarinata* (d'Orbigny 1839), lateral view, 400 µm. **7** *Melonis affinis* (Reuss 1851) lateral view, 100 µm. **8** *Chilostomella oolina* Schwager 1878, lateral view 200 µm. **9** *Cibicides kullenbergi* Parker 1953, 300 µm a. ventral view, b. dorsal view, c. edge view. **10** *Cibicidoides mundula* (Brady, Parker and Jones 1888) 200 µm a. dorsal view, b. ventral view, c. edge view. **11** *Cibicidoides wuellerstorfi* (Schwager 1866), lateral view, 300 µm. **12** *Planulina limbata* Natland 1938, 400 µm a. edge view, b. dorsal view, c. ventral view

Test of medium size, planispiral involute, asymmetrical and compressed, peripheral margin slightly angled, acute and slightly flaring around aperture, umbilical region depressed and filled with bead-like granular material; chambers low, numerous, 10 to 14 in final whorl, rapidly increase in size as added. Sutures depressed curved; wall calcareous, smooth, finely perforate; aperture elongate, narrow, interomarginal with a small equatorial opening at base of apertural face.

Genus *Nonionella* Rhumbler 1949
Nonionella auris (D'Orbigny 1839)
Figure 6, 3a, 3b
Valvulina auris D'Orbigny 1839

Description

Test asymmetrical, slightly trochoid, the spire not raised, periphery broadly rounded chambers 9 to 11, low and broad, very distinct, slightly inflated, the last-formed chamber having an enlarged portion extending over the umbilicus on the ventral side. Sutures distinct, depressed, gently curved; wall smooth, polished, very finely perforate; aperture at the base of the last-formed chamber extending from the periphery ventral, low and broad.

Subfamily PULLENIINAE Schwager 1877
Genus *Pullenia* Parker and Jones 1862
Pullenia bulloides (D'Orbigny 1846)
Figure 6, 4
Nonionina bulloides D'Orbigny 1846

Description

Test free, planispiral, involute, spheroidal, circular in side view, ovate in apertural view. Chambers distinct, inflated, 4–5 in the last whorl. Sutures distinct, radial, flush with surface. Wall calcareous, hyaline, smooth, finely perforate. Aperture a narrow interiomarginal slit extending from one umbilicus to the other, with a thin lip.

Pullenia quinqueloba (Reuss 1851)
Figure 6, 5
Nonionina quinqueloba Reuss, 1851

Description

Test free, planispiral, involute, compressed, slightly lobate, rounded to subrounded in side view. Chambers distinct, usually four to five in the last whorl, gradually increasing in size as added. Sutures distinct, slightly depressed, nearly radial. Wall calcareous, hyaline, smooth, finely perforate. Aperture a narrow interiomarginal slit extending to the umbilicus on either side, with a lip.

Pullenia subcarinata (d'Orbigny 1839)
Figure 6, 6

Nonionina subcarinata D'Orbigny 1839

Description

Test subglobular, involute planispire; somewhat compressed, and slightly lobulate in outline, with a rounded periphery. The (typically) five triangular, inflated chambers in the final whorl increase gradually in size and are separated by depressed, slightly curved, nearly radial sutures. Chamber walls are calcareous, finely perforate and smooth. The primary aperture is a narrow, arcuate, interiomarginal, crescentic slit, extending across the periphery to the umbilical.

Family MELONIDAE Holzmann and Pawlowski 2017
Genus *Melonis* Montfort 1808
Melonis affinis (Reuss 1851)
Figure 6, 7 *Nonionina affinis* Reuss 1851

Description

Test free, planispiral, biconvex, compressed; periphery rounded, nearly circular in side view. Chambers distinct, 10–11 in the last whorl, gradually increasing in size as added. Sutures limbate, flush with surface, gently curved, fused into an imperforate ring around the umbilicus. Wall calcareous, hyaline, smooth; coarsely perforate, except for the imperforate umbilical rim, umbilical flaps and apertural face. Aperture a low arched interiomarginal equatorial slit, extending laterally from one umbilicus to the other.

Melonis pompilioides (Fichtel and Moll 1798)
Figure 8, 7
Nautilus pompilioides Fichtel and Moll 1798

Description

Test free, planispiral, involute, periphery broadly rounded; eight to ten chambers in the final whorl. Wall calcareous, hyaline, smooth, coarsely perforate. Sutures slightly curved, flush with surface. Aperture a low arch that extends to umbilicus on both sides.

Superfamily CHILOSTOMELLOIDEA Brady 1881
Family CHILOSTOMELLIDAE Brady 1881
Subfamily CHILOSTOMELLINAE Brady 1881
Genus *Chilostomella* Reuss 1849
Chilostomella oolina Schwager 1878
Figure 6, 8
Chilostomella oolina Schwager 1878

Description

Test ovate, elongate, rounded, planispiral, involute. There are two strongly embracing chambers per whorl, the last one occupying more than half of the test and the initial ones often in an oblique position relative to the long axis of the test. Chamber walls are thin, smooth, transparent, shiny and finely perforate. The aperture is a narrow, interiomarginal slit, situated approximately two-thirds of the way along the length of the test and sometimes bordered by a small lip.

Superfamily PLANORBULINOIDEA Schwager 1877
Family CIBICIDIDAE Cushman 1927
Subfamily CIBICIDINAE Cushman 1927
Genus *Cibicides* Montfort 1808
Cibicides kullenbergi Parker 1953
Figure 6, 9a, 9b, 9c
Cibicides kullenbergi Parker, 1953

Description

Test free, trochospiral, biconvex, composed of three whorls, periphery acute with a distinct, narrow keel. Chambers distinct, 11–14 in the last-formed whorl, increasing slowly in size, if at all, producing a circular outline with all chambers visible on spiral side. Sutures thin and flush with surface on both sides, curved. Wall calcareous, hyaline, smooth, finely perforate on the involute side, with large punctae rather widely spaced on each chamber on the evolute side. Aperture an interiomarginal equatorial arch, extending along the spiral sutures.

Genus *Cibicidoides* Thalmann 1939
Cibicidoides mundula (Brady, Parker and Jones 1888)
Figure 6, 10a, 10b, 10c
Truncatulina mundula Brady, Parker and Jones 1888

Description

Test free, almost equally biconvex; peripheral margin acutely rounded; chambers numerous, 12–14 in the last-formed coil; sutures oblique dorsally; margins of chambers of clear shell material, raised and thickened in the earlier coils, almost hidings the original surface of the chamber and board even in the last-formed chambers, remainder of the surface distinctly foraminated. Ventral side with the sutures indistinct, with little or no trace of thickenings; walls very coarsely foraminated; aperture a narrow slit extending ventrally from the peripheral margin.

Cibicidoides wuellerstorfi (Schwager 1866)
Figure 6, 11
Anomalina wuellerstorfi Schwager, 1866

Description

Test free, planoconvex; periphery with distinct keel. Sutures strongly recurved on both sides. Wall calcareous, coarsely and densely perforate. Aperture a low arch at the base of the final chamber.

Superfamily PLANORBULINOIDEA Schwager 1877
Family PLANULINIDAE Bermúdez 1952
Genus *Planulina* d'Orbigny 1826
Planulina limbata Natland 1938

Figure 6, 12a, 12b, 12c
Planulina limbata Natland 1938

Description

Test free, discoidal, very low trochospiral of about two whorls, spiral side evolute, umbilical side partially evolute, nine to ten broad, low and arched chambers in the final whorl, septa thick, sutures imperforate, thickened and elevated, strongly curved back at the peripheral margin, periphery truncate, with thick imperforate marginal keel. Wall calcareous, optically radial, finely perforate and with scattered larger pores. Aperture and equatorial and interiomarginal arch with narrow imperforate bordering lip, extending somewhat onto the umbilical folium.

Family CARPENTERIINAE Saidova 1981
Subfamily CARPENTERIINAE Saidova 1981
Genus *Carpenteria* Gray 1858
Carpenteria balaniformis Gray 1858
Figure 7, 1

Description

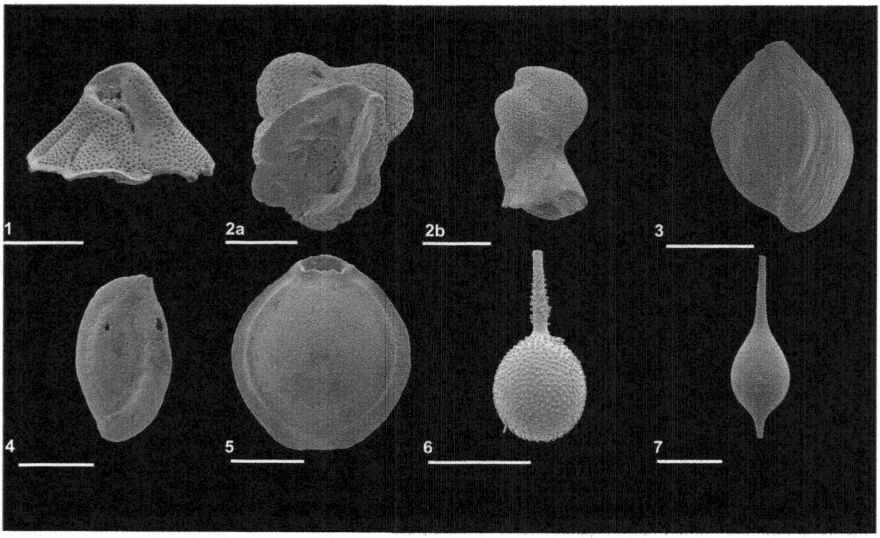

Fig. 7. **1** *Carpenteria balaniformis* Gray 1858, Lateral view, 200 μm. **2** *Rupertina stabilis* (Wallich 1877), a. basal view 300 μm, b. lateral view 500 μm. **3** *Quinqueloculina lamarckiana* D'Orbigny 1839, lateral view, 400 μm. **4** *Quinqueloculina seminula* (Linnaeus 1758), lateral view, 400 μm. **5** *Pyrgo depressa* (D'Orbigny 1826), Lateral view, 200 μm. **6** *Lagena hispidula* Cushman 1913, lateral view, 300 μm. **7** *Procerolagena clavata* (D'Orbigny 1846), Lateral view, 300 μm

Test attached, large, monticular and convex; circumference is deeply lobed, somewhat irregular; early chambers spirally arranged, broad and round at the outer margin, narrow at the umbilical end; later chambers piled up irregularly about a central axis, somewhat inflated. Wall coarsely perforated; aperture a rounded opening at the summit of an erect tubular extension of the end of the central axis.

Superfamily DISCORBINELLOIDEA Sigal 1952
Family PSEUDOPARRELLIDAE Voloshinova 1952
Subfamily PSEUDOPARRELLINAE Voloshinova 1952
Genus *Pseudoparrella* Cushman and Ten Dam 1948
Pseudoparrella sp. 1
Figure 8, 8a, 8b

Description

Test forms a small trochospire; slightly lobulate in outline and unequally biconvex in cross section, with an evolute, slightly convex spiral side, an involute, convex to nearly conical umbilical side, and an acute, lobulate periphery. The five moderately inflated chambers in the last whorl increase gradually in size and are separated by flush, oblique, thickened sutures on the spiral side, and by curved, slightly depressed sutures on the umbilical side. Chamber walls are calcareous, finely perforate and smooth. The primary aperture is an interiomarginal slit extending up the face of the final chamber on the umbilical side.

Superfamily LOXOSTOMATOIDEA Loeblich and Tappan 1962
Family LOXOSTOMATIDAE Loeblich and Tappan 1962 †
Genus *Loxostomum* Ehrenberg 1854 †
Loxostomum minuta (Natland 1938)
Figure 6, 1
Bolivina minuta Natland 1938

Description

Test very compressed, side nearly parallel, periphery angled, flattened, forming a slightly concave side; eight to ten pairs of chambers, rapidly expanding from a subacute initial end; suture distinct, curved, quite oblique; wall thin, very finely perforate; aperture a loop-shaped opening in the top of the last chamber.

Class TUBOTHALAMEA Pawlowski, Holzman and Tyszka 2013
Order MILIOLIDA Delage and Hérouard 1896
Suborder MILIOLINA
Superfamily MILIOLOIDEA Ehrenberg 1839
Family HAUERINIDAE Schwager 1876
Subfamily SIGMOILINITINAE Luczkowska 1974
Genus *Spirosigmoilina* Parr 1942
Spirosigmoilina tenuis (Cžjžek 1848)
Figure 8, 9
Quinqueloculina tenuis Cžjžek 1848

Description

Test about twice as long as wide, narrow, compressed, visible chambers 5 or 6 on the either side, chambers, narrow, rounded, sutures depressed, distinct; wall smooth, aperture exerted, rounded.

Genus *Quinqueloculina* D'Orbigny 1826
Quinqueloculina lamarckiana D'Orbigny 1839
Figure 8, 10
Quinqueloculina lamarckiana D'Orbigny 1839

Description

Test ovate to round in outline triangularly in cross section. Chambers one-half coil in length. Only three chambers visible at the exterior in outside view, three chambers visible form one side and two from that opposite. However, inner construction of transverse section one side and two from that opposite. Wall calcareous, imperforate and porcelaneous, smooth in surface. Aperture ovate, provided with a simple tooth.

Quinqueloculina poeyana D'Orbigny 1839
Figure 7, 3
Quinqueloculina poeyana D'Orbigny 1839

Description

Test two to three time as long as broad; chambers distinct; sutures slightly depressed; periphery rounded; surface ornamented by numerous longitudinal costae, running from the base of the chamber to the apertural lip; apertural in slightly extended, very slightly, if at all, constricted, lip slight; aperture circular, with a single narrow tooth, somewhat slightly bifid at the tip.

Quinqueloculina seminula (Linnaeus 1758)
Figure 7, 4
Quinqueloculina seminula (Linnaeus 1758)

Description

Test elongate, two times as long as broad, ovate in lateral view, ovate in cross section, with rounded periphery; oral end truncated, aboral end inflated, slightly produced; sutures slightly depressed; wall is smooth, polished and glossy; aperture without a neck, arched-shaped with a thickened rim but no lip, provided with a small tooth, often bifid

Subfamily MILIOLINELLINAE Vella 1957
Genus *Pyrgo* Defrance 1824
Pyrgo depressa (D'Orbigny 1826)
Figure 7, 5
Biloculina depressa D'Orbigny 1826

Description

Test free, inflated, subcircular inside view, subovate in transverse section with margin extended and carinate. Test biloculine; chambers inflated. Wall calcareous, porcelaneous, smooth, ornamented with a broad keel that extends completely about the periphery, but possessing a small sinus at posterior end. Aperture terminal, near junction of last two chambers, with a prominent tubular neck, round to ovate, with a bifid tooth

Pyrgo serrata (Bailey 1861)
Figure 8, 11
Biloculina serrata Bailey 1861

Description

Test forms a slightly compressed, biloculine series in the adult form; nearly circular in outline, elliptical in cross section, with a broad, serrate keel. Chamber walls are calcareous, porcelaneous and imperforate. The primary aperture is oval and terminal, with a bifid tooth.

Pyrgo murrhina (Schwager 1866)
Figure 8, 12
Biloculina murrhina Schwager 1866

Description

Test forms a slightly compressed, biloculine series in the adult form; subcircular in outline, and elliptical in cross section, with an acute, extended, keeled periphery. Keel may be smooth or dentate, and is interrupted at a point opposite to the aperture to form a sinus. Chamber walls are calcareous, porcelaneous, smooth and imperforate. The primary aperture is oval, and terminal with a short bifid tooth.

Order LAGENIDA Delage and Hérouard 1896
Superfamily POLYMORPHINOIDEA
Family ELLIPSOLAGENIDAE A. Silvestri 1923
Subfamily ELLIPSOLAGENINAE Silvestri 1923
Genus *Fissurina* Reuss 1850
Fissurina fimbriata (Brady 1881)
Figure 8, 13
Lagena fimbriata Brady 1881

Description

Test pyriform or flask-shaped, broad at the base, compressed; ento or ectosolenian; furnished with a deep vertical wing, encircling the oval base; the wing traversed by parallel tubuli, and sometime fringed at the free margin

Fissurina marginata (Montagu 1803)
Figure 8, 14
Vermiculum marginatum Montagu 1803

Description

Fig. 8. **1** *Lagenammina atlantica* (Cushman 1944), lateral view 100 μm. **2** *Cyclammina cancellata* (Brady 1879), lateral view, 300 μm. **3** *Hoeglundina* sp.1, 500 μm a. dorsal view, b. ventral view. **4** *Protoglobobulimina pupoides* (D'Orbigny 1846), lateral view, 50 μm. **5** *Siphouvigerina hispida* (Schwager 1866), lateral view, 200 μm. **6** *Uvigerina hootsi* Rankin 1934, lateral view, 100 μm. **7** *Melonis pompilioides* (Fichtel and Moll 1798), lateral view, 100 μm. **8** *Pseudoparrella* sp.1, 100 μm a. dorsal view, b. ventral view. **9** *Spirosigmoilina tenuis* (Cžjžek 1848), lateral view, 50 μm. **10** *Quinqueloculina lamarckiana* d'Orbigny 1839, lateral view, 500 μm. **11** *Pyrgo serrata* (Bailey 1861), lateral view, 300 μm. **12** *Pyrgo murrhina* (Schwager 1866), lateral view, 200 μm. **13** *Fissurina fimbriata* (Brady 1881), lateral view, 200 μm. **14** *Fissurina marginata* (Montagu 1803), lateral view, 100 μm. **15** *Fissurina lucida* (Williamson 1848), lateral view, 100 μm. **16** *Lagena striata* (d'Orbigny 1839), lateral view, 100 μm. **17** *Lenticulina limbosa* (Reuss 1863) 300 μm a. lateral view, b. edge view

Test subcircular in contour; with a well-developed keel; central body inflated, highly translucent, finely perforated; inflated area at the base of the produced aperture; bands on each side of the inflated central area, interrupted at the basal and apertural ends of the test. Opaque under the dissecting microscope, whitish on SEM pictures; aperture produced, a long slit with tightly compressed lips.

Fissurina lucida (Williamson 1848)
Figure 8, 15
Entosolenia lucida Williamson 1848

Description

Test pyriform, compressed, with rounded periphery; basal end sometimes with a short projection; wall smooth and opaque under the dissecting microscope, except for the central area, which is clear and translucent.

Superfamily NODOSARIOIDEA Ehrenberg 1838
Family LAGENIDAE Reuss 1862
Genus *Lagena* Walker and Boys 1798
Lagena hispidula Cushman 1913
Figure 7, 6
Lagena hispidula Cushman 1913

Description

Test free, unilocular, circular to ovate in side view, with long apertural neck. Wall calcareous, hyaline, finely perforate, body ornamented with very fine spines, while apertural neck is smooth. Aperture terminal at the end of a long slender neck.

Lagena striata (D'Orbigny 1839)
Figure 8, 16
Oolina striata D'Orbigny 1839

Description

Test free, unilocular, flask-shaped, rounded in cross section. Wall calcareous, hyaline, finely perforate, body ornamented with numerous fine costae oriented from the base to the base of the neck. Aperture terminal, circular, at the end of a long slender neck.

Genus *Procerolagena* Puri 1954
Procerolagena clavata (d'Orbigny 1846)
Figure 7, 7
Oolina clavata D'Orbigny, 1846

Description

Test elongate, fusiform with long delicate neck and well-developed spinose extension of aboral part of test. Circular in transverse section; Wall translucent, thin, smooth; aperture rounded, small, with hyaline lip.

Family VAGINULINIDAE Reuss 1860
Subfamily LENTICULININAE Chapman et al. 1934
Genus *Lenticulina* Lamarck 1804
Lenticulina limbosa (Reuss 1863)
Figure 8, 17a, 17b
Cristellaria (Robulina) limbosa Reuss 1863

Description

Test lenticular, planispiral, bioconvex, involute test. Medium to very large size. Outline subrounded, sharp periphery with narrow keel. Sutures radiate from the large boss in the umbilical area below the recent. Walls calcareous, hyaline.

5 Conclusions

In the Panama Bight, 417 species and 149 genera of benthic foraminifera have been reported. In this zone, the hyaline organisms predominate. The richness of species suggests a latitudinal variation, with the highest values in the continental margin of Ecuador and Colombia and the lowest values in the continental slope of Panama; this variation may be related to the flow of allochthonous species transported by colder currents from the South. The faunal composition corresponds, zoogeographically, to the Panamanian province, and some of these species reported are indicators of environments with high flow of organic matter and high productivity. The new set of data studied by the author allows a more detailed description of the organisms, a future is expected to carry out quantitative analyses (e.g., diversity, abundance, distribution) and expand the information for continental margins in the Colombian Pacific.

Acknowledgements. I want to thank the Instituto Colombiano del Petroleo—Ecopetrol for making this material available for study, and Dr. Daniel Rincon Martínez and M. Sc. Germán David Patarroyo for their advice and constant help throughout the project. Also, I want to thank the biologist Gustavo Torres Angarita, for collaborating in the preparation of samples for optical images, and the Paula Troyon technique for taking the scanning electron microscope images in the Materials Characterization Department, Centro Atómico de Bariloche (CAB).

References

Abu-Zied R, Bantan R et al (2011) Composition, distribution, and taphonomy of nearshore benthic Foraminifera of the Farasan Island, Southern Red Sea, Saudi Arabian. J Foram Res 41:349–362

Altenbach AV, Sarnthein M (1989) Productivity record in benthic Foraminifera. In: Berger WH, Smetacek VS, Wefer G (eds) Productivity of the Oceans: present and past. Dahlem-Konferenzen, Wiley, New York, pp 255–269

Altenbach AV, Pflaumann U et al (1999) Scaling percentages of benthic Foraminifera with flux rates of organic carbon. J Foram Res 29(3):173–185

Armstrong HA, Brasier MD (2008) Microfossils. Blackwell Publishing, London, p 305

Bandy O, Arnal RE (1957) Distribution of recent Foraminifera off west coast of Central America off the west coast of Central America. Bull Am Ass Petroleum Geologist 41:2037–2053

Bandy O, Rodolfo K (1964) Distribution of foraminifera and sediments, Peru-Chile trench area. Deep Sea Res Oceanogr Abstr 11(5):817–837. https://doi.org/10.1016/0011-7471(64)90951-9

Bernhard JM, Sen Gupta BK et al (1997) Benthic Foraminiferal proxy to estimate dysoxic bottom-water oxygen concentrations: Santa Barbara basin, US Pacific continental margin. J Foram Res 27(4):301–310

Betancur M, Martínez I (2001) Productividad y Oxigenación en la Cuenca de Panama, Pacífico colombiano, durante el Pleistoceno tardío - Holoceno: el registro de los Foraminíferos Bentónicos. Geología Colombiana 26:123–151

Betancur M, Martínez I (2003) Foraminíferos bentónicos recientes en sedimentos de fondo de la cuenca de Panamá (Pacífico colombiano), como indicadores de productividad y oxigenación. Boletín de Investigaciones Marinas y Costeras 32:93–123

Boltovskoy E, Gualancañay E (1975) Foraminíferos Bentónicos Actuales de Ecuador. Provincia Esmeraldas. Instituto Oceanográfico Biológico 5:1–56

Boltovskoy E, Wright RC (1976) Recent foraminifera. Springer, La Haya, p 515

Bornmalm L (1997) Taxonomy and paleoecology of late Neogene benthic foraminifera from the Caribbean Sea and eastern Pacific Ocean. Fossils Strata 41:1–96

Burke S, Berger W et al (1993) Benthic Foraminifera in box CORE ERDC 112, Ontong java plateau. J Foram Res 23:19–39

Chavez FP, Strutton PG, Friederich GE, Feely RA, Feldman GC, Foley DG, McPhaden MJ (1999) Biological and chemical response of the Equatorial Pacific Ocean to the 1997–98 El Nin˜o. Science 286:2126–2131

Coulbourn W (1980) Relationship between the distribution of Foraminifera and geologic structures of the Arica Bight. S Am J Paleontol 54(4):696–718

Crouch W, Poag W (1987) Benthic Foraminifera of the Panamanian province: distribution and origins. J Foram Res 17:153–176

Cruz M, De Gonzales M et al (1980) Lista de la fauna sublitoralbentonica del estero salado inferior, Ecuador. Acta oceanográfica del pacifico 1:1–96

Culver SJ, Buzas M (1987) Distribution of Recent benthic foraminifera off the Pacific coast of Mexico and Central America'. Smithson Contrib Mar Sci 30:1–84. https://doi.org/10.5479/si.01960768.30.1

De Gupta AK (2010) Deep-sea faunal provinces and their inferred environments in the Indian Ocean based on distribution of Recent Benthic Foraminifera. Palaeogeogr Palaeoclimatol Palaeoecol 291(3–4):429–442

de Mello e Sousa SH, Passos RF, Fukumoto M, da Silveira I, Figueira L, Koutsoukos EA, Rezende E (2006) Mid-lower bathyal benthic Foraminifera of the Campos Basin, Southeastern Brazilian margin: Biotopes and controlling ecological factors. Mar Micropaleontol 61(1–3):40–57

Eichler B, Pimenta M et al (2014) Living *Bulimina marginata* in the SW Atlantic continental margin: effect of the Subtropical Shelf Front and South Atlantic Central Water. Cont Shelf Res 89:88–92

Figueroa S, Marchant M et al (2005) Foraminíferos Bentónicos Rotalinidos Del Centro Sur de Chile (36° S – 44° N). Gayana 69(2):329–363

Finger KL (1990) Atlas of California Neogene Foraminifera, Cushman foundation for Foraminiferal Research, special publication La Habra California 28:278

Finger KL (1992) Biostratigraphic Atlas of Miocene Foraminifera from The Monterey and Modelo Formations, Central and Southern California. Cushman foundation for Foraminiferal Research, special publication La Habra California 29:179

Forsbergh E (1969) On the climatology, oceanography and fisheries of the Panama Bight. Bull Inter-Am Trop Tuna Comm 14(2):49–385

Fontanier C, Jorissen FJ et al (2002) Live benthic Foraminiferal faunas from the Bay of Biscay: faunal density, composition, and microhabitats. Deep-Sea Res 49:751–785

García-Hansen I, Cortés-Altamirano R et al (2004) La marea roja causada por el dinoflagelado *Alexandrium tamarense* en la costa Pacífica colombiana (2001). Rev Biol Trop 52:59–68

Golik A, Phleger FB (1977) Benthonic Foraminifera from the Gulf of Panama. J Foram Res 7 (2):83–99

Gualancañay E (1983) Foraminíferos bentónicos del golfo de Guayaquil. Acta Oceanográfica del Pacífico 2:589–657

Gualancañay E (1987) Foraminíferos bentónicos de los canales de Jambeli y del Morro en el golfo de Guayaquil. Acta Oceanográfica del Pacífico 4:102–109

Gualancañay E (1989) Distribución de los foraminíferos bentónicos del Golfo de Guayaquil interior y su relación con algunos factores ambientales. Acta Oceanográfica del Pacífico 5:41–48

Gualancañay E (2000) Foraminiferos bentonicos de salinas, ecuador y su relación con el medio. Acta Oceanográfica del Pacífico 10:1–11

Gualancañay E (2007) Distribución zoogeográfica de los géneros *Cibicides* y *Quinqueloculina* en la zona de turbulencia de monteverde. Acta Oceanográfica del Pacífico 14(1):163–167

Gualancañay E (2009) Foraminíferos bentónicos de la bahía de Santa Elena- Ecuador, durante 2006-2007. Acta Oceanográfica del Pacífico 15:15–20

Gualancañay E (2013) Foraminíferos Bentónicos de Manabí, Bahía de Caráquez en la costa ecuatoriana, Durante la Estación lluviosa del 2012. Acta Oceanográfica del Pacífico 18:91–99

Hayward BW, Grenfell HR, Sabaa AT, Hayward CM, Neil H (2006) Ecological distribution of benthic foramini- fera, offshore north-east New Zealand. J Foramin Res 36:332–354

Hermelin O (1989) Pliocene benthic Foraminifera from the ontong-java plateau (Western equatorial Pacific Ocean): faunal response to chaining paleoenviromental. Cushman Found Foram Res 26:1–143

Hess S, Kunt W (1996) Deep-sea benthic Foraminiferal recolonization of the 1991 Mt. Pinabuto ash layer in the South China Sea. Mar Micropaleontol 28(2):171–197

Holboun AE, Henderson AS (2002) Re-illustration and revised taxonomy for selected Deep-Sea Benthic Foraminifera. Paleontología Electronica 4(2):1–36

Jorissen FJ, Stigter HC et al (1995) A conceptual model explaining benthic Foraminiferal microhabitats. Mar Micropaleontol 26:3–15

Lankford RR, Phleger FB (1973) Foraminifera from the nearshore turbulent zone, western North America. J Foram Res 3(3):101–132

Legeckis R (1988) Upwelling off the Gulf of Panama and Papagayo in the tropical pacific during March 1985. J Geophys Res 93:15485–15489

Lei Y, Li T (2016) Atlas of Benthic Foraminifera from China Seas: The Bohai Sea and the Yellow Sea by. Springer, Beijing, p 397

Loeblich AR, Tapan H (1988) Foraminiferal genera and their classification, Van Nostrand Rainhold, New York, 2ª edn, 970 pp

López E (2009) Reporte a bordo de adquisición de piston core, Ecopetrol, 158 pp

Martínez I, Rincón D, Yokoyama Y et al (2006) Foraminifera and coccolithophorid assemblage changes in the Panama Basin during the last deglaciation: response to sea-surface productivity induced by a transient climate change. Palaeogeogr Palaeoclimatol Palaeoecol 234:114–126

Mitchell TP, Wallace JM (1992) The annual cycle in equatorial convection and sea surface temperature. J Clim 5:1140–1156

Murray JW (1991) Ecology and distribution of benthic Foraminifera. In: Lee JJ, Anderson OR (eds) Biology of Foraminifera. Academic Press, New York, pp 221–224

Murray JW (2006) Ecology and applications of Benthic Foraminifera. Cambridge University Press, Cambridge, p 426

Ovsepyan E, Ivanova EV (2009) Benthic Foraminiferal assemblages as indicators of the paleoceanographic conditions in the eastern Equatorial Pacific. Oceanology 49(1):121–129

Parada RC, Tchegliakova N (1990) Foraminíferos y sedimentos de playa blanca, Isla Gorgona. Geología Colombiana 17:227–237

Pascual A, Pujos M (2001) Información suministrada por los foraminíferos del Cuaternario terminal de la plataforma del Pacifico colombiano. Geogaceta 29:87–90

Patarroyo G, Martínez I (2013a) Distribución y preferencias ambientales de los foraminíferos bentónicos de aguas profundas de la cuenca de Panamá, Pacífico Oriental. Caldasia 35:311–324

Patarroyo G, Martínez I (2013b) Foraminíferos bentónicos recientes en las aguas profundas de la cuenca de Panama: ecología y su posible relación con las corrientes. Boletin de Investigaciones marinas y costeras 42:31–55

Pawlowski J (2000) Introduction to the molecular systematics of foraminifera. In: Lee JJ, Hallock P (eds) Advances in the biology of Foraminifera. Micropaleontology 46(supplement 1):1–12

Pawlowski J, Lecroq B (2010) Short rDNA Barcodes for species identification in Foraminifera. J Eukaryot Microbiol 57:197–205

Pawlowski J, Holzmann M et al (2013) New supraordinal classification of Foraminifera: molecules meet morphology. Mar Micropaleontol 100:1–10

Resig J (1981) Biogeography of benthic Foraminifera of the northern Nazca plate and adjacent continental margin. Geol Soc Am Memoir 154:616–665

Restrepo D, Kjerfve B (2000) Water discharge and sediment load from the Western Slopes of the Colombian Andes with focus on Rio San Juan. J Geol 108(1):17–33

Rodriguez-Rubio E, Stuardo J (2002) Variability of photosynthetic pigments in the Colombian Pacific Ocean and its relationship with the wind field using ADEOS-I data. In: Proceedings of the Indian academy of sciences, earth and planetary sciences, vol 111, issue 3, pp 227–236

Rodrìguez-Rubio E, Schneider W (2003) On the seasonal circulation within the Panama Bight derived from satellite observations of wind, altimetry and sea surface temperature. Geophys Res Lett 30(7):1–3

Spezzaferri S, Ruggeberg A et al (2014) Taxonomic notes and ilustration of benthic Foraminifera from cold-water coral ecosystems. Cushman Foundation Special Publication 44:40–140

Scott D, Takayanagi Y et al (2000) Illustration and taxonomic reevaluation of Neogene foraminifera described from Japan. Palaeontologia Electronica 3(2):1–41

Shepherd S, Rathburn E et al (2007) Living Foraminiferal assemblages from the Southern California margin: a comparison of the > 150, 63–150, and > 63 μm fractions. Mar Micropaleontol 65(1–2):54–77

Tomczak M, Godfrey S (1994) Regional oceanography: an introduction. Pergamon, Oxford, p 422

Ujie H (1995) Benthic Foraminifera common in the Bathyal surface sediments of the Ryukyu Island Arc Region, Northwest Pacific. Bull Coll Sci 60:51–111

Wyrtki K (1966) Oceanography of the eastern equatorial Pacific Ocean. Oceanogr Mar Biol Ann Rev 4:33–68

Zapata J, Gutierrez A (1995) Foraminferos Litorales Recientes de Tocopilla, (22° 06' S, 70° 13' W), Chile. Estud Oceanol 14:49–59

Index

G. C. Cusminsky et al. (eds.), *Advances in South American Micropaleontology*, Springer Earth System Sciences, https://doi.org/10.1007/978-3-030-02119-1

Printed by Printforce, the Netherlands